Handbook of Experimental Pharmacology

Volume 176/I

Editor-in-Chief

K. Starke, Freiburg i. Br.

Editorial Board

G.V.R. Born, London
S.P. Duckles, Irvine, CA
M. Eichelbaum, Stuttgart
D. Ganten, Berlin
F. Hofmann, München
W. Rosenthal, Berlin
G. Rubanyi, San Diego, CA

The Vascular Endothelium I

Contributors

F. Antohe, J.D. Catravas, A. Chatterjee, P.A. D'Amore,
A.P. Davenport, C. Dimitropoulou, K. Egan, G.A. FitzGerald,
E.A. Higgs, W.M. Kuebler, J.J. Maguire, A.B. Malik,
L. McCloud, R.D. Minshall, S. Moncada, A.R. Pries,
M. Simionescu, A.M. Suburo, Q.-K. Tran, H. Watanabe,
G. Yetik-Anacak

Editors
Salvador Moncada and Annie Higgs

Prof.
Salvador Moncada
FRCP FRS
The Wolfson Institute
for Biomedical Research
University College London
Gower Street
London WC1E 6BT
UK
E-mail: s.moncada@ucl.ac.uk

Ms.
Annie Higgs
The Wolfson Institute
for Biomedical Research
University College London
Gower Street
London WC1E 6BT
UK
E-mail: a.higgs@ucl.ac.uk

With 42 Figures and 3 Tables

ISSN 0171-2004

ISBN-10 3-540-32966-8 Springer Berlin Heidelberg New York

ISBN-13 978-3-540-32966-4 Springer Berlin Heidelberg New York

This work is subject to copyright. All rights reserved, whether the whole or part of the material is concerned, specifically the rights of translation, reprinting, reuse of illustrations, recitation, broadcasting, reproduction on microfilm or in any other way, and storage in data banks. Duplication of this publication or parts thereof is permitted only under the provisions of the German Copyright Law of September 9, 1965, in its current version, and permission for use must always be obtained from Springer. Violations are liable for prosecution under the German Copyright Law.

Springer is a part of Springer Science+Business Media
springer.com

© Springer-Verlag Berlin Heidelberg 2006

The use of general descriptive names, registered names, trademarks, etc. in this publication does not imply, even in the absence of a specific statement, that such names are exempt from the relevant protective laws and regulations and therefore free for general use.

Product liability: The publishers cannot guarantee the accuracy of any information about dosage and application contained in this book. In every individual case the user must check such information by consulting the relevant literature.

Editor: Simon Rallison, Heidelberg
Desk Editor: Susanne Dathe, Heidelberg
Cover design: *design & production* GmbH, Heidelberg, Germany
Typesetting and production: LE-TEX Jelonek, Schmidt & Vöckler GbR, Leipzig, Germany
Printed on acid-free paper   27/3100-YL - 5 4 3 2 1 0

Preface

It was with great pleasure that I accepted the invitation of Springer to edit this book. My association with the vascular endothelium covers a large part of my scientific career and, as with any good long-standing relationship, it has had moments of great excitement and periods of laborious construction. It has sometimes been difficult but has never given me cause for despondency. Indeed, in the last quarter of a century, research on the vascular endothelium has been very productive and its results have contributed, arguably more than any others, to unravelling the mystery of cardiovascular disease, its origin, its development, its complications and its prevention or treatment once it has developed.

I am very happy that Annie Higgs agreed to join me in this task. Over the years we have collaborated closely and, as always, she has shouldered the brunt of the work and has made sure that things get done to everybody's satisfaction. We have also been fortunate in that the scientists who have made some of the most significant contributions in the field agreed to write chapters; as a result, we have produced two volumes which is a good representation of our knowledge in early 2006. We are, however, aware that the field has expanded beyond all expectation and that there may have been some oversight in the covering of a specific area or some aspect of it. This is compounded by the speed at which knowledge is being generated, with more than 4,100 papers concerning the endothelium published in 2005.

These volumes are organised in such a way that the early chapters discuss the structure, development and function of the normal vascular endothelium. The subsequent chapters consider conditions that lead to disruption of vascular physiology, while the later chapters deal with specific pathologies and their treatment. The final chapter describes various gene-therapy strategies for the treatment of vascular pathologies. Interestingly, although this field of research can now be considered mature, it continues to generate a great deal of new information at a time when some of its fruits are having a direct impact on clinical medicine. This is clearly exemplified in the contents of most of the chapters.

The concept of endothelial dysfunction, although mooted many years ago, has come to the fore and has been very useful in defining a situation which may exist long before the overt signs of vascular diseases can be identified. Although endothelial dysfunction is likely to comprise a variety of disturbances, it is interesting that these days it is almost exclusively measured as a decrease in nitric oxide (NO)-dependent vascular dilatation, either induced by suitable pharmacological agonists or by increases in blood flow. Oxidative stress, which is associated with the genesis of endothelial dysfunction, is a loose term used to define an imbalance between the release of oxygen-derived free radicals and the anti-oxidant systems of the body. Many years ago our work established that reactive oxygen species are important in reducing the local concentrations of both prostacyclin and NO. It is now clear that free radicals also affect other homeostatic systems in the vasculature. However, many things remain to be clarified, especially the origin of oxidative stress in early disease.

The absence of one of these mediators, in this case not NO, but prostacyclin, has been discussed in the scientific and popular press for the past couple of years. The reason is that it is very likely that the cardiovascular side effects which have led to the withdrawal from the market of the anti-inflammatory class of drugs known as COX II inhibitors are due to their inhibitory action on the generation of prostacyclin by the vasculature, leading to a pro-thrombotic situation. The fact that reducing prostacyclin formation in the vasculature leads eventually to cardiovascular events validates the concept we proposed in 1976 that a balance between the generation of thromboxane A_2 by the platelets and prostacyclin by the vessel wall is significant in defining the pro- or anti-thrombotic status of the cardiovascular system. Previously, the only evidence available came from the action of low-dose aspirin which, by inhibiting platelet thromboxane A_2 without affecting prostacyclin, leads to an anti-thrombotic situation. This raises the issue about the status of a cardiovascular system in which both prostacyclin and thromboxane A_2 are inhibited following long-term administration of the classical COX I inhibitors, something which we are only now beginning to address.

The above are just a few considerations that exemplify the problems and challenges that occupy a great deal of our attention today. They show that the vascular endothelium has moved a long way from the "cellophane wrapper" described by early vascular biologists to being recognised as an organ with a variety of functions, some of which, I am sure, remain to be defined. What has yet to be discovered promises to be as exciting and rewarding as that which we already know.

London, S. Moncada
March 2006

List of Contents

Normal Endothelium 1
   A. R. Pries, W. M. Kuebler

Functional Ultrastructure of the Vascular Endothelium:
Changes in Various Pathologies 41
   M. Simionescu, F. Antohe

Development of the Endothelium 71
   A. M. Suburo, P. A. D'Amore

Transport Across the Endothelium:
Regulation of Endothelial Permeability 107
   R. D. Minshall, A. B. Malik

Calcium Signalling in the Endothelium 145
   Q.-K. Tran, H. Watanabe

Eicosanoids and the Vascular Endothelium 189
   K. Egan, G. A. FitzGerald

Nitric Oxide and the Vascular Endothelium 213
   S. Moncada, E. A. Higgs

Angiotensin, Bradykinin and the Endothelium 255
   C. Dimitropoulou, A. Chatterjee, L. McCloud, G. Yetik-Anacak,
   J. D. Catravas

Endothelin 295
   A. P. Davenport, J. J. Maguire

Subject Index 331

Contents of Companion Volume 176/II

Haemostasis 1
   J. Arnout, M. F. Hoylaerts, H. R. Lijnen

Vascular Endothelium and Blood Flow 43
   R. Busse, I. Fleming

Biomechanical Modulation of Endothelial Phenotype:
Implications for Health and Disease 79
   G. García-Cardeña, M. A. Gimbrone Jr.

Leukocyte-Endothelial Interactions in Health and Disease 97
   K. Ley, J. Reutershan

Endothelial Cell Dysfunction, Injury and Death 135
   J. S. Pober, W. Min

Principles and Therapeutic Implications of Angiogenesis,
Vasculogenesis and Arteriogenesis 157
   C. Fischer, M. Schneider, P. Carmeliet

Endothelial Cell Senescence 213
   J. D. Erusalimsky, D. J. Kurz

The Vascular Endothelium in Hypertension 249
   L. E. Spieker, A. J. Flammer, T. F. Lüscher

Vascular Endothelium and Atherosclerosis 285
   P. Libby, M. Aikawa, M. K. Jain

Endothelial Cells and Cancer 307
   L. Nikitenko, C. Boshoff

Gene Therapy: Role in Myocardial Protection 335
   A. S. Pachori, L. G. Melo, V. J. Dzau

Subject Index 351

List of Contributors

Addresses given at the beginning of respective chapters

Antohe, F. , 41

Catravas, J.D. , 255
Chatterjee, A. , 255

D'Amore, P.A. , 71
Davenport, A.P. , 295
Dimitropoulou, C. , 255

Egan, K. , 189

FitzGerald, G.A. , 189

Higgs, E.A. , 213

Kuebler, W.M. , 1

Maguire, J.J. , 295
Malik, A.B. , 107
McCloud, L. , 255
Minshall, R.D. , 107
Moncada, S. , 213

Pries, A.R. , 1

Simionescu, M. , 41
Suburo, A.M. , 71

Tran, Q.-K. , 145

Watanabe, H. , 145

Yetik-Anacak, G. , 255

Normal Endothelium

A. R. Pries (✉) · W. M. Kuebler

Dept. of Physiology, Charité Berlin, Arnimallee 22, 14195 Berlin, Germany
axel.pries@charite.de

1	Central Functional Role of the Endothelium	2
2	Heterogeneity of Endothelial Cells	3
2.1	Heterogeneity Between Different Species	4
2.2	Heterogeneity Between Different Organs	4
2.2.1	Morphological Heterogeneity	4
2.2.2	Specific Vascular Beds	6
2.3	Heterogeneity Between Arterial and Venous Endothelium	13
2.4	Heterogeneity Between Adjacent Endothelial Cells	15
3	Glycocalyx and Endothelial Surface Layer	16
3.1	Endothelial Glycocalyx	16
3.2	Endothelial Surface Layer	20
3.2.1	Composition of the Endothelial Surface Layer	23
3.3	Physiological and Clinical Impact	26
	References	28

Abstract In recent decades, it has become evident that the endothelium is by no means a passive inner lining of blood vessels. This 'organ' with a large surface ($\sim$350 m^2) and a comparatively small total mass ($\sim$110 g) is actively involved in vital functions of the cardiovascular system, including regulation of perfusion, fluid and solute exchange, haemostasis and coagulation, inflammatory responses, vasculogenesis and angiogenesis. The present chapter focusses on two central aspects of endothelial structure and function: (1) the heterogeneity in endothelial properties between species, organs, vessel classes and even within individual vessels and (2) the composition and role of the molecular layer on the luminal surface of endothelial cells. The endothelial lining of blood vessels in different organs differs with respect to morphology and permeability and is classified as 'continuous', 'fenestrated' or 'discontinuous'. Furthermore, the mediator release, antigen presentation or stress responses of endothelial cells vary between species, different organs and vessel classes. Finally there are relevant differences even between adjacent endothelial cells, with some cells exhibiting specific functional properties, e.g. as pacemaker cells for intercellular calcium signals. Organ-specific structural and functional properties of the endothelium are marked in the vascular beds of the lung and the brain. Pulmonary endothelium exhibits a high constitutive expression of adhesion molecules which may contribute to the margination of the large intravascular pool of leucocytes in the lung. Furthermore, the pulmonary microcirculation is less permeable to protein and water flux as compared to large pulmonary vessels. Endothelial cells of the blood-brain barrier exhibit a specialised phenotype with no fenestrations, extensive tight junctions and sparse pinocytotic vesicular transport. This barrier allows a strict control of exchange of solutes and circulating cells between the plasma and the interstitial space. It was observed that average haematocrit levels in muscle capillaries are much lower as compared to systemic haematocrit, and that flow resistance of microvascular beds is higher than expected from in vitro studies of blood rheology. This evidence stimulated the concept of a substantial layer on the luminal endothelial surface (endothelial surface layer, ESL) with a thickness in the range of 0.5–1 µm. In comparison, the typical thickness of the glycocalyx directly anchored in the endothelial plasma membrane, as seen in electron micrographs, amounts to only about 50–100 µm. Therefore it is assumed that additional components, e.g. adsorbed plasma proteins or hyaluronan, are essential in constituting the ESL. Functional consequences of the ESL presence are not yet sufficiently understood and acknowledged. However, it is evident that the thick endothelial surface layer significantly impacts haemodynamic conditions, mechanical stresses acting on red cells in microvessels, oxygen transport, vascular control, coagulation, inflammation and atherosclerosis.

Keywords Heterogeneity · Blood-brain barrier · Pulmonary endothelium · Glycocalyx · Endothelial surface layer

1
Central Functional Role of the Endothelium

For a long time, endothelial cells were considered as a homogeneous population of cells merely forming an inert barrier to separate the vascular space from the interstitium. Florey (1966) challenged these beliefs, pointing out that the endothelium was more than a sheet of nucleated cellophane. About 25 years ago, the ground-breaking investigations on the involvement of the endothelium in regulating vascular smooth muscle tone and coagulation (Moncada et al. 1977; Furchgott 1983; Furchgott and Zawadzki 1980; Palmer et al. 1987, 1988) stressed the fact that the endothelium is not a merely passive barrier. Situated at the interface between blood and tissues, the endothelium plays a central role for critical functions of the cardiovascular system, including regulation of vascular tone, fluid and solute exchange, haemostasis and coagulation, inflammatory responses, vasculogenesis and angiogenesis.

These functions reside in a comparatively 'small organ', albeit with a very large active surface. Based on data established by Mall (1888) on anatomical dimensions of the vascular system in the canine intestine, the total area of the blood/endothelium interface in man can be estimated to be about 350 m^2 (Pries et al. 2000). Depending on the assumed endothelial thickness, this corresponds to a total endothelial mass in the range of only about 110 g (for a thickness of ∼0.3 µm).

The different functional aspects of the endothelium will be addressed in chapters of this book including transport and exchange (R.D. Minshall et al.), regulation of smooth muscle tone (chapters by Q.-K. Tran and H. Watanabe; C. Dimitropoulou et al.; A.P. Davenport and J.J. Maguire; G. García-Cardeña and M.A. Gimbrone, Jr.) and control of haemostasis (J. Arnout et al.). Others will deal with pathophysiological aspects of endothelial function such as hypertension, atherosclerosis, inflammation (J.S. Pober et al.), cancer and metastasis, vasculogenesis and angiogenesis (chapters by C. Fischer et al.; J.D. Erusalimsky and D.J. Kurz).

2
Heterogeneity of Endothelial Cells

Endothelia differ on the basis of their intercellular junctions and can accordingly be classified as 'continuous', 'fenestrated' or 'discontinuous' (Benett et al. 1959). In addition, endothelial cells may differ in terms of morphology, mediator release, antigen presentation or stress responses. Endothelial phenotypes not only differ between species and different organs, but also between consecutive vascular sections. For example, in the kidney, the endothelium is fenestrated in peritubular capillaries, discontinuous in glomerular capillaries and continuous in other regions (Risau 1995). Individual endothelial cells can even differ from the immediately adjacent endothelium, e.g. pacemaker cells that generate interendothelial calcium waves (Ying et al. 1996).

Two putative causes of this heterogeneity have been proposed and are currently a focus of intense study and controversial discussion. The genetic (intrinsic) hypothesis predicts that specific phenotypes are predetermined before endothelial cells migrate from the mesoderm to their specific localisation within the vascular system. This theory is supported by cell lineage studies showing distinct embryonic origins for coronary endothelium and the endocardium (Mikawa and Fischman 1992; Reese et al. 2002). Moreover, arterial and venous endothelial cells express differing profiles of molecules of the ephrin, neuropilin, notch and BMX family early in development, i.e. prior to the onset of circulation and therefore independent of haemodynamic stress (Aird 2003; le Noble et al. 2004).

In contrast, the environmental (extrinsic) hypothesis maintains that site-specific properties of endothelial cells are governed by microenvironmental factors such as soluble mediators, cell-cell and cell-matrix interactions, partial pressures of oxygen or carbon dioxide, or mechanical forces. Plasticity of the endothelium is suggested by transplantation studies in which endothelial cells were shown to adapt to local environmental cues. For example, when avascular tissue from quail brain is transplanted into the coelomic activity of chick embryos, the chick endothelial cells that vascularise the quail brain form a competent blood-brain barrier, whereas when avascular embryonic quail coelomic grafts are transplanted into embryonic chick brain, the chick endothelial cells that invade the mesenchymal tissue grafts form leaky capillaries and venules (Stewart and Wiley 1981). Implantation of astrocytes into the anterior chamber of the eye or into the chick chorioallantoic membrane induces the formation of tight, non-leaky vessels characteristic of the central nervous system (Janzer and Raff 1987). Auricular blood vessels acquire a cardiac endothelial phenotype in the presence of ventricular myocytes (Aird et al. 1997). Moreover, plasticity of endothelial cells is not restricted to organ-specific phenotypes but also applies to arterial-venous differentiation. By implantation of quail arteries and veins into chick embryos, it was shown that arterial endothelial cells can colonise veins and vice versa, and can adapt their gene expression profile accordingly (Moyon et al. 2001a; Othman-Hassan et al. 2001). Therefore, phenotypic heterogeneity of the endothelium is presumably the result of a combination of genetic and environmental factors.

2.1
Heterogeneity Between Different Species

Although endothelia and their ability to react to chemical and physical stimuli are ancestral phenomena present in the different classes of vertebrates (Miller and Vanhoutte 1986), endothelial phenotypes may vary considerably between species in terms of ultrastructure (Rhodin 1968; Higashi et al. 2002), metabolism (Kjellstrom et al. 1987) or signalling mechanisms (Miller and Vanhoutte 1986; Graier et al. 1996). In addition, endothelial cells from different species are heterogeneous at the level of constitutive expression of intracellular as well as cell surface molecules such as the B_1 kinin receptor (Wohlfart et al. 1997), major histocompatibility complex class II antigens (Houser et al. 2004) or selenoproteins (Miller et al. 2002). Like all forms of endothelial heterogeneity, interspecies variability may limit the uncritical transferability of findings from animal experiments or animal cell culture systems to the human situation or vice versa. Moreover, in xenotransplantation, interspecies differences not only contribute to acute vascular rejection reactions (Dorling 2003), but physiological vascular functions within the graft may also differ from vascular responses in the host tissue.

2.2
Heterogeneity Between Different Organs

2.2.1
Morphological Heterogeneity

Endothelial phenotypes within a single organism can be differentiated on the basis of morphology and permeability, as illustrated in Fig. 1. In continuous capillaries, luminal and abluminal plasma membranes fuse only at the tight junctions, which represent the predominant pathway for the exchange of water, glucose, urea and other hydrophilic molecules. Accordingly, the structure of individual tight junctions is the major determinant of vascular permeability in this type of endothelium and accounts, for example, for the tight blood-brain barrier of the brain microcirculation (Ballabh et al. 2004). Fenestrated capillaries are characterised by pores 50–60 nm in diameter which are sealed

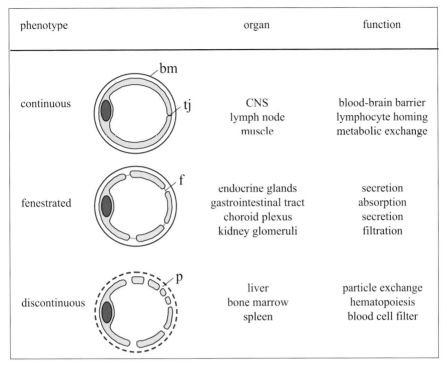

Fig. 1 Different types of endothelial cells, their distribution to different organs and specific functional roles (*bm*, basal membrane; *tj*, tight junction; *f*, fenestrae; *p*, pores)

by a diaphragm. Consistent with their presence at sites of filtration, secretion, and absorption, fenestrated capillaries are more permeable to low-molecular-weight hydrophilic molecules and water (Adeagbo 1997).

Since fenestrated endothelium is located in close proximity to epithelia, interaction between the two cell types has been proposed to trigger differentiation and formation of fenestrae (Risau 1995). Accordingly, co-cultivation of endothelial cells on extracellular matrix derived from a renal epithelial cell line resulted in formation of diaphragmed fenestrations (Milici et al. 1985). Vascular endothelial growth factor (VEGF) may be a relevant paracrine signal in this context. Typically, VEGF is highly expressed in epithelial cells neighbouring fenestrated endothelia (Breier et al. 1992), and VEGF can rapidly induce fenestrae in capillaries in vivo (Roberts and Palade 1995, 1997).

Discontinuous capillaries exhibit large inter- and intracellular gaps 0.1–1 µm in diameter, which are commonly also referred to as fenestrae yet lack a diaphragm. The basal membrane is either absent or involved in the gaps, which are not fixed structures, but can undergo dynamic changes. In hepatic sinusoids, endothelial gaps are clustered in sieve plates that control the exchange of fluids, solutes and macromolecules between the sinusoid and the space of

Disse (Braet and Wisse 2002). Treatment with actin filament-disrupting drugs can induce a substantial and rapid increase in the number of gaps, indicating regulation of the porosity of the endothelial lining by the actin cytoskeleton (Braet et al. 1996). In addition, individual gaps can contract or dilate, depending on the calcium concentration within the liver sinusoidal endothelial cells. Addition of a calcium ionophore induces contraction of endothelial gaps, but chelators of extracellular calcium or calmodulin antagonists suppress the response (Oda et al. 2000). Serotonin-induced contraction of gaps is associated with phosphorylation of myosin light chain kinase, suggesting a crucial role of the calcium-calmodulin-actomyosin cascade in the regulation of gap diameters (Gatmaitan et al. 1996).

2.2.2
Specific Vascular Beds

In addition to morphological differences, endothelial phenotypes from different sites of the vascular bed vary significantly with respect to protein expression and cellular function and are highly adapted to the specific requirements of the individual organ. In addition to the above-mentioned hepatic sinusoidal endothelium, three prominent examples of such organ-specific differentiation are discussed here, i.e. the pulmonary endothelium, the blood-brain barrier, and the high endothelium present in postcapillary venules of peripheral lymph nodes and Peyer's patches.

2.2.2.1
Pulmonary Endothelium

The pulmonary microcirculation is unique in that it accommodates 100% of the cardiac output, maintains a low pressure and resistance system, and facilitates exchange of blood gases with the ambient air. As with almost every vascular bed, endothelial cells in the lung express a unique repertoire of genes and gene products, including lung endothelial cell adhesion molecule-1 expressed exclusively in the pulmonary circulation (Zhu et al. 1991; Elble et al. 1997), endothelial-specific molecule-1 (Lassalle et al. 1996; Bechard et al. 2000) and DANCE (developing arteries and neural crest EGF-like) (Jean et al. 2002) present in lung, kidney and the gastrointestinal tract or the spleen, respectively, or membrane dipeptidase, which is predominantly expressed in lung and kidneys (Rajotte and Ruoslahti 1999). By infusion of radiolabelled antibodies, Panes and co-workers (1995) determined regional differences in the constitutive expression of intercellular adhesion molecule-1 (ICAM-1), which mediates the adhesion of circulating leucocytes by binding to β_2-integrins. Radioactivity in the lung exceeded values from other organs by more than 30-fold, and even after correction for vascular surface area, ICAM-1 expression was most prominent in the lung (Panes et al. 1995). Accordingly, ICAM-targeting can be applied successfully for drug delivery to the pulmonary endothelium (Murciano et al. 2003). Similar to ICAM-1, expression of the adhesion molecule P-selectin is highest in the lung as compared to other organs (Eppihimer et al. 1996) and is predominantly localised at microvascular bifurcations (Kuebler et al. 1999). The vitronectin receptor $\alpha_v\beta_3$ integrin mediates adhesion of circulating platelets (Gawaz et al. 1997), but its expression is generally considered to be confined to proliferating and tumour vessels (Brooks et al. 1994). However, in the lung, $\alpha_v\beta_3$ integrin is constitutively present on both the luminal and the abluminal face of the microvascular endothelium (Singh et al. 2000). The high constitutive expression of adhesion molecules on the pulmonary endothelium may be closely linked to the large pool of intravascular leucocytes which are physiologically marginated in pulmonary arterioles, venules and particularly the dense capillary network of the lung (Kuebler et al. 1994, 1997). Considering the high exchange rate between the alveolar space and the ambient environment, a pro-inflammatory endothelial phenotype in the lung may be regarded as a phylogenetically beneficial defence mechanism (Kuebler and Goetz 2002).

In addition to the expression of surface markers, pulmonary endothelial cells exhibit unique functional attributes, including signal transduction and barrier properties. However, these functional differences exist not only between the endothelium of the pulmonary and the systemic circulation, but also between different blood vessel types in the lung. Studies combining vascular casting and electron microscopy suggest that lung macrovascular endothelium is derived from the pulmonary truncus by angiogenesis, whereas the microvascular endothelium is derived from blood islands formed through vasculogenesis in the mesenchyme of the embryonic lung before the in-growth of the pulmonary artery (deMello et al. 1997; deMello and Reid 2000). The specific phenotype of lung macro- and microvascular endothelial cells is even preserved when cells are isolated and cultured under identical conditions. It was shown that microvascular endothelial cells still express more vascular endothelial (VE)-cadherin and less endothelial nitric oxide synthase (eNOS) than endothelial cells isolated from the pulmonary artery (Stevens et al. 2001), and both phenotypes can be differentiated based on the binding of various lectins (King et al. 2004). These data support the notion that endothelial heterogeneity is not the sole result of environmental factors, but is also in part attributable to programmed determinants. This heterogeneity is also evident at the functional level. For example, infusion of the plant alkaloid thapsigargin, which activates calcium entry through store-operated calcium channels, causes perivascular oedema in pre- and postcapillary vessels, but does not alter capillary barrier function (Chetham et al. 1999). Segmental measurements of hydraulic conductivity confirmed that the pulmonary microcirculation is more restrictive to protein and water flux than is the macrocirculation (Parker and Yoshikawa 2002). Furthermore, lung microvascular endothelial cells grow faster than their macrovascular counterparts (Stevens 2002). This site-specific functional heterogeneity can be partly attributed to phenotypically distinct signal transduction cascades. As compared to lung macrovascular endothelial cells, the magnitude of store-operated calcium entry is substantially reduced in endothelial cells of the pulmonary microcirculation (Stevens et al. 1997; Kelly et al. 1998). Furthermore, lung microvascular endothelial cells possess an intrinsic capacity to preserve intracellular cyclic adenosine monophosphate (cAMP) concentrations (Stevens et al. 1999) which enhances their barrier function (Stevens et al. 2001). The larger store-operated calcium response in macrovascular cells may be functionally linked to a shorter coupling distance between the apical plasmalemma and the endoplasmic reticulum in endothelial cells of the pulmonary artery as compared to lung microvascular endothelial cells (King et al. 2004). Dynamic rearrangements of endothelial microtubules and the actin cytoskeleton may control the intracellular distribution of the endoplasmic reticulum and have therefore been implicated in this scenario (Wu et al. 2001). Of note, the segmental distribution of growth and permeability responses is exactly opposite in the bronchial circulation, in which macrovascular endothelial cells grow faster and exhibit a more restrictive barrier function, illustrating again the site-specificity of the endothelial phenotype (Moldobaeva and Wagner 2002).

2.2.2.2
Blood-Brain Barrier

The endothelium of the cerebral microvasculature at the interface between blood and the central nervous system (CNS) exhibits specific protective properties that strictly regulate the infiltration of plasma components and circulating cells (Fig. 2).

For this purpose, endothelial cells of the blood-brain barrier differ from other endothelial phenotypes by the absence of fenestrations, the formation of extensive tight junctions and sparse pinocyotic vesicular transport (Reese and Karnovsky 1967; Brightman and Reese 1969). Tight junctions consist of three integral membrane proteins, namely, claudin, occludin and junction adhesion molecules, and several cytoplasmic accessory proteins including ZO-1, ZO-2, ZO-3, cingulin and others. In endothelial tight junctions forming the blood-brain barrier, claudins-1, -3 and -5 have been described (Morita et al. 1999; Liebner et al. 2000; Wolburg et al. 2003) as well as expression of occludin (Papadopoulos et al. 2001), junctional adhesion molecule (JAM-1) and peripheral zonula occludens protein (ZO-1) (Dobrogowska and Vorbrodt 2004; Vorbrodt and Dobrogowska 2004). Expression of occludin is much higher in brain endothelial cells compared to non-neuronal tissues, suggesting that occludin may be a regulatory protein reducing paracellular permeability (Hirase et al. 1997). This notion is supported by several facts. First, the expression of occludin inversely correlates with the leak of contrast dye in human brain tumours (Papadopoulos et al. 2001). Second, brain oedema formation following intracarotid infusion of hyperosmotic arabinose solution is closely associated

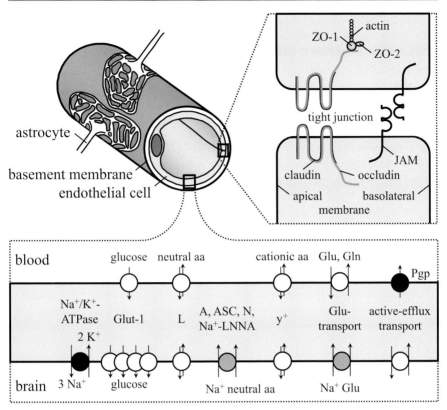

Fig. 2 Schematic drawing of the blood-brain barrier. *Upper left panel*: Perivascular astrocytic end feet forming "rosette"-like structures on the abluminal brain capillary surface. *Upper right panel*: Enlarged view of transmembrane and associated intracellular proteins of interendothelial tight junctions. Claudin, occludin and junctional adhesion molecule (*JAM*) are the transmembrane proteins, and peripheral zonula occludens proteins (*ZOs*, cingulin and others) are cytoplasmic proteins which link claudin to the cytoskeleton. *Lower panel*: Enlarged view of important transport mechanisms of the blood-brain barrier. Na^+-K^+-ATPase is predominantly localised at the abluminal membrane, but may also be present on the luminal surface (Manoonkitiwongsa et al. 2000). The Na^+-independent glucose transporter Glut-1 is expressed in ~4-fold greater abundance on the abluminal as compared to the luminal membrane (Farrell and Pardridge 1991). Na^+-dependent (*grey*) and -independent (*white*) transport systems regulate influx and efflux of amino acids (*aa*) across the blood-brain barrier. Active efflux transport involves the sequential action of an energy-independent carrier and an energy-dependent (*black*) transporter such as P-glycoprotein (Pgp)

with reduced expression of occludin and its spatial disorganisation from the junctional complexes (Dobrogowska and Vorbrodt 2004). Third, differential expression of occludin may also account for regional heterogeneities in the function of the blood-nerve barrier.

The dorsal root ganglion of the peripheral nervous system consists of a nerve fibre-rich area with a relatively tight blood-nerve barrier and a cell body-rich area with considerable leakage. Of note, endothelial cells in the tight nerve fibre-rich area express occludin in addition to claudin-5, whereas those in the cell body-rich area express claudin-5 but no occludin (Hirakawa et al. 2004). In addition to occludin, claudin-5 has been implicated in the regulation of paracellular conductance. Expression of claudin-5 in a renal epithelial cell line increased transepithelial resistance fivefold and selectively decreased the permeability to monovalent cations (Wen et al. 2004). Claudin-5-deficient mice have morphologically normal brain microvessels with no signs of bleeding or oedema, but a size-selective loosening of the blood-brain barrier against molecules of less than 800 daltons (Nitta et al. 2003). However, in two inflammatory disorders, experimental autoimmune encephalitis and human glioblastoma multiforme, formation of brain oedema was solely associated with a selective loss of claudin-3, whereas other tight junction proteins remained unchanged (Wolburg et al. 2003). Hence, differential expression of tight junction molecules may specifically regulate blood-brain barrier function in response to different stimuli. Selective ion permeability may be mediated by different claudins forming paracellular pores or channels (Tsukita et al. 2001). Likewise, removal of single claudins may cause a size-dependent increase in permeability by activation of mechanisms that mediate size-selective paracellular diffusion, e.g. by association with occludin (Matter and Balda 2003).

The presence of a well-developed system of tight junctions is probably responsible for the high degree of polarisation of the brain capillary endothelium which is required for the directed transport of solutes between the blood and the nervous system (Joo 1996). Biochemical studies of brain capillary endothelial cells resulted in the identification of two plasma membrane fractions, a light luminal fraction containing alkaline phosphatase and γ-glutamyl transpeptidase, and a heavier abluminal fraction containing Na^+-K^+-ATPase and 5′-nucleotidase (Betz et al. 1980). Whereas small lipophilic substances such as O_2 and CO_2 easily diffuse across the blood-brain barrier, small polar solutes, including the brain's primary metabolic substrate glucose, require specific carriers. Brain capillary endothelial cells express high amounts of the sodium-independent glucose transporter Glut-1 (Mueckler et al. 1985; Pardridge et al. 1990). Asymmetric distribution of Glut-1, as well as other cytoplasmic and membrane-bound enzymatic processes, warrants energy-independent glucose transport at the blood-brain barrier and thus allows the brain to meet its high metabolic demand in the face of varying plasma glucose levels (Farrell and Pardridge 1991; McAllister et al. 2001).

Brain capillary endothelial cells also exhibit an extensive set of amino acid transporters, including the Na^+-independent systems L and y^+ for large neutral and cationic amino acids (Stoll et al. 1993; Sanchez del Pino et al. 1995). In addition, several Na^+-dependent carriers such as system A (Betz and Goldstein 1978), system ASC (Hargreaves and Pardridge 1988), system N (Lee et al. 1998), and system Na^+-LNAA (O'Kane and Hawkins 2003) facilitate the transport of neutral amino acids at the blood-brain barrier. The exclusive location of these carriers in the abluminal membrane of brain endothelial cells may contribute importantly to the maintenance of low amino acid concentrations in the cerebrospinal fluid, which-with the exception of glutamine-are approximately 10% of those in plasma (McGale et al. 1977). Some of these Na^+-dependent transport systems are regulated by oxoproline, an intracellular product of γ-glutamyl amino acids which are formed through the transfer of the γ-glutamyl moiety from extracellular glutathione to acceptor amino acids at the luminal membrane of the endothelium (Orlowski and Meister 1970). The γ-glutamyl cycle may thus regulate the exit of amino acids from brain to blood, thereby protecting the brain against elevated amino acid levels (Lee et al. 1996). Furthermore, specific carriers mediate the efflux of potentially toxic metabolites from the CNS. Active extrusion of glutamine and glutamate from the brain via Na^+-dependent transport systems on the abluminal membrane and facilitative transport on the luminal side may provide an essential mechanism for removal of nitrogen and nitrogen-rich amino acids from brain (Lee et al. 1998). mdr1a P-glycoprotein is an energy-dependent efflux carrier at the luminal membrane of brain endothelial cells which transports a wide variety of low-molecular-weight molecules out of the brain to the circulation and confers the multidrug-resistance phenotype on brain capillaries (Thiebaut et al. 1989). Disruption of the mdr1a P-glycoprotein gene results in elevated drug levels in, and decreased drug elimination from, the brain (Schinkel et al. 1994). Hence, active efflux transporters play a major role in protecting the brain from xenobiotics and are currently a major target for interventional therapies aimed at increasing drug delivery to the brain (Pardridge 2003).

Regulation of blood-brain barrier integrity is considered to depend on juxtaposed astrocytes (Davson and Oldendorf 1967). The endfeet of astrocytic glia form a lacework of fine lamellae closely apposed to the outer surface of the endothelium. This structural arrangement facilitates astrocytic-endothelial communication and warrants free diffusion between the endothelium and the brain parenchyma (Kacem et al. 1998). In cell culture, cerebral endothelial cells lose their blood-brain barrier characteristics, but maintain them when co-cultured with astrocytes or in the presence of astrocyte-conditioned media (Prat et al. 2001; Abbott 2002). Transitory focal astrocyte loss in the inferior colliculus by intraperitoneal administration of 3-chloropropanediol results in a loss of occludin, claudin-5 and ZO-1 from the sites of tight junction complexes which correlated with focal vascular leak of high molecular weight markers (Willis et al. 2004). Tight junction protein expression returns when astrocytes repopulate the lesion. In contrast, removal of astrocytes from a co-culture with brain endothelial cells increases the permeability of the endothelial monolayer, yet does not result in visible changes of the molecular composition of endothelial tight junctions (Hamm et al. 2004). Although the reasons underlying these contradicting results in vitro and in vivo remain to be elucidated, both studies suggest that direct astrocyte-endothelial contact or paracrine release of short-range diffusible factors by glial cells determine the cerebral endothelial phenotype.

2.2.2.3
Endothelium of High Endothelial Venules

The lymphatic microvasculature plays a central role in the homing of naive B- and T-lymphocytes, which emigrate from the blood through high endothelial venules (HEVs) of peripheral lymphatic tissues and then recirculate through efferent lymph and the thoracic duct back to blood (Gowans and Knight 1964; Marchesi and Gowans 1964) until they find their cognate antigen (Cahill et al. 1976). HEVs of the peripheral lymph nodes facilitate this route of lymphocyte traffic by specific morphological and functional properties. The plump, almost cuboidal endothelial cells are linked by discontinuous, 'spot-welded' tight junctions (Anderson and Shaw 1993), which presumably facilitate the passage of large numbers of emigrating lymphocytes (Girard and Springer 1995). Most importantly, HEVs constitutively and exclusively express a group of adhesion molecules facilitating lymphocyte homing. The first step of this adhesion/emigration cascade is the tethering and rolling of naive T and B lymphocytes along the wall of HEVs. This process is mediated by interaction of L-selectin (CD62L) expressed on the lymphocyte with O-linked glycosylated carbohydrate moieties (Warnock et al. 1998). These ligands, collectively termed the peripheral lymph node addressin (PNAd), were identified by the monoclonal antibody MECA-79, which stains all HEVs within lymphoid tissues yet does not interact with postcapillary venules or large vessels in spleen, thymus or non-lymphoid tissues (Streeter et al. 1988; Michie et al. 1993). MECA-79 prevents lymphocyte adhesion to HEVs in vitro and inhibits lymphocyte emigration through HEVs in vivo (Streeter et al. 1988; Michie et al. 1993). L-selectin ligands in HEVs contain fucose, sialic acid and sulphate and include several HEV glycoproteins such as glycosylation-dependent cell adhesion molecule 1 (GlyCAM-1) (Lasky et al. 1992), CD34 (Baumheter et al. 1993) and mucosal addressin cell adhesion molecule 1 (MadCAM-1) (Berg et al. 1993). None of these glycoproteins is specific for HEV, e.g. MadCAM-1 is expressed in cultured brain-derived endothelial cells (Berg et al. 1993), and CD34 is widely expressed on endothelial cells in most organs (Puri et al. 1995). Therefore, binding of MECA-79 and L-selectin-mediated lymphocyte homing crucially depend upon post-translational modifications of these glycoproteins.

Lymph node-specific sulphation (van Zante et al. 2003; Uchimura et al. 2004) and O-linked glycosylation (Smith et al. 1996; Lowe 2002) of the carbohydrate moieties are required for efficient binding of L-selectin, and the tissue-specific role of glycosyl- and sulphotransferases is currently a topic of intense study. In addition, some lymph node venules seem to express L-selectin ligands that are not MECA-79 reactive (M'Rini et al. 2003; van Zante et al. 2003). HEVs constitutively express the CC-chemokine ligand 21 (CCL21), which binds to the CC chemokine receptor 7 (CCR7) on T cells, resulting in the activation of T cell integrins, and thus facilitating firm arrest of rolling T cells (Campbell et al. 1998; Gunn et al. 1998; Stein et al. 2000). A second CCR7 agonist, CCL19, is not expressed by HEVs, but by lymphatic endothelium and interstitial cells in the lymph node. However, HEV may activate rolling T cells by transcytosis and luminal expression of CCL19 (Baekkevold et al. 2001). Furthermore, HEV expression of CXCL12 and interaction with CXCR4 may contribute to T and B cell homing (Okada et al. 2002).

Interruption of afferent lymphatic flow results in partial loss of the characteristic HEV morphology and vascular addressin expression, suggesting that local environmental factors may at least partially regulate the specialisation of the HEV phenotype (Mebius et al. 1991, 1993).

2.3
Heterogeneity Between Arterial and Venous Endothelium

Endothelial cells from arterial and venous vascular sites differ in terms of morphology as well as function. Endothelial cells in terminal arterioles are generally elongated, reaching a width-to-length ratio of 1:6.8 in rat tracheal mucosa, whereas capillary (1:4.7) and particularly venular (1:2.4) endothelial cells are rounder (McDonald 1994). Endothelial-dependent relaxations are generally more pronounced in arteries than in corresponding veins (Seidel and LaRochelle 1987). Since most veins respond well to nitrovasodilators, the heterogeneity of endothelium-dependent responses appears to be determined by the endothelium rather than by the smooth muscle (De Mey and Vanhoutte 1982). At the microvascular level, shear stress-induced, nitric oxide (NO)-mediated dilatation is more pronounced in arterioles compared to venules of the porcine epicardium (Kuo et al. 1991, 1993). This is consistent with a higher basal expression of NO synthase in arteriolar vs venular endothelium (Nichols et al. 1994). Leucocyte-endothelial interactions, on the other hand, are predominantly confined to the venular compartment (Cohnheim 1867) and only present in arterioles in severe tissue injury (Mayrovitz et al. 1980). An exception is the pulmonary microcirculation, in which leucocyte rolling and adhesion are not uncommon in arterioles, yet less pronounced than in venules (Kuebler et al. 1994). The preferential venous distribution of leucocyte-endothelial interactions was long attributed to higher flow velocities and thus to higher shear rates in arterioles, preventing the binding of leucocyte adhesion molecules to their endothelial ligands. Although high shear rates reduce leucocyte rolling in venules, reduced shear rates do not cause leucocyte rolling in arterioles, indicating that arteriolar and venular endothelial phenotypes differ with respect to adhesion molecule expression (Ley and Gaehtgens 1991). Indeed, most adhesion molecules, including P-selectin, E-selectin and ICAM-1, are found to be exclusively or preferentially expressed on venular endothelium (Cotran et al. 1986; McEver et al. 1989; Iigo et al. 1997).

Differences in haemodynamic factors were generally held responsible for the development of these different phenotypes. In vitro studies suggest that endothelial ICAM-1 expression may be upregulated by reduced shear stress (Nagel et al. 1994), a finding that is in accordance with the fact that pulmonary arterioles have low shear rates, express ICAM-1 and show considerable interaction of circulating leucocytes with the endothelium (Kuebler et al. 1994; Sato et al. 2000). The notion that haemodynamic factors govern the differential arterial and venous endothelial phenotypes is also supported by the observation that endothelial cells of vein grafts transplanted into the arterial system undergo morphological and cytoskeletal changes characteristic for arterial endothelium (Yoshida and Sugimoto 1996). However, the recent discovery of molecules that are specifically expressed in arterial or venous endothelial cells during early development prior to the onset of circulation has challenged this view.

Arterial endothelial cells of chick, mouse and zebrafish selectively express ephrin-B2 (Wang et al. 1998; Adams et al. 1999), neuropilin 1 (Herzog et al. 2001), Bmx tyrosine kinase (Rajantie et al. 2001) and members of the Notch signalling pathway including Notch 3, DLL4 and gridlock (Shutter et al. 2000; Zhong et al. 2000; Villa et al. 2001). Other gene products expressed predominantly in arterial endothelium include tyrosine phosphatase-µ and endothelial per-arylhydrocarbon receptor-nuclear translocator-SIM domain protein-1 (EPAS-1) (Tian et al. 1997; Bianchi et al. 1999). On the other hand, the ephrin-B2-receptor EphB4 is specific for the venous endothelium (Wang et al. 1998; Gerety et al. 1999), and several other molecules-including neuropilin 2 and the angiopoietin receptor tie-2-are preferentially expressed in veins in avian embryos (Herzog et al. 2001; Moyon et al. 2001b). The fact that these expression profiles are evident even before the output of the first embryonic heart beat suggests that segment-specific expression of gene products is genetically predetermined and may regulate arterial-venous differentiation, patterning and cell fate (Wang et al. 1998; Torres-Vazquez et al. 2003). Accordingly, cultured endothelial cells of arterial origin differ substantially from those of the venous circulation and maintain phenotypical differences, e.g. in protein synthesis in the absence of haemodynamic stress (Wagner et al. 1988).

In spite of preset genetic programs, endothelial cells show a high degree of plasticity and are able to adjust their genetic make-up, depending on local haemodynamics (le Noble et al. 2005). This was recently demonstrated by elegant flow manipulation experiments in the chick embryo yolk sac (le Noble et al. 2004). After ligation of the right vitelline artery, part of the arterial system was perfused in a retrograde manner, thus forming a new venular tree. Arterial venularisation resulted in rapid downregulation of the arterial markers ephrin-B2 and neuropilin 1, followed by a subsequent upregulation of the venous markers neuropilin 2 and Tie-2. Hence, different segmental endothelial phenotypes seem to originate from the combined effects of genetic imprinting and endothelial plasticity in response to haemodynamic factors.

2.4
Heterogeneity Between Adjacent Endothelial Cells

Recently, Ying and co-workers reported that even endothelial cells immediately adjacent to each other may exhibit differential signalling and functional properties. Using fluorescence imaging in isolated-perfused lungs, they identified a specific subset of endothelial cells in pulmonary microvessels called pacemaker cells (Fig. 3) with the unique ability to spontaneously generate oscillations of the intracellular calcium concentration (Kuebler et al. 2002; Ying et al. 1996).

Calcium oscillations can be communicated to adjacent non-pacemaker cells, thus generating interendothelial calcium waves travelling along the microvas-

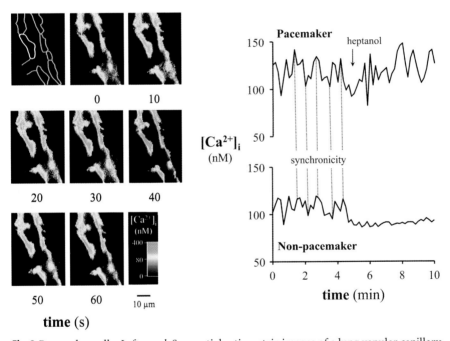

Fig. 3 Pacemaker cells. *Left panel*: Sequential ratiometric images of a lung venular capillary with endothelial cells loaded with the calcium-sensitive dye fura-2. A schematic drawing (*top left*) outlines an endothelial pacemaker cell (*white*) located at the vessel bifurcation and adjacent non-pacemaker cells (*blue*). Images taken at 10-s intervals and colour-coded for the intracellular calcium concentration ($[Ca^{2+}]_i$) show the spontaneous generation of a calcium oscillation in the pacemaker cell and its propagation along the vascular wall. *Right panel*: Tracings of the intracellular calcium concentration ($[Ca^{2+}]_i$) in an endothelial pacemaker (*top*) and an adjacent non-pacemaker cell (*bottom*). Calcium oscillations are synchronous in both cells, but delayed by ~10 s in non-pacemaker as compared to pacemaker cells. The gap junction uncoupler heptanol blocks interendothelial propagation, but not generation of pacemaker oscillations. Methods and experimental protocol as described in Kuebler et al. (1999) and Ying et al. (1996)

cular wall at a speed of approximately 5 μm/s. Pacemaker cells are preferentially located at microvascular branch points. As compared to adjacent non-pacemaker cells, they exhibit increased vesicular trafficking and expression of P-selectin as well as a higher density of mitochondria (Kuebler et al. 1999; Parthasarathi et al. 2002). However, the molecular basis underlying the unique ability of pacemaker cells to generate calcium oscillations remains to be elucidated. Of note, intercellular calcium waves are absent in lung capillaries of rats with congestive heart failure, which are simultaneously characterised by an impaired control of the vascular lumen (Kuebler 2005). Hence, pacemaker-generated intercommunication between adjacent endothelial cells may play an important role in co-ordinating spatial and temporal signalling in the lung vasculature and homogenise changes in tone or permeability.

3
Glycocalyx and Endothelial Surface Layer

3.1
Endothelial Glycocalyx

About 60 years ago, Danielli (1940) and Chambers and Zweifach (1947) introduced the concept of a thin non-cellular layer on the endothelial surface and in the inter-endothelial clefts (endocapillary layer) to explain the results of studies on endothelial permeability. Since then, many studies have investigated specific molecules residing in the endothelial membrane and have shown that the endothelial plasma-membrane is decorated by a large variety of extracellular domains of membrane-bound molecules. This coat includes glycolipids, glycoproteins, and proteoglycans and constitutes the endothelial glycocalyx in a strict sense (Fig. 4).

Most electron microscopic studies indicate the presence of a glycocalyx with a thickness in the range of 20 nm (Luft 1966; Ito 1974). However, fixation methods for electron microscopy are likely to lead to a collapse of gel-like surface structures with a high water content (Sims and Horne 1993). This led to the search for methods more capable of visualising the thickness of the glycocalyx in situ (Baldwin and Winlove 1984; Sims et al. 1991; Clough and Moffitt 1992; Rostgaard and Qvortrup 1997). The glycocalyx reported from these studies showed an average thickness ranging from about 60 to 110 nm, in line with the assumed length of typical glycoproteins and proteoglycans.

In these studies, larger projections into the vascular lumen (Baldwin and Winlove 1984; Sims et al. 1991; Clough and Moffitt 1992) and filamentous plugs composed of 20–40 filaments with a length of about 350 nm on the surface of endothelial fenestrae (Rostgaard and Qvortrup 1997) were also reported. Such structures may reflect hyaluronan (hyaluronic acid) anchored in the endothelial plasma membrane, according the concept of Duling and co-workers (Henry and Duling 1999; Platts et al. 2003), or molecules bound reversibly and dynamically to glycoproteins and proteoglycans of the glycocalyx (Pries et al. 2000). While it is not possible based on the available information to distinguish between the different options, it appears helpful in order to avoid ambiguity and confusion to restrict the term 'endothelial glycocalyx' to the glycolipids, glycoproteins, and proteoglycans which are integrated in the endothelial plasma membrane (Pries et al. 2000; Hjalmarsson et al. 2004). In contrast, the much thicker zone on the endothelial surface which exhibits mechanical and physicochemical properties which differ from those of the free-flowing plasma (which includes the glycocalyx proper) was named the 'endothelial surface layer' (see Sect. 3.2).

Prominent examples of molecular components of the glycocalyx are cell adhesion molecules involved in immune reactions and inflammatory processes, e.g. selectins and integrins (Springer 1990, 1995; Ley 1996; Esmon et al. 1999; Pries et al. 2000; Hjalmarsson et al. 2004) and components of the coagulation/fibrinolysis system, e.g. tissue factor or plasminogen (Shih and Hajjar 1993; Rao and Pendurthi 1998). Despite the large amount of information available on individual molecules, not much is known on the quantitative composition of the glycocalyx, e.g. the relative number of different molecule classes or the number of molecules per surface area. In a recent study, Squire and co-workers (2001) used computer-assisted analysis of electron micrograph images to analyse the structural arrangement of molecules in the glycocalyx. They describe a three-dimensional fibrous meshwork with a fibre diameter of about 10–12 nm and characteristic spacing between fibres of about 20 nm. They also report that fibres may be arranged in clusters with a common inter-cluster spacing of about 100 nm. According to a model analysis (Weinbaum et al. 2003), this arrangement is in line with the barrier functions of the endothelium.

Probes like cationised ferritin, colloidal gold and a number of different lectins have provided information on the distribution of specific chemical moieties on the endothelial surface. Leabu et al. (1987) reported a rather homogeneous coating with cationic basic residues (e.g. amino groups) for rabbit aorta and coronaries. In contrast, anionic sites were not distributed homogeneously. About one-third of these were constituted by neuraminidase-cleavable sialic acids. Results from critical electrolyte staining experiments indicated a major contribution of carboxyl groups to surface charge while sulphate groups were also present (Haldenby et al. 1994). Lectin binding experiments have demonstrated the presence of a variety of saccharide components including sialyl [i.e. N(O)-acetylneuramin (muramin) acid], mannosyl and galactosyl residues, as well as N-acetylglucosamine and N-acetylgalactosamine (Milici and Porter 1991; Noble et al. 1996; Thurston et al. 1996).

Usually, the term 'glycoprotein' is reserved for those glycoconjugates in which the carbohydrate side chains are short (about 2 to 15 sugar residues) and branched (Montreuil et al. 1986; Leabu et al. 1987). Glycosylation of proteins is a very regular event, and most of the known proteins at the endothelial surface (e.g. selectins, integrins, members of the immunoglobulin superfamily, etc.) belong to this type. The carbohydrate components in this extremely variable class of molecules are dominated by mannose, galactose, N-acetylgalactosamine, glucose, N-acetylglucosamine and fucose, the charge being provided mainly by sialyl residues.

In contrast, the more strictly defined class of proteoglycans is characterised by long (about 200 sugar residues, stretched length about 80 nm) and unbranched side chains. Of the proteoglycans associated with endothelial cells, 50%–90% are heparan sulphate proteoglycans (HSPGs) (Bauersachs et al. 1997; Rosenberg et al. 1997) in which a varying number of heparan sulphate (HS) glycosaminoglycan (GAG) side chains are attached to the core protein (Fig. 4). The core proteins present at the luminal side of endothelial cells belong to the syndecan or glypican families. The transmembrane proteins syndecan-1, -2 and -4 with molecular weights of 33, 23 and 22 kDa have a short, highly conserved cytoplasmatic tail which contains four tyrosine residues at fixed positions (Bernfield et al. 1992; Rosenberg et al. 1997) and may activate protein kinase C upon homo-oligomerisation in a variety of cellular reactions. The variable extracellular domain exhibits a dibasic cleavage site at which the proteoglycan may be detached from the cell surface by proteases. GAG chains are attached to 3–5 specific sites which mostly exhibit Ser-Gly(Ala)-X-Gly(Ala) motifs.

Glypicans 1 to 4 exhibit molecular weights ranging from 57 to 69 kDa and are attached to the membrane by a glycosylphosphatidylinositol (GPI) anchor. Therefore, they can be released from the endothelial surface by phospholipase

Fig. 4 The glycocalyx. *Upper panel*: Schematic drawing of typical components of the endothelial glycocalyx. A glycoprotein and two types of heparan sulphate proteoglycans (HSPG) belonging to the syndecan and glypican families are shown. Glycoproteins (e.g. selectins, integrins, members of the immunoglobulin superfamily) are characterised by short and branched carbohydrate side chains, while proteoglycans exhibit long unbranched side chains. *Lower panel*: Protein core with a typical insertion sequence and basic structure of the attached glycosaminoglycan chain. The multiple disaccharide units of this chain are partially modified by specific epimerisation and sulphatation, resulting in typical sulphated oligosaccharide motifs of the heparan sulphate type (*lower half*) separated by unchanged regions. The pentasaccharide shown represents the minimal specific binding site for antithrombin III (ATIII), and is thus crucial for the anticoagulatory properties of the glycocalyx. The stretched length of typical glycosaminoglycan side chains is in the order of 80 nm. The *red arrow* shows a typical cleavage site for heparinase (heparin lyase, EC 4.2.2.7). Given at the *bottom* is the basic structural unit of hyaluronic acid (stretched length up to several micrometers) which also belongs to the glycosaminoglycan (GAG) family. Endothelial cells produce hyaluronic acid (Suzuki et al. 2003), which may also be adsorbed from the plasma (Saegusa et al. 2002) to endothelial surface receptors (McCourt et al. 1999; Nandi et al. 2000). *gly*, glycine; *ala*, alanine; *ser*, serine; *Xyl*, xylose; *Gal*, galactose; *GlcA*, glucuronic acid; *IdoA*, iduronic acid; *IodA 2S*, 2-O-sulphated iduronic acid; *GlcNAc*, N-acetylglucosamine; *GlcNAc 6S*, 6-O-sulphated N-acetylglucosamine; *GlcNS 3S*, 3-O-sulphated N-sulphated glucosamine; *GlcNS 3S 6S*, 3-O- and 6-O-sulphated N-sulphated glucosamine

Normal Endothelium

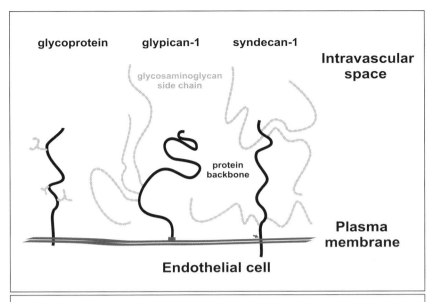

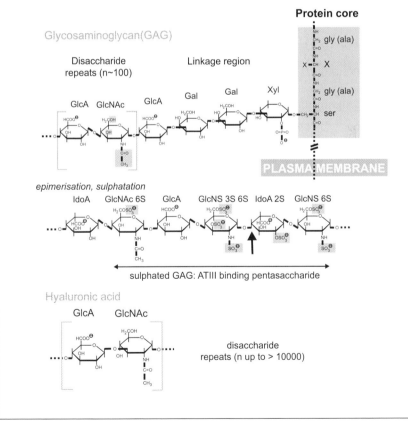

activity. Their extracellular region, with 3–6 GAG attachment sites, has a compact tertiary structure stabilised by 14 invariant cysteine residues (Rosenberg et al. 1997). Based on the molecular weights and the corresponding number of amino acids for the above-mentioned proteins (about 200 to 600), the stretched length of the protein core would range from about 70 to 210 nm. Taking into account the secondary and tertiary structures, the effective length of these molecules is in the same range as the thickness of the glycocalyx as seen in the electron microscope, i.e. 50–100 nm.

Glycosaminoglycans are bound to the respective attachment sites, preferably with a typical tetrasaccharide (GlcA-Gal-Gal-Xyl) acceptor sequence. For syndecans, the specificity for HS is about 60%, the remaining sites being linked to chondroitin sulphate (CS). The HS specificity of glypican is nearly 100% (Rosenberg et al. 1997). Complex and not completely defined mechanisms involving a number of highly specialised enzymes determine the structure of the HS side chains. The process starts by the addition of a repetitive chain of about 100 disaccharide units (GlcA-GlcNAc) to the initial linkage tetrasaccharide. Based on the number of saccharides, a length of about 80 nm can be estimated for a typical HS side chain. The resulting copolymer is then modified by a combination of epimerisation and sulphonisation, resulting in a large variety of HS motifs with different functional properties.

The above-mentioned results indicate the presence of a fairly dense macromolecular coat on the endothelial surface with a thickness of about 50–100 nm, consisting of numerous members of the very diverse class of glycoproteins and a considerable amount of HSPGs (about 10^5–10^6 per cell). This coat is characterised by a significant amount of negative charges at terminal sialyl residues (glycoproteins) and in the sulphated domains of HS side chains (proteoglycans).

3.2
Endothelial Surface Layer

Experimental data on haematocrit levels in the microcirculation and on flow resistance in microvessels led to the concept that the endothelial surface is covered with a stationary layer (endothelial surface layer, ESL) which is much thicker than the glycocalyx described above (Klitzman and Duling 1979; Klitzman and Johnson 1982; Desjardins and Duling 1987; Duling and Desjardins 1987; Pries et al. 1990, 1994). In order to explain the observations, this layer was assumed to exclude red cells and not to allow significant axial flow.

Measurements of the volume fraction of red blood cells ('micro-haematocrit' or 'tube haematocrit', H_T) in capillaries yielded values much lower than the respective systemic haematocrit. For capillaries of the hamster cremaster muscle, Klitzman and Duling (1979) reported H_T values of only about 10% at a systemic haematocrit (H_{SYS}) of 53%, i.e. $H_T/H_{SYS} = 0.19$. This exceeds by far the haematocrit reduction which is to be expected on the basis of the Fahraeus effect

(Albrecht et al. 1979) which describes the relationship between the micro-haematocrit, H_T and the discharge haematocrit H_D (which would be obtained if the blood flowing through the tube was collected at the outflow end; Goldsmith et al. 1989). The reduction of H_T relative to H_D results from the fact that red cells preferentially travel in axial regions of the microvessels and thus the mean red cell velocity (V_{rbc}) exceeds the mean blood velocity (V_{mean}). According to experimental data, the H_T/H_D level for capillary-sized vessels at typical discharge haematocrits varies between about 0.6 and 0.8 (Albrecht et al. 1979; Barbee and Cokelet 1971).

Thus the Fahraeus effect cannot explain the observed H_T/H_{SYS} levels if it is assumed that the discharge haematocrit does not differ substantially from the systemic haematocrit. Distribution of plasma flow and red cell flow in microvascular networks leads to a reduction of mean discharge haematocrit relative to the systemic haematocrit (network Fahraeus effect; Pries et al. 1986). However, this effect is relatively small. In contrast, a stationary layer on the endothelial surface from which red cells are excluded could explain much stronger reductions of micro-haematocrit values. Accordingly, Klitzman, Duling and Desjardins (Klitzman and Duling 1979; Duling and Desjardins 1987) hypothesised that a slow-moving plasma layer with a thickness in the order of 1–1.2 µm was responsible for the low capillary haematocrits observed in their studies.

Additional evidence for the presence of a thick stationary layer on the endothelial surface came from measurements and predictions of flow distribution and flow resistance in microvascular networks (Pries et al. 1990, 1994, 1997). Mathematical flow models based on observed network structures were used to predict flow velocities in individual vessel segments. If values for apparent viscosity as derived from in vitro studies with blood-perfused glass tubes (Pries et al. 1992) were used, the predictions did not agree with experimental observations. However, experimental data could be reconciled with the theoretical predictions if a stationary plasma layer on the endothelial surface with a thickness of about 1.1 µm was assumed (Pries et al. 1994). The same was true for the comparison of experimental determinations of pressure drop across the complete microvascular networks with respective model simulations (Fig. 5): Based on in vitro viscosity findings, an overall pressure drop of only about 24 mmHg was predicted. Assuming the presence of a thick endothelial surface layer (1.1 µm) the pressure drop increased to 54 mmHg, close to the experimental value of 62 mmHg (Pries et al. 1994).

For both the haematocrit and the flow/resistance-based approach, it was shown that discrepancies between experimental findings and theoretical expectations not assuming a thick endothelial surface layer could be reduced by microinfusion of heparinase, which cleaves sugar side chains from proteoglycans and thus partially degrades the cell-bound glycocalyx. Heparinase treatment led to an increase of micro-haematocrit (Desjardins and Duling 1990) and to a decrease in microvascular flow resistance (Pries et al. 1997).

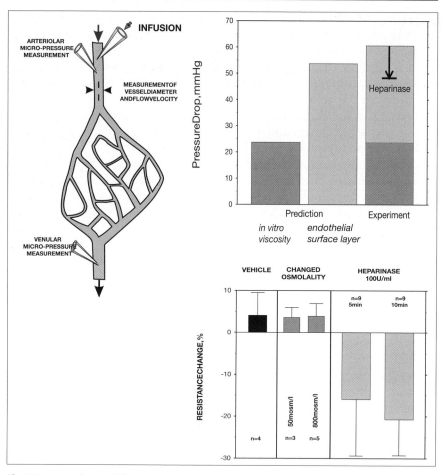

Fig. 5 Pressure drop and flow resistance in microvascular networks. The schematic drawing on the *left* shows the experimental set-up. Pressures were measured in the feeding arteriole and the draining venule of the network. In addition, the volume flow into the network was derived from flow velocity and diameter of the feeding arteriole to allow calculation of flow resistance. The *upper right panel* gives values for pressure differences calculated for three networks of the rat mesentery with a mathematical flow simulation using the in vitro viscosity law (*left bar*) and the in vivo viscosity law (*middle bar*) including the assumption of an endothelial surface layer. (Modified after Pries et al. 1994). The *right bar* gives the mean pressure drop determined by micropuncture before and after microinfusion of heparinase, which cleaves carbohydrate side chains from the glycocalyx. The *lower right panel* shows changes in flow resistance (mean±SE) upon micro-infusion of fluids with different osmolality (Pries et al. 1998b) and heparinase. The degradation of the endothelial surface layer due to heparinase infusion led to a resistance decrease of up to about 20%. According to model simulations, this corresponds to an average reduction in layer thickness of about 0.55 µm (Pries et al. 1997). (Modified after Pries et al. 1997)

These findings were consistent with a significant reduction in the thickness of the endothelial surface layer by about 0.5–1 μm. Direct evidence for the presence of an ESL was obtained by experiments in which the free-flowing plasma was fluorescently stained by fluorescein isothiocyanate (FITC)-dextran (Vink and Duling 1996). In intravital investigations of hamster cremaster capillaries, Vink and Duling found that the width of the labelled plasma column was 0.8–1.0 μm smaller than the diameter of the vessel ('anatomical diameter'), as judged from the estimated position of the endothelial cell surface. This corresponds to a layer thickness of about 0.4–0.5 μm.

Recently, the introduction of microparticle image velocimetry (μPIV) (Smith et al. 2003; Long et al. 2004) to intravital microscopy opened the possibility of directly assessing the hydrodynamically effective thickness of the ESL in medium-sized microvessels (∼20–60 μm). For venules with diameters between 30 and 50 μm the ESL thickness reported varied between about 0.5 and 0.8 μm. From the results of a clinical study with double tracer infusion (indocyanine green, stained autologous erythrocytes), Rehm et al. (2001) estimated the pre-treatment total body volume of the ESL to be in the range of 720 ml. For a total endothelial surface area of about 350 m^2, this corresponds to an average layer thickness of about 2 μm, showing again that the ESL is much thicker than the glycocalyx constituted by molecules directly bound to the endothelial plasma membrane.

The presence of a thick layer on endothelial cells will have a powerful effect on the microhaemodynamics at the endothelial surface. According to the current concepts of the ESL, shear stress is transmitted to the endothelial cell surface by the cell-bound molecules of the glycocalyx, while fluid shear stresses on endothelial cell membranes are minimal (Secomb et al. 2001a). The mechanotransduction in endothelial cells has been explained using a model with two signalling pathways in response to fluid shear stress (Thi et al. 2004) related to the torque effected on the endothelial cell via anchoring points of the glycocalyx, and to focal adhesions and stress fibres.

The mechanical properties of the layer furthermore lead to a strong attenuation of fast fluctuations in shear stress on the endothelial surface (Secomb et al. 2001a). The presence of the ESL tends to smooth the inner capillary surface and thus reduces the importance of capillary irregularities for flow resistance. It will also attenuate fast changes in shear forces experienced by red blood cells traversing irregular capillaries (Secomb et al. 2002), which in turn may increase red cell survival.

3.2.1
Composition of the Endothelial Surface Layer

Up to now, the precise chemical, structural and physical properties of the endothelial surface layer are not known. However, the available studies contain a number of observations which hint at certain properties and components:

- The layer is substantially degraded by treatment with heparinase (Desjardins and Duling 1990; Pries et al. 1997; Vogel et al. 2000), indicating that side chains of HS proteoglycans play a major role in the integrity of the ESL.
- The accessibility of the layer for macromolecules is substantially altered by hyaluronidase treatment (Henry and Duling 1999), suggesting a central involvement of hyaluronic acid in the ESL composition.
- The thickness of the layer is modified by changes in the plasma composition due to infusion of artificial fluids (Vink and Duling 1996; Pries et al. 1998a; Rehm et al. 2001; Long et al. 2004). Thus, components of the ESL seem to be in a dynamic exchange with the free-flowing plasma.
- The layer excludes flowing red cells but not white cells or stationary red cells (Vink and Duling 1996). After passage of a white cell, the layer recovers after about 1 s.
- The difficulties in visualising the layer, e.g. by changes in the refractive index together with theoretical analyses of its mechanical properties (Damiano 1998; Secomb et al. 1998; Weinbaum et al. 2003), indicate that the layer probably consists of a very dilute matrix with a concentration of macromolecules not very much higher than that of free-flowing plasma.

These observations led to the generation of conceptual models for the composition of the ESL (Pries et al. 2000; Platts et al. 2003) which are represented in Fig. 6. The further development and application of new experimental approaches, such as µPIV (Smith et al. 2003; Long et al. 2004), and new imaging

Fig. 6 Concepts for the composition of the endothelial surface layer. *Left*: The endothelial surface layer with a total thickness of up to a micrometer is composed of two zones. The glycocalyx proper, i.e. the comparatively thin (50–100 nm) region on the endothelial surface is dominated by molecules (glycoproteins and proteoglycans) bound directly to the endothelial plasma membrane. A much thicker layer, consisting of a complex three-dimensional array of soluble plasma components possibly including a variety of proteins, glycosaminoglycans and hyaluronan, is attached to the glycocalyx. Components of this layer are dynamically exchanged with the flowing plasma. The thickness and composition of the surface layer depend on the plasma composition, the local haemodynamic conditions and the functional state of the endothelium (Modified according to Pries et al. 2000). *Right*: A different concept was proposed by Duling and co-workers. Here, hyaluronan (hyaluronic acid) produced in the endothelial plasma membrane, or bound to it, plays a more important role. The entire layer is labelled 'glycocalyx' which, however, may exhibit different properties at different distances from the endothelial cell. (Modified after Platts et al. 2003)

Normal Endothelium

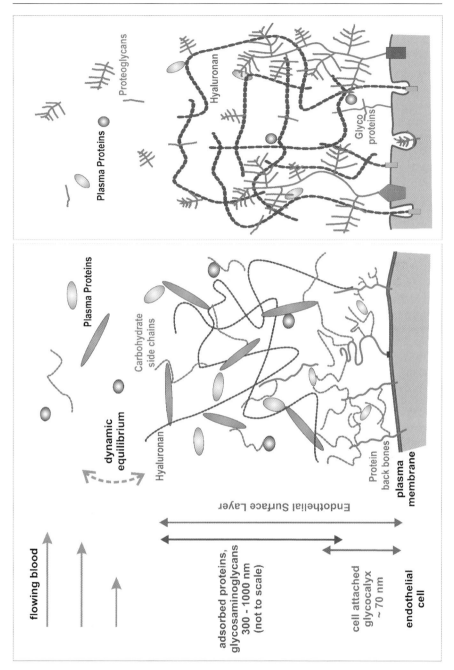

techniques, such as multiphoton imaging and second harmonic imaging, are needed to distinguish between the different concepts and, more importantly, to allow an analysis of the dynamic changes in composition and properties of the ESL in different functional states and pathophysiological conditions.

3.3
Physiological and Clinical Impact

The glycocalyx and ESL constitute the first line of the blood/tissue interface and are thus involved in a substantial number of physiological and pathophysiological processes (Pries et al. 2000), including many of the functionally relevant aspects of the endothelium addressed in other chapters of this book.
These include:

- Transport along vessels due to the effects of the ESL on haematocrit and flow resistance (Klitzman and Duling 1979; Desjardins and Duling 1987; 1990; Pries et al. 1997)
- Mechanical stress on blood cells and the endothelium due to the damping mechanical properties of the ESL (Damiano 1998; Secomb et al. 2001b; 2002; Thi et al. 2004)
- Regulation of vascular tone due to the effect of the ESL on mechanosensitivity of the endothelium (Secomb et al. 2001a; Thi et al. 2004)
- Exchange across the endothelium and control of tissue fluid content (oedema) due to the central role of the ESL and the glycocalyx proper for permeability of the vessel wall for different substances (Henry and Duling 1999; Squire et al. 2001; Dull et al. 2003; van den Berg et al. 2003; Ueda et al. 2004; Rehm et al. 2004)
- Coagulation due to (1) the physicochemical properties of ESL and glycocalyx proper, (2) the presence of specific receptors and activators and (3) its content of heparan sulphate proteoglycans (HSPGs) with anticoagulatory potency (Benedict et al. 1994; Lijnen and Collen 1997; Platts and Duling 2004)
- Blood cell/endothelium interaction and inflammation due to the influence it exerts on the presentation or accessibility of specific adhesion molecules (Zhao et al. 2001; Mulivor and Lipowsky 2002; Constantinescu et al. 2003)
- Angiogenesis and angioadaptation due to the strong mutual interaction of ESL and glycocalyx components, especially HSPGs, with the production, localisation and activity of growth factors (Klagsbrun 1992; Brown et al. 1996; Sasaki et al. 1999; Iozzo and San Antonio 2001; Pieper et al. 2002)
- Cancer and metastasis due to the control the ESL exerts on angiogenesis and on the interaction of embolised tumour cells with the endothelium

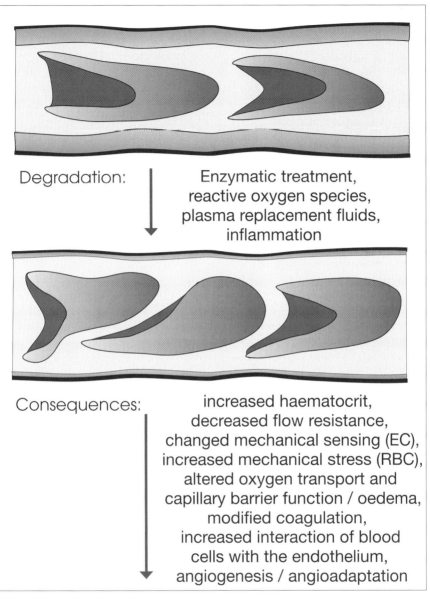

Fig. 7 The thickness of the endothelial surface can be reduced by experimental measures, e.g. enzymatic treatment, but also in the context of pathophysiological events (e.g. inflammation or accumulation of reactive oxygen species) as well as during therapeutic interventions (e.g. infusion of artificial plasma replacement fluids). Reduced thickness of the ESL, in turn, will have significant corollaries on haemodynamic and functional parameters. (Modified after Pries et al. 1997)

(Kishibe et al. 2000; Sanderson 2001; Liu et al. 2002; Kim et al. 2003; Qiao et al. 2003; Xu et al. 2003; Reiland et al. 2004)

In turn, a number of physiological, pathophysiological and therapeutic mechanisms may influence the thickness, composition and integrity of the ESL (Desjardins and Duling 1990; Adamson and Clough 1992; Ward and Donnelly 1993; Beresewicz et al. 1998; Pries et al. 1998a; Constantinescu et al. 2001; Rehm et al. 2001; Platts et al. 2003; Platts and Duling 2004; Fig. 7). Such factors include oxidised low-density lipoproteins, adenosine, growth factors (since, for example, fibroblast growth factor-2 and transforming growth factor-β1 increase the expression of HSPGs), hypoxia, ischaemia-reperfusion, changes in plasma composition (for example, by infusion of artificial plasma replacement fluids) and enzymes degrading ESL or glycocalyx components.

References

Abbott NJ (2002) Astrocyte-endothelial interactions and blood-brain barrier permeability. J Anat 200:629–638

Adams RH, Wilkinson GA, Weiss C, et al (1999) Roles of ephrinB ligands and EphB receptors in cardiovascular development: demarcation of arterial/venous domains, vascular morphogenesis, and sprouting angiogenesis. Genes Dev 13:295–306

Adamson RH, Clough G (1992) Plasma proteins modify the endothelial cell glycocalyx of frog mesenteric microvessels. J Physiol (Lond) 445:473–486

Adeagbo AS (1997) Endothelium-derived hyperpolarizing factor: characterization as a cytochrome P450 1A-linked metabolite of arachidonic acid in perfused rat mesenteric prearteriolar bed. Am J Hypertens 10:763–771

Aird WC (2003) Endothelial cell heterogeneity. Crit Care Med 31:S221-S230

Aird WC, Edelberg JM, Weiler-Guettler H, Simmons WW, Smith TW, Rosenberg RD (1997) Vascular bed-specific expression of an endothelial cell gene is programmed by the tissue microenvironment. J Cell Biol 138:1117–1124

Albrecht KH, Gaehtgens P, Pries AR, Heuser M (1979) The Fahraeus effect in narrow capillaries (i.d. 3.3 to 11.0 µm). Microvasc Res 18:33–47

Anderson AO, Shaw S (1993) T cell adhesion to endothelium: the FRC conduit system and other anatomic and molecular features which facilitate the adhesion cascade in lymph node. Semin Immunol 5:271–282

Baekkevold ES, Yamanaka T, Palframan RT, et al (2001) The CCR7 ligand elc (CCL19) is transcytosed in high endothelial venules and mediates T cell recruitment. J Exp Med 193:1105–1112

Baldwin AL, Winlove CP (1984) Effects of perfusate composition on binding of ruthenium red and gold colloid to glycocalyx of rabbit aortic endothelium. J Histochem Cytochem 32:259–266

Ballabh P, Braun A, Nedergaard M (2004) The blood-brain barrier: an overview: structure, regulation, and clinical implications. Neurobiol Dis 16:1–13

Barbee JH, Cokelet GR (1971) The Fahraeus effect. Microvasc Res 3:6–16

Bauersachs J, Popp R, Fleming I, Busse R (1997) Nitric oxide and endothelium-derived hyperpolarizing factor: formation and interactions. Prostaglandins Leukot Essent Fatty Acids 57:439–446

Baumheter S, Singer MS, Henzel W, et al (1993) Binding of L-selectin to the vascular sialomucin CD34. Science 262:436–438

Bechard D, Meignin V, Scherpereel A, et al (2000) Characterization of the secreted form of endothelial-cell-specific molecule 1 by specific monoclonal antibodies. J Vasc Res 37:417–425

Benedict CR, Pakala R, Willerson JT (1994) Endothelial-dependent procoagulant and anticoagulant mechanisms. Recent advances in understanding. Tex Heart Inst J 21:86–90

Benett HS, Luft JH, Hampton JC (1959) Morphological classifications of vertebrate blood capillaries. Am J Physiol 196:381–390

Beresewicz A, Czarnowska E, Maczewski M (1998) Ischemic preconditioning and superoxide dismutase protect against endothelial dysfunction and endothelium glycocalyx disruption in the postischemic guinea-pig hearts. Mol Cell Biochem 186:87–97

Berg EL, McEvoy LM, Berlin C, Bargatze RF, Butcher EC (1993) L-selectin-mediated lymphocyte rolling on MAdCAM-1. Nature 366:695–698

Bernfield M, Kokenyesi R, Kato M, et al (1992) Biology of the syndecans: a family of transmembrane heparan sulfate proteoglycans. Annu Rev Cell Biol 8:365–393

Betz AL, Goldstein GW (1978) Polarity of the blood-brain barrier: neutral amino acid transport into isolated brain capillaries. Science 202:225–227

Betz AL, Firth JA, Goldstein GW (1980) Polarity of the blood-brain barrier: distribution of enzymes between the luminal and antiluminal membranes of brain capillary endothelial cells. Brain Res 192:17–28

Bianchi C, Sellke FW, Del Vecchio RL, Tonks NK, Neel BG (1999) Receptor-type proteintyrosine phosphatase mu is expressed in specific vascular endothelial beds in vivo. Exp Cell Res 248:329–338

Braet F, Wisse E (2002) Structural and functional aspects of liver sinusoidal endothelial cell fenestrae: a review. Comp Hepatol 1:1

Braet F, De Zanger R, Jans D, Spector I, Wisse E (1996) Microfilament-disrupting agent latrunculin A induces and increased number of fenestrae in rat liver sinusoidal endothelial cells: comparison with cytochalasin B. Hepatology 24:627–635

Breier G, Albrecht U, Sterrer S, Risau W (1992) Expression of vascular endothelial growth factor during embryonic angiogenesis and endothelial cell differentiation. Development 114:521–532

Brightman MW, Reese TS (1969) Junctions between intimately apposed cell membranes in the vertebrate brain. J Cell Biol 40:648–677

Brooks PC, Clark RA, Cheresh DA (1994) Requirement of vascular integrin alpha v beta 3 for angiogenesis. Science 264:569–571

Brown MD, Egginton S, Hudlicka O, Zhou AL (1996) Appearance of the capillary endothelial glycocalyx in chronically stimulated rat skeletal muscles in relation to angiogenesis. Exp Physiol 81:1043–1046

Cahill RN, Frost H, Trnka Z (1976) The effects of antigen on the migration of recirculating lymphocytes through single lymph nodes. J Exp Med 143:870–888

Campbell JJ, Hedrick J, Zlotnik A, Siani MA, Thompson DA, Butcher EC (1998) Chemokines and the arrest of lymphocytes rolling under flow conditions. Science 279:381–384

Chambers R, Zweifach BW (1947) Intercellular cement and capillary permeability. Physiol Rev 27:436–463

Chetham PM, Babal P, Bridges JP, Moore TM, Stevens T (1999) Segmental regulation of pulmonary vascular permeability by store-operated Ca(2+) entry. Am J Physiol 276:L41-L50

Clough G, Moffitt H (1992) Immunoperoxidase labelling of albumin at the endothelial cell surface of frog mesenteric microvessels. Int J Microcirc Clin Exp 11:345–358

Cohnheim J (1867) Über Entzündung und Eiterung. Virchows Arch Patholo Anat Physiol Klin Med 40:1–79

Constantinescu AA, Vink H, Spaan JA (2001) Elevated capillary tube hematocrit reflects degradation of endothelial cell glycocalyx by oxidized LDL. Am J Physiol 280:H1051-H1057

Constantinescu AA, Vink H, Spaan JA (2003) Endothelial cell glycocalyx modulates immobilization of leukocytes at the endothelial surface. Arterioscler Thromb Vasc Biol 23:1541–1547

Cotran RS, Gimbrone MA Jr, Bevilacqua MP, Mendrick DL, Pober JS (1986) Induction and detection of a human endothelial activation antigen in vivo. J Exp Med 164:661–666

Damiano ER (1998) The effect of the endothelial-cell glycocalyx on the motion of red blood cells through capillaries. Microvasc Res 55:77–91

Danielli JF (1940) Capillary permeability and edema in the perfused frog. J Physiol (Lond) 98:109–129

Davson H, Oldendorf WH (1967) Symposium on membrane transport. Transport in the central nervous system. Proc R Soc Med 60:326–329

De Mey JG, Vanhoutte PM (1982) Heterogeneous behavior of the canine arterial and venous wall. Importance of the endothelium. Circ Res 51:439–447

deMello DE, Reid LM (2000) Embryonic and early fetal development of human lung vasculature and its functional implications. Pediatr Dev Pathol 3:439–449

deMello DE, Sawyer D, Galvin N, Reid LM (1997) Early fetal development of lung vasculature. Am J Respir Cell Mol Biol 16:568–581

Desjardins C, Duling BR (1987) Microvessel hematocrit: measurement and implications for capillary oxygen transport. Am J Physiol 252:H494-H503

Desjardins C, Duling BR (1990) Heparinase treatment suggests a role for the endothelial cell glycocalyx in regulation of capillary hematocrit. Am J Physiol 258:H647-H654

Dobrogowska DH, Vorbrodt AW (2004) Immunogold localization of tight junctional proteins in normal and osmotically-affected rat blood-brain barrier. J Mol Histol 35:529–539

Dorling A (2003) Are anti-endothelial cell antibodies a pre-requisite for the acute vascular rejection of xenografts? Xenotransplantation 10:16–23

Duling BR, Desjardins C (1987) Capillary hematocrit-what does it mean? News Physiol Sci 2:66–69

Dull RO, Dinavahi R, Schwartz L, et al (2003) Lung endothelial heparan sulfates mediate cationic peptide-induced barrier dysfunction: a new role for the glycocalyx. Am J Physiol Lung Cell Mol Physiol 285:L986-L995

Elble RC, Widom J, Gruber AD, et al (1997) Cloning and characterization of lung-endothelial cell adhesion molecule-1 suggest it is an endothelial chloride channel. J Biol Chem 272:27853–27861

Eppihimer MJ, Wolitzky B, Anderson DC, Labow MA, Granger DN (1996) Heterogeneity of expression of E- and P-selectins in vivo. Circ Res 79:560–569

Esmon CT, Fukudome K, Mather T, et al (1999) Inflammation, sepsis, and coagulation. Haematologica 84:254–259

Farrell CL, Pardridge WM (1991) Blood-brain barrier glucose transporter is asymmetrically distributed on brain capillary endothelial lumenal and ablumenal membranes: an electron microscopic immunogold study. Proc Natl Acad Sci U S A 88:5779–5783

Furchgott RF (1983) Role of endothelium in responses of vascular smooth muscle. Circ Res 53:558–573

Furchgott RF, Zawadzki JV (1980) The obligatory role of endothelial cells in the relaxation of arterial smooth muscle by acetylcholine. Nature 288:373–376

Gatmaitan Z, Varticovski L, Ling L, Mikkelsen R, Steffan AM, Arias IM (1996) Studies on fenestral contraction in rat liver endothelial cells in culture. Am J Pathol 148:2027–2041

Gawaz M, Neumann FJ, Dickfeld T, et al (1997) Vitronectin receptor (alpha(v)beta3) mediates platelet adhesion to the luminal aspect of endothelial cells: implications for reperfusion in acute myocardial infarction. Circulation 96:1809–1818

Gerety SS, Wang HU, Chen ZF, Anderson DJ (1999) Symmetrical mutant phenotypes of the receptor EphB4 and its specific transmembrane ligand ephrin-B2 in cardiovascular development. Mol Cell 4:403–414

Girard JP, Springer TA (1995) High endothelial venules (HEVs): specialized endothelium for lymphocyte migration. Immunol Today 16:449–457

Florey L (1996) The endothelial cell. Br Med J 5512:487–490

Goldsmith HL, Cokelet GR, Gaehtgens P (1989) Robin Fahraeus: evolution of his concepts in cardiovascular physiology. Am J Physiol 257:H1005-H1015

Gowans JL, Knight EJ (1964) The route of re-circulation of lymphocytes in the rat. Proc R Soc Lond B Biol Sci 159:257–282

Graier WF, Holzmann S, Hoebel BG, Kukovetz WR, Kostner GM (1996) Mechanisms of L-NG nitroarginine/indomethacin-resistant relaxation in bovine and porcine coronary arteries. Br J Pharmacol 119:1177–1186

Gunn MD, Tangemann K, Tam C, Cyster JG, Rosen SD, Williams LT (1998) A chemokine expressed in lymphoid high endothelial venules promotes the adhesion and chemotaxis of naive T lymphocytes. Proc Natl Acad Sci U S A 95:258–263

Haldenby KA, Chappell DC, Winlove CP, Parker KH, Firth JA (1994) Focal and regional variations in the composition of the glycocalyx of large vessel endothelium. J Vasc Res 31:2–9

Hamm S, Dehouck B, Kraus J, et al (2004) Astrocyte mediated modulation of blood-brain barrier permeability does not correlate with a loss of tight junction proteins from the cellular contacts. Cell Tissue Res 315:157–166

Hargreaves KM, Pardridge WM (1988) Neutral amino acid transport at the human blood-brain barrier. J Biol Chem 263:19392–19397

Henry CB, Duling BR (1999) Permeation of the luminal capillary glycocalyx is determined by hyaluronan. Am J Physiol 277:H508-H514

Herzog Y, Kalcheim C, Kahane N, Reshef R, Neufeld G (2001) Differential expression of neuropilin-1 and neuropilin-2 in arteries and veins. Mech Dev 109:115–119

Higashi N, Ueda H, Yamada O, et al (2002) Micromorphological characteristics of hepatic sinusoidal endothelial cells and their basal laminae in five different animal species. Okajimas Folia Anat Jpn 79:135–142

Hirakawa H, Okajima S, Nagaoka T, Kubo T, Takamatsu T, Oyamada M (2004) Regional differences in blood-nerve barrier function and tight-junction protein expression within the rat dorsal root ganglion. Neuroreport 15:405–408

Hirase T, Staddon JM, Saitou M, et al (1997) Occludin as a possible determinant of tight junction permeability in endothelial cells. J Cell Sci 110:1603–1613

Hjalmarsson C, Johansson BR, Haraldsson B (2004) Electron microscopic evaluation of the endothelial surface layer of glomerular capillaries. Microvasc Res 67:9–17

Houser SL, Benjamin LC, Wain JC, Madsen JC, Allan JS (2004) Constitutive expression of major histocompatibility complex class II antigens in pulmonary epithelium and endothelium varies among different species. Transplantation 77:605–607

Iigo Y, Suematsu M, Higashida T, et al (1997) Constitutive expression of ICAM-1 in rat microvascular systems analyzed by laser confocal microscopy. Am J Physiol 273:H138-H147

Iozzo RV, San Antonio JD (2001) Heparan sulfate proteoglycans: heavy hitters in the angiogenesis arena. J Clin Invest 108:349–355

Ito S (1974) Form and function of the glycocalyx on free cell surfaces. Philos Trans R Soc Lond B Biol Sci 268:55–66

Janzer RC, Raff MC (1987) Astrocytes induce blood-brain barrier properties in endothelial cells. Nature 325:253–257

Jean JC, Eruchalu I, Cao YX, Joyce-Brady M (2002) DANCE in developing and injured lung. Am J Physiol Lung Cell Mol Physiol 282:L75-L82

Joo F (1996) Endothelial cells of the brain and other organ systems: some similarities and differences. Prog Neurobiol 48:255–273

Kacem K, Lacombe P, Seylaz J, Bonvento G (1998) Structural organization of the perivascular astrocyte endfeet and their relationship with the endothelial glucose transporter: a confocal microscopy study. Glia 23:1–10

Kelly JJ, Moore TM, Babal P, Diwan AH, Stevens T, Thompson WJ (1998) Pulmonary microvascular and macrovascular endothelial cells: differential regulation of Ca2+ and permeability. Am J Physiol 274:L810-L819

Kim H, Xu GL, Borczuk AC, et al (2003) The heparan sulfate proteoglycan GPC3 is a potential lung tumor suppressor. Am J Respir Cell Mol Biol 29:694–701

King J, Hamil T, Creighton J, et al (2004) Structural and functional characteristics of lung macro- and microvascular endothelial cell phenotypes. Microvasc Res 67:139–151

Kishibe J, Yamada S, Okada Y, et al (2000) Structural requirements of heparan sulfate for the binding to the tumor-derived adhesion factor/angiomodulin that induces cord-like structures to ECV-304 human carcinoma cells. J Biol Chem 275:15321–15329

Kjellstrom BT, Ortenwall P, Risberg B (1987) Comparison of oxidative metabolism in vitro in endothelial cells from different species and vessels. J Cell Physiol 132:578–580

Klagsbrun M (1992) Mediators of angiogenesis: the biological significance of basic fibroblast growth factor (bFGF)-heparin and heparan sulfate interactions. Semin Cancer Biol 3:81–87

Klitzman B, Duling BR (1979) Microvascular hematocrit and red cell flow in resting and contracting striated muscle. Am J Physiol 237:H481-H490

Klitzman B, Johnson PC (1982) Capillary network geometry and red cell distribution in hamster cremaster muscle. Am J Physiol 242:H211-H219

Kuebler WM (2005) Pressure-induced inflammatory signaling in lung endothelial cells. In: Bhattacharya J (ed) Cell signaling in vascular inflammation. Humana Press, Totowa, pp 61–71

Kuebler WM, Goetz AE (2002) The marginated pool. Eur Surg Res 34:92–100

Kuebler WM, Kuhnle GE, Groh J, Goetz AE (1994) Leukocyte kinetics in pulmonary microcirculation: intravital fluorescence microscopic study. J Appl Physiol 76:65–71

Kuebler WM, Kuhnle GE, Groh J, Goetz AE (1997) Contribution of selectins to leucocyte sequestration in pulmonary microvessels by intravital microscopy in rabbits. J Physiol 501:375–386

Kuebler WM, Ying X, Singh B, Issekutz AC, Bhattacharya J (1999) Pressure is proinflammatory in lung venular capillaries. J Clin Invest 104:495–502

Kuebler WM, Ying X, Bhattacharya J (2002) Pressure-induced endothelial Ca(2+) oscillations in lung capillaries. Am J Physiol Lung Cell Mol Physiol 282:L917-L923

Kuo L, Chilian WM, Davis MJ (1991) Interaction of pressure- and flow-induced responses in porcine coronary resistance vessels. Am J Physiol 261:H1706-H1715

Kuo L, Arko F, Chilian WM, Davis MJ (1993) Coronary venular responses to flow and pressure. Circ Res 72:607–615

Lasky LA, Singer MS, Dowbenko D, et al (1992) An endothelial ligand for L-selectin is a novel mucin-like molecule. Cell 69:927–938
Lassalle P, Molet S, Janin A, et al (1996) ESM-1 is a novel human endothelial cell-specific molecule expressed in lung and regulated by cytokines. J Biol Chem 271:20458–20464
le Noble F, Moyon D, Pardanaud L, et al (2004) Flow regulates arterial-venous differentiation in the chick embryo yolk sac. Development 131:361–375
le Noble F, Fleury V, Pries A, Corvol P, Eichmann A, Reneman RS (2005) Control of arterial branching morphogenesis in embryogenesis: go with the flow. Cardiovasc Res 65:619–628
Leabu M, Ghinea N, Muresan V, Colceag J, Hasu M, Simionescu N (1987) Cell surface chemistry of arterial endothelium and blood monocytes in the normolipidemic rabbit. J Submicrosc Cytol 19:193–208
Lee WJ, Hawkins RA, Peterson DR, Vina JR (1996) Role of oxoproline in the regulation of neutral amino acid transport across the blood-brain barrier. J Biol Chem 271:19129–19133
Lee WJ, Hawkins RA, Vina JR, Peterson DR (1998) Glutamine transport by the blood-brain barrier: a possible mechanism for nitrogen removal. Am J Physiol 274:C1101-C1107
Ley K (1996) Molecular mechanisms of leukocyte recruitment in the inflammatory process. Cardiovasc Res 32:733–742
Ley K, Gaehtgens P (1991) Endothelial, not hemodynamic, differences are responsible for preferential leukocyte rolling in rat mesenteric venules. Circ Res 69:1034–1041
Liebner S, Fischmann A, Rascher G, et al (2000) Claudin-1 and claudin-5 expression and tight junction morphology are altered in blood vessels of human glioblastoma multiforme. Acta Neuropathol (Berl) 100:323–331
Lijnen HR, Collen D (1997) Endothelium in hemostasis and thrombosis. Prog Cardiovasc Dis 39:343–350
Liu D, Shriver Z, Qi Y, Venkataraman G, Sasisekharan R (2002) Dynamic regulation of tumor growth and metastasis by heparan sulfate glycosaminoglycans. Semin Thromb Hemost 28:67–78
Long DS, Smith ML, Pries AR, Ley K, Damiano ER (2004) Microviscometry reveals reduced blood viscosity and altered shear rate and shear stress profiles in microvessels after hemodilution. Proc Natl Acad Sci U S A 101:10060–10065
Lowe JB (2002) Glycosylation in the control of selectin counter-receptor structure and function. Immunol Rev 186:19–36
Luft JH (1966) Fine structure of capillary and endocapillary layer as revealed by ruthenium red. Fed Proc 25:1773–1783
M'Rini C, Cheng G, Schweitzer C, et al (2003) A novel endothelial L-selectin ligand activity in lymph node medulla that is regulated by alpha(1,3)-fucosyltransferase-IV. J Exp Med 198:1301–1312
Mall JP (1888) Die Blut- und Lymphwege im Dünndarm des Hundes. Abh Math-Phys Kl Koniglich Saech Gessellschaft Wiss 24:151–189
Manoonkitiwongsa PS, Schultz RL, Wareesangtip W, Whitter EF, Nava PB, McMillan PJ (2000) Luminal localization of blood-brain barrier sodium, potassium adenosine triphosphatase is dependent on fixation. J Histochem Cytochem 48:859–865
Marchesi VT, Gowans JL (1964) The migration of lymphocytes through endothelium of venules in lymph nodes: an electron microscope study. Proc R Soc Lond B Biol Sci 159:283–290
Matter K, Balda MS (2003) Holey barrier: claudins and the regulation of brain endothelial permeability. J Cell Biol 161:459–460

Mayrovitz HN, Tuma RF, Wiedeman MP (1980) Leukocyte adherence in arterioles following extravascular tissue trauma. Microvasc Res 20:264–274

McAllister MS, Krizanac-Bengez L, Macchia F, et al (2001) Mechanisms of glucose transport at the blood-brain barrier: an in vitro study. Brain Res 904:20–30

McCourt PA, Smedsrod BH, Melkko J, Johansson S (1999) Characterization of a hyaluronan receptor on rat sinusoidal liver endothelial cells and its functional relationship to scavenger receptors. Hepatology 30:1276–1286

McDonald DM (1994) Endothelial gaps and permeability of venules in rat tracheas exposed to inflammatory stimuli. Am J Physiol 266:L61-L83

McEver RP, Beckstead JH, Moore KL, Marshall-Carlson L, Bainton DF (1989) GMP-140, a platelet alpha-granule membrane protein, is also synthesized by vascular endothelial cells and is localized in Weibel-Palade bodies. J Clin Invest 84:92–99

McGale EH, Pye IF, Stonier C, Hutchinson EC, Aber GM (1977) Studies of the interrelationship between cerebrospinal fluid and plasma amino acid concentrations in normal individuals. J Neurochem 29:291–297

Mebius RE, Streeter PR, Breve J, Duijvestijn AM, Kraal G (1991) The influence of afferent lymphatic vessel interruption on vascular addressin expression. J Cell Biol 115:85–95

Mebius RE, Dowbenko D, Williams A, Fennie C, Lasky LA, Watson SR (1993) Expression of GlyCAM-1, an endothelial ligand for L-selectin, is affected by afferent lymphatic flow. J Immunol 151:6769–6776

Michie SA, Streeter PR, Bolt PA, Butcher EC, Picker LJ (1993) The human peripheral lymph node vascular addressin. An inducible endothelial antigen involved in lymphocyte homing. Am J Pathol 143:1688–1698

Mikawa T, Fischman DA (1992) Retroviral analysis of cardiac morphogenesis: discontinuous formation of coronary vessels. Proc Natl Acad Sci U S A 89:9504–9508

Milici AJ, Porter GA (1991) Lectin and immunolabeling of microvascular endothelia. J Electron Microsc Tech 19:305–315

Milici AJ, Furie MB, Carley WW (1985) The formation of fenestrations and channels by capillary endothelium in vitro. Proc Natl Acad Sci U S A 82:6181–6185

Miller S, Walker SW, Arthur JR, et al (2002) Selenoprotein expression in endothelial cells from different human vasculature and species. Biochim Biophys Acta 1588:85–93

Miller VM, Vanhoutte PM (1986) Endothelium-dependent responses in isolated blood vessels of lower vertebrates. Blood Vessels 23:225–235

Moldobaeva A, Wagner EM (2002) Heterogeneity of bronchial endothelial cell permeability. Am J Physiol Lung Cell Mol Physiol 283:L520-L527

Moncada S, Herman AG, Higgs EA, Vane JR (1977) Differential formation of prostacyclin (PGX or PGI2) by layers of the arterial wall. An explanation for the anti-thrombotic properties of vascular endothelium. Thromb Res 11:323–344

Montreuil J, Bouquelet S, Debray H, Fournet B, Spik G, Strecker G (1986) Glycoproteins. In: Chaplin MF, Kennedy JF (eds) Carbohydrate analysis. A practical approach. IRL Press, Washington, pp 143–203

Morita K, Sasaki H, Furuse M, Tsukita S (1999) Endothelial claudin: claudin-5/TMVCF constitutes tight junction strands in endothelial cells. J Cell Biol 147:185–194

Moyon D, Pardanaud L, Yuan L, Breant C, Eichmann A (2001a) Plasticity of endothelial cells during arterial-venous differentiation in the avian embryo. Development 128:3359–3370

Moyon D, Pardanaud L, Yuan L, Breant C, Eichmann A (2001b) Selective expression of angiopoietin 1 and 2 in mesenchymal cells surrounding veins and arteries of the avian embryo. Mech Dev 106:133–136

Mueckler M, Caruso C, Baldwin SA, et al (1985) Sequence and structure of a human glucose transporter. Science 229:941–945

Mulivor AW, Lipowsky HH (2002) Role of glycocalyx in leukocyte-endothelial cell adhesion. Am J Physiol Heart Circ Physiol 283:H1282-H1291

Murciano JC, Muro S, Koniaris L, et al (2003) ICAM-directed vascular immunotargeting of antithrombotic agents to the endothelial luminal surface. Blood 101:3977–3984

Nagel T, Resnick N, Atkinson WJ, Dewey CF Jr (1994) Gimbrone MA, Jr. Shear stress selectively upregulates intercellular adhesion molecule-1 expression in cultured human vascular endothelial cells. J Clin Invest 94:885–891

Nandi A, Estess P, Siegelman MH (2000) Hyaluronan anchoring and regulation on the surface of vascular endothelial cells is mediated through the functionally active form of CD44. J Biol Chem 275:14939–14948

Nichols K, Staines W, Rubin S, Krantis A (1994) Distribution of nitric oxide synthase activity in arterioles and venules of rat and human intestine. Am J Physiol 267:G270-G275

Nitta T, Hata M, Gotoh S, et al (2003) Size-selective loosening of the blood-brain barrier in claudin-5-deficient mice. J Cell Biol 161:653–660

Noble LJ, Mautes AE, Hall JJ (1996) Characterization of the microvascular glycocalyx in normal and injured spinal cord in the rat. J Comp Neurol 376:542–556

O'Kane RL, Hawkins RA (2003) Na+-dependent transport of large neutral amino acids occurs at the abluminal membrane of the blood-brain barrier. Am J Physiol Endocrinol Metab 285:E1167-E1173

Oda M, Han JY, Yokomori H (2000) Local regulators of hepatic sinusoidal microcirculation: recent advances. Clin Hemorheol Microcirc 23:85–94

Okada T, Ngo VN, Ekland EH, et al (2002) Chemokine requirements for B cell entry to lymph nodes and Peyer's patches. J Exp Med 196:65–75

Orlowski M, Meister A (1970) The gamma-glutamyl cycle: a possible transport system for amino acids. Proc Natl Acad Sci U S A 67:1248–1255

Othman-Hassan K, Patel K, Papoutsi M, Rodriguez-Niedenfuhr M, Christ B, Wilting J (2001) Arterial identity of endothelial cells is controlled by local cues. Dev Biol 237:398–409

Palmer RM, Ferrige AG, Moncada S (1987) Nitric oxide release accounts for the biological activity of endothelium-derived relaxing factor. Nature 327:524–526

Palmer RMJ, Ashton DS, Moncada S (1988) Vascular endothelial cells synthesize nitric oxide from L-arginine. Nature 333:664–666

Panes J, Perry MA, Anderson DC, et al (1995) Regional differences in constitutive and induced ICAM-1 expression in vivo. Am J Physiol 269:H1955-H1964

Papadopoulos MC, Saadoun S, Woodrow CJ, et al (2001) Occludin expression in microvessels of neoplastic and non-neoplastic human brain. Neuropathol Appl Neurobiol 27:384–395

Pardridge WM (2003) Blood-brain barrier drug targeting: the future of brain drug development. Mol Interv 3:90–105, 51

Pardridge WM, Boado RJ, Farrell CR (1990) Brain-type glucose transporter (GLUT-1) is selectively localized to the blood-brain barrier. Studies with quantitative western blotting and in situ hybridization. J Biol Chem 265:18035–18040

Parker JC, Yoshikawa S (2002) Vascular segmental permeabilities at high peak inflation pressure in isolated rat lungs. Am J Physiol Lung Cell Mol Physiol 283:L1203-L1209

Parthasarathi K, Ichimura H, Quadri S, Issekutz A, Bhattacharya J (2002) Mitochondrial reactive oxygen species regulate spatial profile of proinflammatory responses in lung venular capillaries. J Immunol 169:7078–7086

Pieper JS, Hafmans T, van Wachem PB, et al (2002) Loading of collagen-heparan sulfate matrices with bFGF promotes angiogenesis and tissue generation in rats. J Biomed Mater Res 62:185–194

Platts SH, Duling BR (2004) Adenosine A3 receptor activation modulates the capillary endothelial glycocalyx. Circ Res 94:77–82

Platts SH, Linden J, Duling BR (2003) Rapid modification of the glycocalyx caused by ischemia-reperfusion is inhibited by adenosine A2A receptor activation. Am J Physiol Heart Circ Physiol 284:H2360-H2367

Prat A, Biernacki K, Wosik K, Antel JP (2001) Glial cell influence on the human blood-brain barrier. Glia 36:145–155

Pries AR, Ley K, Gaehtgens P (1986) Generalization of the Fahraeus principle for microvessel networks. Am J Physiol 251:H1324-H1332

Pries AR, Secomb TW, Gaehtgens P, Gross JF (1990) Blood flow in microvascular networks- experiments and simulation. Circ Res 67:826–834

Pries AR, Neuhaus D, Gaehtgens P (1992) Blood viscosity in tube flow: dependence on diameter and hematocrit. Am J Physiol 263:H1770-H1778

Pries AR, Secomb TW, Gessner T, Sperandio MB, Gross JF, Gaehtgens P (1994) Resistance to blood flow in microvessels in vivo. Circ Res 75:904–915

Pries AR, Secomb TW, Jacobs H, Sperandio M, Osterloh K, Gaehtgens P (1997) Microvascular blood flow resistance: role of endothelial surface layer. Am J Physiol 273:H2272-H2279

Pries AR, Secomb TW, Sperandio M, Gaehtgens P (1998a) Blood flow resistance during hemodilution: effect of plasma composition. Cardiovasc Res 37:225–235

Pries AR, Secomb TW, Gaehtgens P (1998b) Structural adaptation and stability of microvascular networks: theory and simulations. Am J Physiol 275:H349-H360

Pries AR, Secomb TW, Gaehtgens P (2000) The endothelial surface layer. Pflugers Arch 440:653–666

Puri KD, Finger EB, Gaudernack G, Springer TA (1995) Sialomucin CD34 is the major L-selectin ligand in human tonsil high endothelial venules. J Cell Biol 131:261–270

Qiao D, Meyer K, Mundhenke C, Drew SA, Friedl A (2003) Heparan sulfate proteoglycans as regulators of fibroblast growth factor-2 signaling in brain endothelial cells. Specific role for glypican-1 in glioma angiogenesis. J Biol Chem 278:16045–16053

Rajantie I, Ekman N, Iljin K, et al (2001) Bmx tyrosine kinase has a redundant function downstream of angiopoietin and vascular endothelial growth factor receptors in arterial endothelium. Mol Cell Biol 21:4647–4655

Rajotte D, Ruoslahti E (1999) Membrane dipeptidase is the receptor for a lung-targeting peptide identified by in vivo phage display. J Biol Chem 274:11593–11598

Rao LV, Pendurthi UR (1998) Tissue factor on cells. Blood Coagul Fibrinolysis 9 Suppl 1:S27-S35

Reese DE, Mikawa T, Bader DM (2002) Development of the coronary vessel system. Circ Res 91:761–768

Reese TS, Karnovsky MJ (1967) Fine structural localization of a blood-brain barrier to exogenous peroxidase. J Cell Biol 34:207–217

Rehm M, Haller M, Orth V, et al (2001) Changes in blood volume and hematocrit during acute preoperative volume loading with 5% albumin or 6% hetastarch solutions in patients before radical hysterectomy. Anesthesiology 95:849–856

Rehm M, Zahler S, Lotsch M, et al (2004) Endothelial glycocalyx as an additional barrier determining extravasation of 6% hydroxyethyl starch or 5% albumin solutions in the coronary vascular bed. Anesthesiology 100:1211–1223

Reiland J, Sanderson RD, Waguespack M, et al (2004) Heparanase degrades syndecan-1 and perlecan heparan sulfate: functional implications for tumor cell invasion. J Biol Chem 279:8047–8055

Rhodin JA (1968) Ultrastructure of mammalian venous capillaries, venules, and small collecting veins. J Ultrastruct Res 25:452–500

Risau W (1995) Differentiation of endothelium. FASEB J 9:926-933

Roberts WG, Palade GE (1995) Increased microvascular permeability and endothelial fenestration induced by vascular endothelial growth factor. J Cell Sci 108:2369-2379

Roberts WG, Palade GE (1997) Neovasculature induced by vascular endothelial growth factor is fenestrated. Cancer Res 57:765-772

Rosenberg RD, Shworak NW, Liu J, Schwartz JJ, Zhang L (1997) Heparan sulfate proteoglycans of the cardiovascular system. Specific structures emerge but how is synthesis regulated? J Clin Invest 100:S67-S75

Rostgaard J, Qvortrup K (1997) Electron microscopic demonstrations of filamentous molecular sieve plugs in capillary fenestrae. Microvasc Res 53:1-13

Saegusa S, Isaji S, Kawarada Y (2002) Changes in serum hyaluronic acid levels and expression of CD44 and CD44 mRNA in hepatic sinusoidal endothelial cells after major hepatectomy in cirrhotic rats. World J Surg 26:694-699

Sanchez del Pino MM, Peterson DR, Hawkins RA (1995) Neutral amino acid transport characterization of isolated luminal and abluminal membranes of the blood-brain barrier. J Biol Chem 270:14913-14918

Sanderson RD (2001) Heparan sulfate proteoglycans in invasion and metastasis. Semin Cell Dev Biol 12:89-98

Sasaki T, Larsson H, Kreuger J, et al (1999) Structural basis and potential role of heparin/heparan sulfate binding to the angiogenesis inhibitor endostatin. EMBO J 18:6240-6248

Sato N, Suzuki Y, Nishio K, et al (2000) Roles of ICAM-1 for abnormal leukocyte recruitment in the microcirculation of bleomycin-induced fibrotic lung injury. Am J Respir Crit Care Med 161:1681-1688

Schinkel AH, Smit JJ, van Tellingen O, et al (1994) Disruption of the mouse mdr1a P-glycoprotein gene leads to a deficiency in the blood-brain barrier and to increased sensitivity to drugs. Cell 77:491-502

Secomb TW, Hsu R, Pries AR (1998) A model for red blood cell motion in glycocalyx-lined capillaries. Am J Physiol 274:H1016-H1022

Secomb TW, Hsu R, Pries AR (2001a) Effect of the endothelial surface layer on transmission of fluid shear stress to endothelial cells. Biorheology 38:143-150

Secomb TW, Hsu R, Pries AR (2001b) Motion of red blood cells in a capillary with an endothelial surface layer: effect of flow velocity. Am J Physiol 281:H629-H636

Secomb TW, Hsu R, Pries AR (2002) Blood flow and red blood cell deformation in nonuniform capillaries: effects of the endothelial surface layer. Microcirculation 9:189-196

Seidel CL, LaRochelle J (1987) Venous and arterial endothelia: different dilator abilities in dog vessels. Circ Res 60:626-630

Shih GC, Hajjar KA (1993) Plasminogen and plasminogen activator assembly on the human endothelial cell. Proc Soc Exp Biol Med 202:258-264

Shutter JR, Scully S, Fan W, et al (2000) Dll4, a novel Notch ligand expressed in arterial endothelium. Genes Dev 14:1313-1318

Sims DE, Horne MM (1993) Non-aqueous fixative preserves macromolecules on the endothelial cell surface: an in situ study. Eur J Morphol 32:59-64

Sims DE, Westfall JA, Kiorpes AL, Horne MM (1991) Preservation of tracheal mucus by nonaqueous fixative. Biotech Histochem 66:173-180

Singh B, Fu C, Bhattacharya J (2000) Vascular expression of the alpha(v)beta(3)-integrin in lung and other organs. Am J Physiol Lung Cell Mol Physiol 278:L217-L226

Smith ML, Long DS, Damiano ER, Ley K (2003) Near-wall micro-PIV reveals a hydrodynamically relevant endothelial surface layer in venules in vivo. Biophys J 85:637-645

Smith PL, Gersten KM, Petryniak B, et al (1996) Expression of the alpha(1,3)fucosyltransferase Fuc-TVII in lymphoid aggregate high endothelial venules correlates with expression of L-selectin ligands. J Biol Chem 271:8250–8259

Springer TA (1990) Adhesion receptors of the immune system. Nature 346:425–434

Springer TA (1995) Traffic signals on endothelium for lymphocyte recirculation and leukocyte emigration. Annu Rev Physiol 57:827–872

Squire JM, Chew M, Nneji G, Neal C, Barry J, Michel C (2001) Quasi-periodic substructure in the microvessel endothelial glycocalyx: a possible explanation for molecular filtering? J Struct Biol 136:239–255

Stein JV, Rot A, Luo Y, et al (2000) The CC chemokine thymus-derived chemotactic agent 4 (TCA-4, secondary lymphoid tissue chemokine, 6Ckine, exodus-2) triggers lymphocyte function-associated antigen 1-mediated arrest of rolling T lymphocytes in peripheral lymph node high endothelial venules. J Exp Med 191:61–76

Stevens T (2002) Bronchial endothelial cell phenotypes and the form:function relationship. Am J Physiol Lung Cell Mol Physiol 283:L518-L519

Stevens T, Fouty B, Hepler L, et al (1997) Cytosolic Ca2+ and adenylyl cyclase responses in phenotypically distinct pulmonary endothelial cells. Am J Physiol 272:L51-L59

Stevens T, Creighton J, Thompson WJ (1999) Control of cAMP in lung endothelial cell phenotypes. Implications for control of barrier function. Am J Physiol 277:L119-L126

Stevens T, Rosenberg R, Aird W, et al (2001) NHLBI workshop report: endothelial cell phenotypes in heart, lung, and blood diseases. Am J Physiol Cell Physiol 281:C1422-C1433

Stewart PA, Wiley MJ (1981) Developing nervous tissue induces formation of blood-brain barrier characteristics in invading endothelial cells: a study using quail-chick transplantation chimeras. Dev Biol 84:183–192

Stoll J, Wadhwani KC, Smith QR (1993) Identification of the cationic amino acid transporter (System y+) of the rat blood-brain barrier. J Neurochem 60:1956–1959

Streeter PR, Rouse BT, Butcher EC (1988) Immunohistologic and functional characterization of a vascular addressin involved in lymphocyte homing into peripheral lymph nodes. J Cell Biol 107:1853–1862

Suzuki K, Yamamoto T, Usui T, Suzuki K, Heldin P, Yamashita H (2003) Expression of hyaluronan synthase in intraocular proliferative diseases: regulation of expression in human vascular endothelial cells by transforming growth factor-beta. Jpn J Ophthalmol 47:557–564

Thi MM, Tarbell JM, Weinbaum S, Spray DC (2004) The role of the glycocalyx in reorganization of the actin cytoskeleton under fluid shear stress: a "bumper-car" model. Proc Natl Acad Sci U S A 101:16483–16488

Thiebaut F, Tsuruo T, Hamada H, Gottesman MM, Pastan I, Willingham MC (1989) Immunohistochemical localization in normal tissues of different epitopes in the multidrug transport protein P170: evidence for localization in brain capillaries and crossreactivity of one antibody with a muscle protein. J Histochem Cytochem 37:159–164

Thurston G, Baluk P, Hirata A, McDonald DM (1996) Permeability-related changes revealed at endothelial cell borders in inflamed venules by lectin binding. Am J Physiol 271:H2547-H2562

Tian H, McKnight SL, Russell DW (1997) Endothelial PAS domain protein 1 (EPAS1), a transcription factor selectively expressed in endothelial cells. Genes Dev 11:72–82

Torres-Vazquez J, Kamei M, Weinstein BM (2003) Molecular distinction between arteries and veins. Cell Tissue Res 314:43–59

Tsukita S, Furuse M, Itoh M (2001) Multifunctional strands in tight junctions. Nat Rev Mol Cell Biol 2:285–293

Uchimura K, Kadomatsu K, El Fasakhany FM, et al (2004) N-acetylglucosamine 6-O-sulfotransferase-1 regulates expression of L-selectin ligands and lymphocyte homing. J Biol Chem 279:35001-35008

Ueda A, Shimomura M, Ikeda M, Yamaguchi R, Tanishita K (2004) Effect of glycocalyx on shear-dependent albumin uptake in endothelial cells. Am J Physiol Heart Circ Physiol 287:H2287-H2294

van den Berg BM, Vink H, Spaan JA (2003) The endothelial glycocalyx protects against myocardial edema. Circ Res 92:592-594

van Zante A, Gauguet JM, Bistrup A, Tsay D, von Andrian UH, Rosen SD (2003) Lymphocyte-HEV interactions in lymph nodes of a sulfotransferase-deficient mouse. J Exp Med 198:1289-1300

Villa N, Walker L, Lindsell CE, Gasson J, Iruela-Arispe ML, Weinmaster G (2001) Vascular expression of Notch pathway receptors and ligands is restricted to arterial vessels. Mech Dev 108:161-164

Vink H, Duling BR (1996) Identification of distinct luminal domains for macromolecules, erythrocytes, and leukocytes within mammalian capillaries. Circ Res 79:581-589

Vogel J, Sperandio M, Pries AR, Linderkamp O, Gaehtgens P, Kuschinsky W (2000) Influence of the endothelial glycocalyx on cerebral blood flow in mice. J Cereb Blood Flow Metab 20:1571-1578

Vorbrodt AW, Dobrogowska DH (2004) Molecular anatomy of interendothelial junctions in human blood-brain barrier microvessels. Folia Histochem Cytobiol 42:67-75

Wagner WH, Henderson RM, Hicks HE, Banes AJ, Johnson G Jr (1988) Differences in morphology, growth rate, and protein synthesis between cultured arterial and venous endothelial cells. J Vasc Surg 8:509-519

Wang HU, Chen ZF, Anderson DJ (1998) Molecular distinction and angiogenic interaction between embryonic arteries and veins revealed by ephrin-B2 and its receptor Eph-B4. Cell 93:741-753

Ward BJ, Donnelly JL (1993) Hypoxia induced disruption of the cardiac endothelial glycocalyx: implications for capillary permeability. Cardiovasc Res 27:384-389

Warnock RA, Askari S, Butcher EC, von Andrian UH (1998) Molecular mechanisms of lymphocyte homing to peripheral lymph nodes. J Exp Med 187:205-216

Weinbaum S, Zhang X, Han Y, Vink H, Cowin SC (2003) Mechanotransduction and flow across the endothelial glycocalyx. Proc Natl Acad Sci U S A 100:7988-7995

Wen H, Watry DD, Marcondes MC, Fox HS (2004) Selective decrease in paracellular conductance of tight junctions: role of the first extracellular domain of claudin-5. Mol Cell Biol 24:8408-8417

Willis CL, Leach L, Clarke GJ, Nolan CC, Ray DE (2004) Reversible disruption of tight junction complexes in the rat blood-brain barrier, following transitory focal astrocyte loss. Glia 48:1-13

Wohlfart P, Dedio J, Wirth K, Scholkens BA, Wiemer G (1997) Different B1 kinin receptor expression and pharmacology in endothelial cells of different origins and species. J Pharmacol Exp Ther 280:1109-1116

Wolburg H, Wolburg-Buchholz K, Kraus J, et al (2003) Localization of claudin-3 in tight junctions of the blood-brain barrier is selectively lost during experimental autoimmune encephalomyelitis and human glioblastoma multiforme. Acta Neuropathol (Berl) 105:586-592

Wu S, Sangerman J, Li M, Brough GH, Goodman SR, Stevens T (2001) Essential control of an endothelial cell ISOC by the spectrin membrane skeleton. J Cell Biol 154:1225-1233

Xu X, Quiros RM, Maxhimer JB, et al (2003) Inverse correlation between heparan sulfate composition and heparanase-1 gene expression in thyroid papillary carcinomas: a potential role in tumor metastasis. Clin Cancer Res 9:5968–5979

Ying X, Minamiya Y, Fu C, Bhattacharya J (1996) Ca2+ waves in lung capillary endothelium. Circ Res 79:898–908

Yoshida K, Sugimoto K (1996) Morphological and cytoskeletal changes in endothelial cells of vein grafts under arterial hemodynamic conditions in vivo. J Electron Microsc (Tokyo) 45:428–435

Zhao Y, Chien S, Weinbaum S (2001) Dynamic contact forces on leukocyte microvilli and their penetration of the endothelial glycocalyx. Biophys J 80:1124–1140

Zhong TP, Rosenberg M, Mohideen MA, Weinstein B, Fishman MC (2000) gridlock, an HLH gene required for assembly of the aorta in zebrafish. Science 287:1820–1824

Zhu DZ, Cheng CF, Pauli BU (1991) Mediation of lung metastasis of murine melanomas by a lung-specific endothelial cell adhesion molecule. Proc Natl Acad Sci U S A 88:9568–9572

Functional Ultrastructure of the Vascular Endothelium: Changes in Various Pathologies

M. Simionescu (✉) · F. Antohe

Institute of Cellular Biology and Pathology "Nicolae Simionescu", 8, B.P. Hasdeu Street, P.O. Box 35-14, Bucharest, Romania
Maya.Simionescu@icbp.ro

1	Introduction	42
2	The Many Functions of the Endothelium Enable It to Be Called the Chief Governor of Body Homeostasis	43
3	The Endothelial Cell Structure Supports the Cell Functions	44
3.1	Plasma Membrane and Associated Structures	44
3.2	Endothelial Cell Organelles as Functional Instruments	46
3.2.1	Plasmalemmal Vesicles/Caveolae: Not One, but Many Classes	46
3.2.2	Transendothelial Channels: A Close Relative of Caveolae	50
3.2.3	Fenestrae: The Smallest Polarised Membrane Microdomain	51
3.2.4	Uncoated and Coated Pits, Coated Vesicles	51
3.2.5	Cytoskeleton	52
3.3	Intercellular Junctions: Cross-talk Between Cells	53
4	Basal Lamina	54
5	Innate Phenotypic Heterogeneity of the Vascular Endothelium	56
6	Endothelial Cell Receptors as Operational Tools	57
6.1	Receptors for Vasoactive Mediators	57
6.2	Receptors for Plasma Proteins	57
6.2.1	Receptors for Metalloproteins	57
6.2.2	Insulin Receptors	58
6.2.3	Receptors for Lipoproteins	58
6.2.4	Receptors for Albumin	58
7	Mechanisms of Endothelial Sorting of Molecules	59
7.1	Endocytosis	59
7.2	Transcytosis	61
8	Changes of the Vascular Endothelium in Different Pathologies	61
8.1	The Endothelial Cell Is the Key Player in All Stages of Atherosclerosis	62
8.1.1	Changes in the Endothelium During the Pre-lesional Stage of Atherosclerosis	62
8.1.2	Alterations in the Endothelium at the Lesional Stage of Atherosclerosis	63
8.2	Diabetes Induces Modifications of Endothelial Cells	64
8.2.1	Alterations of Large Vessel Endothelial Cells	65

8.2.2 Modifications of Capillary Endothelial Cells 65
8.3 Alterations of the Endothelium in Alzheimer's Disease 66

9 Conclusion: Lessons from a Brave Cell 66

References 67

Abstract Biology has revealed that form follows function or function creates the organ. Translating this law at the cellular level, we may say that the ultrastructure follows function or function creates the ultrastructure. The vascular endothelium is an accurate illustration of this rule due to its numerous and many-sided functions carried out by highly specialised cells, structurally equipped for their tasks. Occupying a strategic position between the blood and tissues, the endothelial cell (EC) tightly monitors the transport of plasma molecules, employing bidirectional receptor-mediated and receptor-independent transcytosis and endocytosis, regulates the vascular tone, synthesises and secretes a large variety of factors, and is implicated in the regulation of cell cholesterol, lipid homeostasis, signal transduction, immunity, inflammation and haemostasis. Ultrastructurally, besides the common set of organelles, the characteristic features of the ECs are the particularly high number of vesicles (caveolae) endowed with numerous receptors, transendothelial channels, the specialised plasma membrane microdomains of distinct chemistry, and characteristic intercellular junctions. In addition, by virtue of their number ($\sim 6 \times 10^{13}$), aggregated mass (~ 1 kg), large surface area ($\sim 7,000$ m^2) and distribution throughout the body, the ECs can perform all the assumed functions. The vascular endothelium, with its broad spectrum of paracrine, endocrine and autocrine functions, can be regarded as a multifunctional organ and chief governor of body homeostasis. The ECs exists in a high-risk position. The cells react progressively to aggressive factors, at first by modulation of the constitutive functions (permeability, synthesis), followed by EC dysfunction (loss, impairment or new functions); if the insults persist (in time or intensity), cell damage and death ultimately occur. In conclusion, the ECs are daring cells that have the functional–structural attributes to adapt to the ever-changing surrounding milieu, to use innate mechanisms to confront and defend against insults and to monitor and maintain the body's homeostasis.

Keywords Endothelium · Ultrastructure · Functions · Transcytosis · Pathology

1
Introduction

Multicellular organisms operate on the principle of division of labour, a rigorously controlled operation to assist the proper functioning of all cells, tissues and organs. The cardiovascular system is the distributor of oxygen, nutrients, hormones and other essential molecules throughout the body, and thus the normal life of each and every cell crucially depends on the appropriate regulatory mechanisms that operate at the level of the blood vessels. Endothelial cells (ECs) line all vessels, and constitute the only interface between the plasma and the interstitial fluid as well as the underlying cells. Although distributed throughout the body, the aggregated mass of all ECs is quite large, being equal to that of the liver ($\sim 1,000$ g), and corresponding to a sizeable surface of approximately 7,000 m^2. The vascular endothelium is representative of the premise that the function creates the structure and that the structure supports the function.

2
The Many Functions of the Endothelium Enable It to Be Called the Chief Governor of Body Homeostasis

Due to their strategic position and large surface area, the ECs have assumed a great variety of functions, including the control of exchanges of molecules between the plasma and the interstitial fluid (transcytosis), the regulation of vascular tone, the synthesis and secretion of various molecules, the presentation of histocompatibility antigens (immunity), the control of smooth muscle cell (SMC) proliferation, and the maintenance of the proper balance between pro- and anticoagulant factors that ensure the blood fluidity (Fig. 1). Moreover, in response to various stimuli, ECs synthesise and release a large

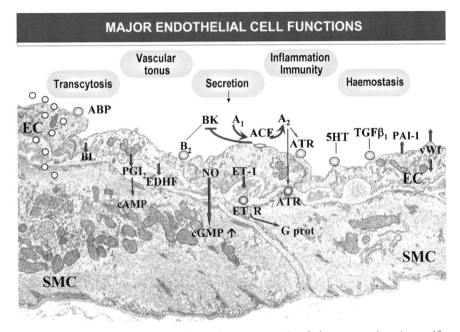

Fig. 1 The endothelial cells (*EC*) (1) monitor transcytosis of plasma proteins via specific receptors (e.g. ABP, albumin binding proteins), (2) maintain vascular tone by secreting prostacyclin (*PGI$_2$*), endothelium-derived hyperpolarising factor (*EDHF*), nitric oxide (*NO*) and endothelin (*ET$_1$*) that acts on the endothelin receptors (*ET$_A$R*) of the smooth muscle cell (*SMC*), as well as angiotensin-converting enzyme (*ACE*) that converts angiotensin I (*A$_1$*) to angiotensin II (*A$_2$*) concomitantly with the inactivation of bradykinin (*BK*), (3) synthesises components of the basal lamina (*BL*) and (4) is implicated in immunity, inflammation and haemostasis. *ATR*, angiotensin I receptors; *5HT*, 5-hydroxytryptamine; *TGFβ1*, transforming growth factor; *PAI-1*, plasminogen activation inhibitor; × 15,000. (Reprinted with permission from Simionescu et al. 2004)

number of vasoactive substances, cytokines, adhesion molecules, endothelins and other factors; therefore, the vascular endothelium is considered to be the largest endocrine organ in the body. Given these complex activities, the vascular endothelium as a whole can be regarded as a multifunctional organ that has a broad spectrum of paracrine, endocrine and autocrine functions. The numerous tasks of the ECs make them collectively the chief supervisor and monitor of body homeostasis, being able to maintain the equilibrium between the main body fluids (the plasma, interstitial fluid and lymph) and the proper functioning of each cell under physiological conditions.

3
The Endothelial Cell Structure Supports the Cell Functions

The vascular endothelium is a type of simple squamous epithelium of mesodermal origin. The ECs are polygonal in shape (10–15 mm wide and 25–50 mm long), generally orientated along the long axis of the vessels (due to the effect of shear stress) and they number approximately 6×10^{13} cells for the entire vasculature. Interposed between two different fluid compartments, the ECs are polarized cells, having a luminal front facing the plasma and an abluminal front, bathed by the interstitial fluid. The polarity is manifested by a distinct protein composition of the apical and basolateral plasmalemma (Muller and Gimbrone 1986) and the regulated secretion of molecules to either the luminal or the abluminal blood front. The intercellular tight junctions impede the diffusion of molecules between the apical and basolateral membrane, thus contributing to the maintenance of cell polarity. The EC apical plasmalemma expresses specific receptors for several plasma molecules, such as vasoactive agents, hormones, procoagulant, anticoagulant and fibrinolytic factors, carrier proteins and lipoproteins (Fig. 1). Although ECs were once viewed as an inert cellophane barrier, progress in cell biology has led to the discovery of characteristic EC structures—such as the plasmalemmal vesicles (Palade 1953), currently named caveolae (small caves), and transendothelial channels (Simionescu et al. 1975a)—and microdomains of the plasmalemma (Simionescu et al. 1981) and numerous membrane receptors.

3.1
Plasma Membrane and Associated Structures

The endothelial plasmalemma is a complex mosaic of proteins, glycoproteins and glycolipids embedded in a lipid bilayer. The ectodomains of the membrane components form the glycocalyx (30–50 nm thick), made up primarily of glycosaminoglycans, oligosaccharide moieties of glycoproteins and glycolipids and sialoconjugates (for review see Simionescu 1979; A.R. Pries and W.M. Kuebler, volume I). The blood–endothelial interface is composed of the plasmalemma proper and the temporarily associated plasma proteins (immunoglobulin, fibrinogen, albumin, α-2-macroglobulin) as well as enzymes

Functional Ultrastructure of the Vascular Endothelium

such as angiotensin-converting enzyme (ACE) and lipoprotein lipase (LPL), whose distribution varies according to the vascular bed: i.e. ACE is well represented in lung capillaries, whereas LPL occurs in large vessel endothelia (Simionescu 1991). The endothelial plasmalemma exhibits membrane-associated microdomains, namely plasmalemmal vesicles, transendothelial channels, fenestrae, coated pits and coated vesicles (Fig. 2). Among these,

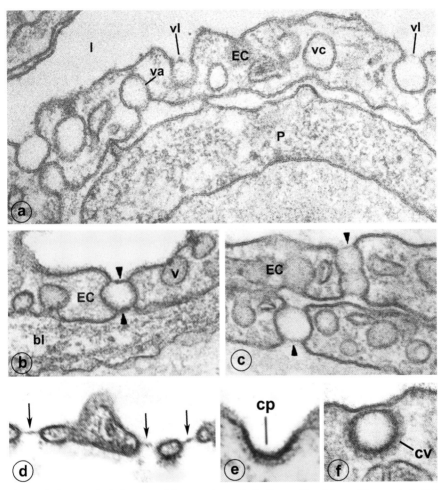

Fig. 2 a–f Characteristic features of the vascular endothelium. **a** The thin capillary endothelial cell (*EC*) accommodates a large number of vesicles open to the luminal front (*vl*), abluminal front (*va*), or enclosed within the cytoplasm (*vc*). **b, c** It has transendothelial channels made up of one or two caveolae (*arrow heads*); **d** diaphragmed fenestrae (*arrows*) in fenestrated capillaries; **e** coated pits (*cp*); and **f** coated vesicles (*cv*). *l*, lumen; *p*, pericyte; *bl*, basal lamina. **a, b, c** × 80,000; **d** × 50,000; **e, f** × 140,000. [Reprinted with permission from Simionescu 1991 (part **a**); Simionescu et al. 2002 (parts **b** and **c**)]

caveolae are the most characteristic structure of the ECs; they appear in direct continuity with either the luminal or abluminal plasma membrane, or are enclosed within the cytoplasm (Fig. 2a). As in other cells, the chemical composition of EC plasmalemma confers a net negative surface charge. Interestingly, the membrane of caveolae, channels and their associated diaphragms (lacking sulphate and/or sialate groups) are devoid of strong anionic sites, a feature that led to the assumption that vesicles represent a preferential pathway for the transport of plasma proteins, most of which are anionic (reviewed in Simionescu and Simionescu 1991).

3.2
Endothelial Cell Organelles as Functional Instruments

Like all eukaryotic cells, the ECs are provided with the common set of organelles mostly gathered in the paranuclear zone. The Golgi complex, endoplasmic reticulum, mitochondria, multivesicular bodies, endosomes and lysosomes and in particular caveolae are present in various numbers of copies as a function of the state of the cell. A characteristic of non-capillary endothelia are the Weibel–Palade bodies (WPB), which are membrane-bound rod-shaped granules, 3–4 mm long, containing several parallel tubes (15 nm diameter) embedded in a dense matrix (Weibel and Palade 1964). The role of WPB is to store and discharge (when needed) the von Willebrand factor (vWF) either to the plasma, or within the vessel wall where vWF has a major role in inflammation.

These EC organelles are operational in the synthesis of basal lamina, extracellular matrix (collagen IV, fibronectin and proteoglycans), vWF, and relaxing and contracting factors such as NO, prostacyclins, endothelins and angiotensin II (Fig. 1). These molecules are differentially sorted and secreted to the luminal, abluminal, or both endothelial fronts. The EC synthetic capacity is profoundly altered in pathological conditions.

3.2.1
Plasmalemmal Vesicles/Caveolae: Not One, but Many Classes

First described in ECs by Palade, plasmalemmal vesicles, a common feature of many cell types, are particularly numerous in the vascular endothelium (10,000–15,000/cell) and especially in capillary ECs (e.g. heart, lung, muscle), with the exception of the brain capillaries (Palade 1953). One can safely assume that the number of EC caveolae varies according to the vascular bed involved. Caveolae appear as spherical vesicles (60–70 nm diameter), either in direct continuity with the apical or basolateral plasmalemma (thus almost doubling the EC surface) or free within the cytoplasm (Fig. 2); sometimes two or more vesicles fuse together. The vesicles open to the EC surface through a neck (10–40 nm diameter) often spanned by a thin ($\sim$7 nm) diaphragm provided with a central knob. As detected by freeze-fracture technique of filipin-treated ECs, the vesicular neck (and the fenestral opening) is surrounded by a peristomal ring of sterols (Fig. 3), assumed to function in the phase separation and the preservation of the sharp bend between the caveolae membrane and the plasmalemma (Simionescu et al. 1983).

Caveolae are dynamic structures that in the process of transcytosis undergo frequent fission and fusion with the plasmalemma. Molecules involved in the vesicle formation, fission, docking and fusion with the target membrane are the vesicular SNAP receptor (vSNARE), synaptobrevin (VAMP)-2, monomeric and trimeric GTPases, annexins II and VI, N-ethyl maleimide-sensitive fusion factor (NSF) and its attachment protein, SNAP (Schnitzer et al. 1995). In the fission process, the large GTPase, dynamin (also associated with clathrin-coated vesicles), oligomerises around the neck of caveolae, a process that requires GTP hydrolysis (Oh et al. 1998). The intracellular movement of caveolae is facilitated by interaction with the cytoskeletal proteins, such as actin, myosin, gelsolin, spectrin and dystrophin (Lisanti et al. 1994). For specific docking, the endothelial caveolar VAMP-2 interacts with the complementary target membrane tSNARE localised on the acceptor membranes (McIntosh and Schnitzer 1999).

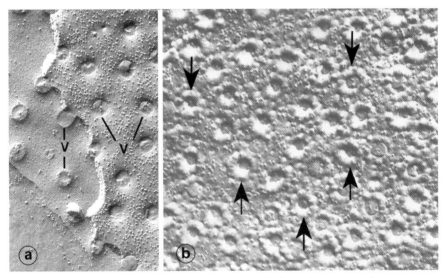

Fig. 3 a,b Freeze-fracture image of a capillary endothelium depicting **a** the opening of vesicles (*v*) to the plasmalemma and **b** the presence of rings of cholesterol (*arrows*) around the vesicular neck as detected by incubation of cells with filipin, which forms specific filipin-sterol complexes. **a** and **b** × 60,000. (Reprinted with permission from Simionescu et al. 1983)

Caveolae Constitute Chemically Distinct Microdomains of the EC Plasmalemma

Caveolins are the marker proteins of endothelial (and other) vesicles. Caveolin-1 is a non-conventional membrane-spanning protein having both the N and C termini towards the cytoplasm and a single hydrophobic region; it appears as a highly ordered homo-oligomer of 14–16 monomers or, upon interaction with caveolin-2, forms stable high-molecular-mass hetero-oligomers. Caveolin-1 binds many proteins via its scaffolding domain that acts as a "master regulator" of signalling molecules and may also be the site that regulates cellular Ca^{2+} concentration and Ca^{2+}-dependent signal transduction (Minshall et al. 2003).

Caveolin-1 binds cholesterol and is critical in the transport of cholesterol from the site of synthesis to the plasmalemma. In addition, caveolin-1 is involved in cholesterol efflux from the cells, a process that implies an association between high-density lipoprotein (HDL) and SR-BI (scavenger receptor class B type I), located in caveolae. Due to the dual function of SR-BI, caveolae are

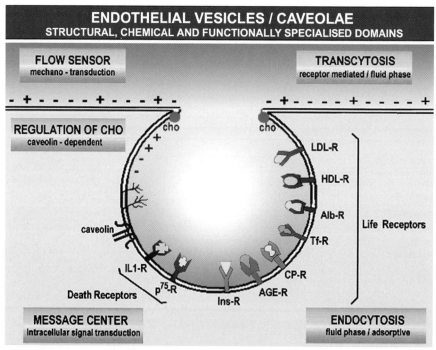

Fig. 4 Diagram depicting the main functions attributed to caveolae and the receptors present in the caveolar membrane. Note the distribution of cholesterol (*cho*) around the caveolar neck and the receptors for LDL (*LDL-R*), HDL (*HDL-R*), albumin (*Alb-R*), transferrin (*Tf-R*), ceruloplasmin (*CP-R*), advance glycation end products (*AGE-R*), insulin (*Ins-R*)—termed "life receptors" (essential in maintaining tissue homeostasis)—as well as IL1-R and p^{75}-R—named "death receptors" (involved in apoptosis). The distribution of anionic sites is prevalent on the EC membrane, but not on the vesicle membrane. (Reprinted with permission from Simionescu et al. 2002)

also important sites for cholesterol uptake, a process regulated by caveolin-1 (reviewed in Razani et al. 2002).

Numerous receptors involved in the transport of plasma proteins and signalling have been identified within the caveolar membrane; they include receptors for plasma proteins (Fig. 4), epidermal growth factor (EGF), platelet-derived growth factor (PDGF), endothelin, CD36, interleukin (IL)-1 and P^{75}, as well as G protein-coupled receptors and inositol triphosphate receptors (reviewed in Simionescu et al. 2002; Schnitzer et al. 1995).

The lipid composition of caveolae consists mainly of cholesterol and sphingolipids (sphingomyelin and glycosphingolipid). Cholesterol has a support function in that it creates the frame into which other molecules are inserted.

We have found differences between the chemical composition of the capillary EC plasma membrane and the caveolar membrane. In contrast to the plasmalemma, the caveolae lack strong anionic sites (Fig. 5a) of low pK_a, sialo-

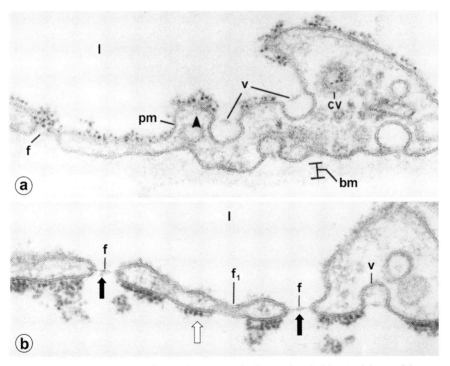

Fig. 5 a,b Microdomains of different charges on the luminal and abluminal front of the endothelial cells as revealed by decoration with cationised ferritin. **a** Note the even distribution of anionic sites on the luminal plasma membrane (*pm*), the heavily labelled fenestral diaphragm (*f*) and the absence of anionic sites on vesicles (*v*) and their diaphragms. **b** On the abluminal front, the plasmalemma is similarly decorated (*white arrow*) and the vesicles (*v*) lack anionic sites, but the abluminal front of fenestral diaphragms is devoid of anionic sites (*black arrows*). *l*, lumen; *bm*, basement membrane; *cv*, coated vesicle; × 120,000. [Reprinted with permission from Simionescu et al. 1981 (part **a**) and 1982 (part **b**)]

conjugates and proteoglycans, and are enriched in *N*-acetylglucosaminyl and galactosyl residues (Simionescu et al. 1982a); the aortic EC caveolae contain a higher concentration of Ca^{2+}-ATPase and some specific glycoproteins and are enriched in palmitoleic and stearic acids (Gafencu et al. 1998).

Caveolae Represent a Functionally Distinct Microdomain Studies based on tracers of various dimension, chemistry and shape (injected in vivo) have indicated that a fraction of vesicles transport plasma molecules, operating either as separate shuttling units or as channels formed by one or more caveolae opening simultaneously to the luminal and abluminal plasmalemma (Fig. 2b, c). For the process of transendothelial transport, Simionescu coined the term "transcytosis" in 1979, an appellation that was further extended to all epithelial cells; the concept was broadened to indicate that the transcytotic mechanisms imply fluid phase, non-specific adsorptive, or receptor-mediated transcytosis. Employing native plasma molecules such as albumin or low-density lipoproteins (LDL), we have found that caveolae function in transcytosis of these proteins across the ECs, employing either receptor-mediated, or receptor-independent transcytosis. Interestingly, caveolae take up cholesterol-carrying LDL via a dual process: by receptor-mediated endocytosis for use by the cell itself and by transcytosis for use by the underlying cells (reviewed in Simionescu and Simionescu 1991).

Other vesicles that are present in close proximity or open to the intercellular space constitutively contain platelet endothelial cell adhesion molecule (PECAM) which, during leucocyte transmigration, establishes a homophilic interaction with the PECAM expressed on the leucocyte membrane, thus assisting cell diapedesis through the junction (reviewed in Dejana 2004).

Based on the above data, one can safely predict that, by virtue of their distinct chemistry, caveolae comprise not one but several distinct classes, with well-defined functions. One fraction of the caveolae is devised to carry out endocytosis, others execute fluid phase, adsorptive or receptor-mediated transcytosis, and others are implicated in cholesterol and lipid homeostasis, signal transduction or leucocyte diapedesis.

3.2.2
Transendothelial Channels: A Close Relative of Caveolae

Vesicles fuse between themselves, so that sometimes a single, or a chain of two or three, fused vesicles open simultaneously on both EC fronts to form a channel that spans the cell (Fig. 2b,c), a feature common in fenestrated capillaries but also demonstrated in continuous capillaries. The formation of transendothelial channels may be facilitated by the high density of vesicles, the extreme attenuation of the ECs and the existence of a large number of vesicles open to both cell fronts (Simionescu et al. 1975a). It is assumed that channels represent a highly dynamic structure, a transient, hydrophilic pathway that forms as an adaptation to temporary local needs or in response to a pathological condition (e.g. ischaemia, inflammation).

3.2.3
Fenestrae: The Smallest Polarised Membrane Microdomain

Fenestrae are round openings (~70 nm diameter) that connect the two EC fronts. They may or may not be spanned by a diaphragm and are characteristic of capillaries of the intestinal mucosa, pancreas, endocrine glands that are termed fenestrated capillaries (Fig. 2d). The diaphragms are thin, lipid-free structures provided with a central knob (15 nm) from which spokes radiate and anchor into a polygonal rim with wedge-shaped spaces in between the spokes (Bearer and Orci 1976). Diaphragms are lacking in liver sinusoids and glomerular capillary ECs, whereas in adrenal cortex capillaries, diaphragmed fenestrae coexist with large aperture-free openings.

Interestingly, the chemistry of the fenestral diaphragm varies on its two aspects, namely the luminal face exposes heparan sulphate proteoglycans and heparin (strong anionic residues) and receptors for wheat germ agglutinin (choriocapillaries) (Pino 1986), whereas their abluminal facet is devoid of anionic sites (Fig. 5a, b). Thus, the fenestrae represent the smallest polarised subcellular component of the EC surface (Simionescu et al. 1982b).

Structural and biochemical data support the concept that vesicles, channels and fenestrae are interrelated structures. Ultrastructurally, stages indicative of fusion and fission of vesicles and the formation of channels are often seen; fenestrae may be considered a collapsed channel. In addition, caveolae, channels and fenestrae have in common a ring of cholesterol at their openings and an endothelium-specific structural protein, PV1 (a rod-like protein of 60 kDa), reported to be involved in the formation of the diaphragm (Stan et al. 2004).

3.2.4
Uncoated and Coated Pits, Coated Vesicles

In addition to open vesicles, the EC plasmalemma is endowed with shallow invaginations—the uncoated pits (Fig. 6b)—that by "en face" images of the EC cytoplasmic surface appear to have a distinctive striated coat (Rothberg et al. 1992). The relationship between uncoated pits and caveolae, and whether they represent vesicle precursors, is not yet known.

Coated pits and coated vesicles similar to those found in other epithelial cells (~120 nm diameter) have a geodesic basketwork of clathrin on their cytoplasmic aspect (Fig. 2e, f). With some exceptions (hepatic sinusoids, intestinal, pancreatic and adrenal fenestrated capillaries), the frequency of coated pits/coated vesicle in the EC is relatively small by comparison with the number of caveolae. The coated vesicles are endowed with a high density of anionic sites that contrast with that of the caveolae membrane and diaphragms.

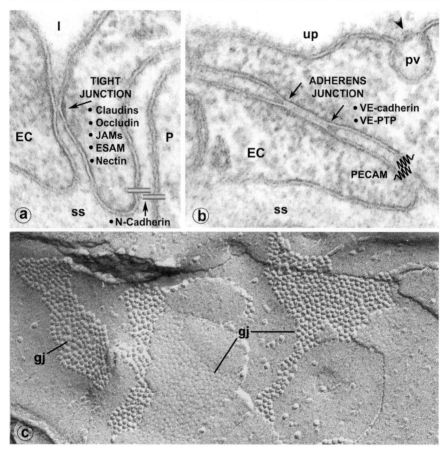

Fig. 6 a–c Junctions between neighbouring microvascular endothelial cells (*EC*) and some of the constituent molecules. **a** Thin section of a capillary tight junction. **b** A capillary adherens junction. **c** A freeze-fracture image of an arteriole EC gap junction (*gj*). Note in part **b** an uncoated pit (*up*) and a vesicle (*pv*) spanned by a diaphragm (*arrowhead*). *l*, lumen; *ss*, subendothelial space; *p*, pericyte; **a, b** × 180,000; **c** × 60,000. (Parts **a** and **b** reprinted with permission from Simionescu 1991)

3.2.5
Cytoskeleton

Direct exposure of the ECs to the plasma requires continuous adaptation to the ever-changing haemodynamic stress and blood pressure. In addition, the EC has to respond rapidly to the chemical signals received either from the blood or host tissue. These modulations of the EC are serviced, in part, by the contractile cytoskeleton, whose major components are actin, myosin II, tropomyosin, α-actinin and actin-binding proteins, such as fodrin, gelsolin, protein 4.1, filamin, vinculin, talin, vimentin and non-muscle caldesmon, that represent a large proportion of the cell proteins (reviewed in Drenckhahn and Ness 1997). All these cytoskeleton components operate, singly or in concert, for (1) the adhesion of the EC to the substratum, (2) the integrity of intercellular junctions, (3) the scaffolding of plasmalemma, (4) immobilisation of membrane proteins and (5) changes of the cell shape in response to shear stress.

3.3
Intercellular Junctions: Cross-talk Between Cells

Along the cardiovascular system, adjacent ECs are connected by various types of junctions, made up of intramembranous specific proteins linked to the cytoskeleton proteins.

Endothelial junctions guarantee the separation between the blood and the interstitial fluid, maintain the cell polarity and lining integrity, ensure contact inhibition, and play a role in remodelling and angiogenesis.

In ECs, the main types of junctions are: (1) tight junctions (zonula occludens) that seal completely the intercellular spaces, (2) adherent junctions (zonula adherens) that together with the former maintain the cell polarity and integrity (Fig. 6a, b), and (3) gap (communicating) junctions (macula communicans) (Fig. 6c). Syndesmos (complexus adhaerentes), an equivalent of epithelial desmosomes, was detected in the lymphatic endothelium at the level at which desmoplakin co-distributes with vascular endothelial (VE)-cadherin/cadherin-5 (Schmelz and Franke 1993).

The molecules detected at the level of tight junctions are members of the claudin family, occludin, junctional adhesion molecules, endothelial cell-selective adhesion molecule (ESAM) and, on their cytoplasmic aspect, ZO-1 (zonula occludens-1), ZO-2 (zonula occludens-2), calcium/calmodulin-dependent serine protein kinase, afadin, partitioning defective-3 (PAR3) and multi-PDZ-domain protein-1 (MUPP1).

As for the molecules characteristic of adherens junctions, the EC expresses VE-cadherin, which can be associated with VE-PTP (vascular endothelial protein tyrosine phosphatase), E-cadherin (in brain endothelium) and neuronal N-cadherin which, because of its extra-junctional location, mediates binding of ECs to pericytes and other neighbouring cells. Many components of the tight and adherens junctions form complexes with catenins and associate with cytoskeleton proteins, zyxin, moesin and others (reviewed in Dejana 2004).

The constituents of gap junctions are a family of proteins, connexins, of which Co43, Co40 and Co37 have been identified in the EC. Clusters of 20-nm (diameter) transmembrane hydrophilic channels (connexons) function in the transfer of ions and small molecules between adjoining cells and warrant the metabolic and electrotonic coupling between neighbouring ECs (homotypic communication), as well as between ECs and the underlining SMCs (heterotypic communication). The organisation of EC junctions varies along the vasculature. In large arteries, complex occluding, adherens and numerous gap junctions link the neighbouring ECs. In addition, ECs are in direct contact and communication with the underlying SMCs via myoendothelial junctions that are essential in the coupling and signal transmission between these cells; the presence of gap junctions between SMCs ensures the transmission of signals from one SMC to neighbouring SMCs, the result of which is the co-ordinated response of the vessel wall to extravascular stimuli and the regulation of the vascular tone. In veins, the composition of EC junctions is similar, but their organisation is less elaborate and the frequency of gap junctions is lower (Simionescu et al. 1975b).

Arterioles exhibit the most elaborate system of junctions, consisting of a combination of occluding and intercalated gap junctions, an association that ensures strong cell-to-cell adhesion, sealing of the intercellular spaces and communication between cells; there are also myoendothelial junctions.

The capillary endothelium is characterised by the presence of occluding junctions only; morphologically distinct gap junctions are absent.

Postcapillary venule ECs exhibit loosely organised tight junctions, of which roughly 30% are open to a gap of approximately 60 nm. The existence of loosely organised junctions, the presence of numerous mast cells along the venules and the uncovering of high-affinity histamine receptors, principally localised in the parajunctional regions of venular ECs (Heltianu et al. 1982), may account for the rapid response of this segment of the vasculature to vasomediators, which renders these vessels the preferential site for plasma and cell extravasation (Fig. 7). Disruption of EC junctions has major effects on vascular homeostasis, with severe consequences in vascular diseases.

4
Basal Lamina

The EC rests on a thin basal lamina, the molecules of which are synthesised and secreted by the cells themselves. The chemical composition of the basal lamina consists mainly of type IV and type V collagen, laminin, entactin (nidogen) and heparan sulphate proteoglycans (Sage et al. 1983). The EC basal lamina encloses the pericytes in capillaries and small venules, whereas in other vessels, basal lamina separates the ECs from the underlying SMCs; however, focal points of direct contact via myoendothelial junctions allow direct cross-talk between ECs and SMCs. Pericytes are smooth muscle-like cells wrapped around the EC, having important metabolic, signalling and mechanical roles (Sims 2000). Pericytes exchange information with ECs by direct contact and/or by releasing cytokines such as tumour necrosis factor (TNF)-β and other soluble factors that are potent inhibitors of EC growth and promote vasoconstriction by up-regulating endothelin-1 and down-regulating inducible NO synthesis (Martin et al. 2000). Recent data demonstrate the role of pericytes in vascular morphogenesis (Betsholtz et al. 2005). Pericyte alteration or degeneration is linked directly to microangiopathies in diabetes, hypertension, microvascular

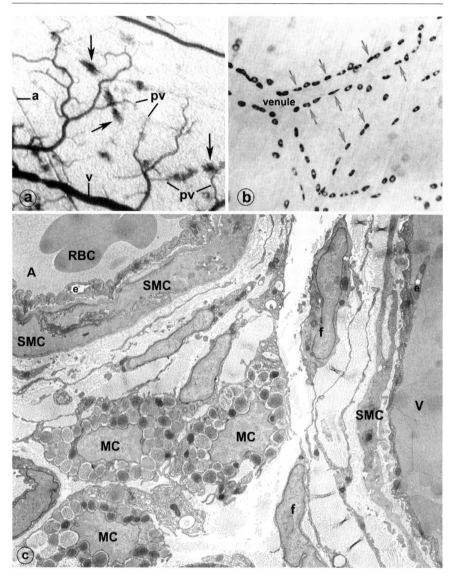

Fig. 7 a Microvasculature of the hamster cheek pouch showing histamine-induced leakage of plasma (*arrows*) at the level of postcapillary venules (*pv*) only. *v*, venules; *a*, arteriole. **b** A similar area of the microvasculature stained with toluidine blue reveals the high density of mast cells (*arrows*) bordering the venules. **c** Electron microscopy depicting the close proximity of mast cells (*MC*) to a venule (*V*) and arteriole (*A*). *SMC*, smooth muscle cell; *RBC*, red blood cell; *f*, fibroblast. **a** × 60; **b** × 120; **c** × 40,000. (Part **b** used with permission from Antohe et al. 1989)

vasculitis in the brain and retina, and possibly inappropriate calcification of blood vessels.

5
Innate Phenotypic Heterogeneity of the Vascular Endothelium

Under the influence of the local environment and the specific needs of the neighbouring cells, the apparently similar ECs have undergone segmental differentiation expressed as a considerable site-specific phenotypic heterogeneity, a feature that explains the varied cell response to normal or aggressive stimuli.

Endothelial Heterogeneity Along the Vasculature As a function of the vessel or the tissue in which it resides, the structural-functional differentiation of the EC consists of variation in shape, proliferative capacity, frequency of WPB, response to vasoactive factors, expression of surface molecules, secretory capacity, reaction to changes in shear stress, organisation of junctions and basic cellular constituents (frequency of caveolae, channels, fenestrae and their diaphragms). These dynamic modulations are expressed in large phenotypic variants for the ECs of large vessels versus microvessels (arterioles, capillaries, venules) and, among the latter, differences within the same class. For instance, based on the EC structure, the capillaries have been classified as continuous (having caveolae only, e.g. the heart and the majority of blood vessels), fenestrated (provided with caveolae and diaphragmed fenestrae, e.g. visceral organs) and discontinuous, characterised by the presence of caveolae and large ($\sim$100 nm) gaps (e.g. haematopoietic tissues and liver). In addition, significant modulations exist within the same type of capillary endothelium, e.g. there are subtypes of continuous capillaries, the extremes being at one end the brain capillaries (with rare caveolae and very tight junctions) and at the other end the myocardial capillaries (with a high number of caveolae and comparatively few tight intercellular junctions). Moreover, within the microvascular endothelium, the intercellular sealing and the cell-to-cell communication is more complex in arterioles than in capillaries and venules; the latter exhibit loosely organised junctions, a feature that has great implications in pathological processes (inflammation, thrombosis).

There is also an antigenic heterogeneity of the vascular endothelium: capillary ECs express major histocompatibility complex (MHC) classes I and II, intercellular adhesion molecule (ICAM) and the monocyte/endothelial marker OKM5, suggesting that capillaries are the site of antigen presentation and the immune attack and response. By contrast, these molecules are almost undetectable on large-vessel ECs that in turn express vWF and endothelial leucocyte adhesion molecule-1 (ELAM-1) (Page et al. 1992).

Phenotypic Heterogeneity Within a Single EC Defines Differentiated Microdomains
In the capillary ECs, we have reported the presence of strong anionic sites (heparan sulphate) on coated pits and the luminal aspect (only) of fenestral

diaphragms, whereas they are almost absent on the membrane of caveolae and their diaphragms. In contrast, the caveolae, transendothelial channels and their diaphragms are particularly rich in β-D-galactose and β-N-acetylglucosamine (Simionescu et al. 1982b). The lung alveolar capillary ECs have a thin avesicular zone and a vesicular zone, the former being postulated to be associated with the gas exchanges.

The heterogeneity of ECs explains the blood vessel-specific reactivity and pathology. Thus, lymphocytes emigrate from the vasculature only via the specialised high-endothelial venules (localised in lymphoid tissues), atherosclerotic plaques develop in specific, arterial lesion-prone areas, vascular leakage occurs in venules and thrombosis occurs in veins.

6
Endothelial Cell Receptors as Operational Tools

6.1
Receptors for Vasoactive Mediators

ECs have receptors for histamine (Antohe et al. 1986), serotonin (5HT) (Shepro and Dunham 1986), bradykinin, thrombin (Haselton et al. 1992) and leukotriene C_4, and respond to these soluble mediators by retracting from one another, thereby increasing the permeability of the monolayer. Upon binding to the EC receptors, these mediators increase the cytosolic free calcium and induce the contraction of the cytoskeleton and the opening of the intercellular junctions in specific segments of the vasculature, i.e. postcapillary venules.

Histamine receptors (predominantly the H2 type) were reported on the ECs of all microvessels, but their frequency is particularly high in postcapillary venules (Heltianu et al 1982). As already stated in postcapillary venules, the presence of histamine receptors, loose EC junctions (∼30%) and mast cells explains the fast response of these vessels to insults such as inflammatory mediators. The mast cells lining the venules (Fig. 7b, c) secrete upon request histamine and vasoactive substances contained within their cytoplasmic granules. Moreover, the mast cell plasmalemma also expresses histamine H2 receptors that modulate histamine release by negative feedback (Antohe et al 1989). In the pathogenesis of atherosclerosis, the histamine/cytokine network regulates inflammatory and immune responses.

6.2
Receptors for Plasma Proteins

6.2.1
Receptors for Metalloproteins

Receptors for transferrin, the plasma iron-carrying glycoprotein, were detected in brain capillaries functioning in receptor-mediated transcytosis of transferrin at the level of the blood–brain barrier (BBB). In the liver sinusoidal ECs, transferrin receptors are restricted to coated pits. Data exist showing that the iron-transferrin complexes are transcytosed across the EC in an intact form. Transferrin is desialylated within the EC and released into the space of Disse from where it is taken up via asialoglycoprotein receptors by the hepatocytes: iron is retained by the cell and transferrin is recycled.

Receptors for ceruloplasmin, a multi-functional copper-containing glycoprotein, are located in the coated pits and vesicles of the liver sinusoidal ECs. The pathway of ceruloplasmin is similar to that of transferrin, being desialylated within the EC and taken up by hepatocyte asialoglycoprotein receptors.

6.2.2
Insulin Receptors

Insulin receptors mediate the metabolic and growth action of insulin. Upon binding to its receptors, insulin initiates a cascade of events and activates multiple signalling pathways in the EC. In large vessel ECs, insulin is taken up by receptor-mediated endocytosis, whereas in capillaries the intact molecule is transported to the target cells by specific transcytotic receptors that also represent a rate-limiting step of the process.

6.2.3
Receptors for Lipoproteins

Cholesterol-carrying lipoproteins (LDL) are taken up by the EC by receptor-independent and receptor-mediated endocytosis and transcytosis. The LDL-receptors (LDL-R) are localised in caveolae, coated pits and coated vesicles. ECs are supplied by receptor-mediated endocytosis with the cholesterol needed for their own use, a process that leads to down-regulation of endogenous biosynthesis of cholesterol. Upon LDL binding, the LDL-R translocate preferentially from the apical to the abluminal plasmalemma, a condition that facilitates the transport of cholesterol-carrying LDL to the underlying cells and tissues (Antohe et al 1999). ECs also possess scavenger receptors that take up oxidised LDL from the circulation.

6.2.4
Receptors for Albumin

Receptors for albumin have been identified in the myocardium, lung, adipose tissue and ECs of microvessels and large vessels. In the EC, the receptor comprises two proteins of approximately 18 and 31 kDa, is located in caveolae and binds albumin specifically and with high affinity (Ghitescu et al. 1986). An albumin receptor of a different molecular weight (60 kDa, albondin) was also found in caveolae of continuous endothelia; its activation stimulates the

Src protein tyrosine kinase signalling pathway, and this may regulate the transcytosis of albumin across the ECs (Tiruppathi et al. 1997). The relationship between the 18, 31 and 60 kDa peptides remains to be established.

Since albumin is the main plasma protein and the carrier of numerous molecules (free fatty acids, thyroid and steroid hormones, bilirubin, bile acids, drugs), the controlled transport of the protein to the right destination and the correct concentration is a prerequisite for the normal functioning of all tissues and, on a larger scale, body homeostasis. There are data to indicate that albumin receptors may selectively discriminate between native and modified albumin and between different ligands bound to albumin. Thus, we have suggested a dual role for albumin receptors, namely as a "docking" protein that recognises and carries molecules bound to albumin (e.g. free fatty acids) and as a "transcytotic" receptor that binds and transports albumin across the EC (Simionescu and Simionescu 1991).

7
Mechanisms of Endothelial Sorting of Molecules

The biochemical and functional attributes of the EC endows it with the ability to sort and direct permeant molecules to the right destination. Caveolae, channels, coated pits and vesicles are equipped to take up and transport macromolecules within the EC by endocytosis, or across the cell by transcytosis, processes that are tightly regulated as a function of the state of the host tissue. Some molecules are endocytosed to be used by the cell itself or to be removed from the plasma, others are transcytosed to reach the underlying cells, whereas others undergo both processes (Fig. 8). There is a precise destination for any given molecule; caveolae are a common denominator in endocytosis and transcytosis.

7.1
Endocytosis

In the EC, endocytosis occurs either via a non-specific (fluid phase or adsorptive process) or by specific receptor-mediated mechanisms.

Non-specific fluid phase endocytosis is dependent on the plasma concentration of a given molecule, whereas adsorptive endocytosis is characteristic for molecules that bind electrostatically to the cell surface. The destination of the endocytosed vesicles is dependent on the molecule involved, but generally the endosomal/lysosomal compartment is the final destination.

Receptor-mediated endocytosis commonly involves specific binding sites localised in coated pits/coated vesicles or caveolae that direct the molecules to different stations before reaching the endosomal/lysosomal compartment; this pathway applies to LDL, beta very low density lipoproteins (β-VLDL), in-

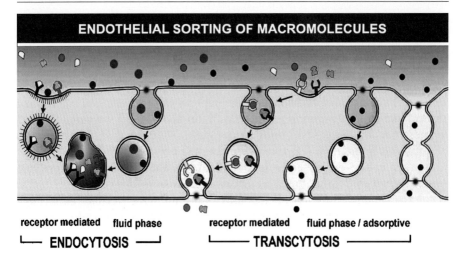

Fig. 8 Diagrammatic representation of the mechanisms involved in the endothelial sorting of macromolecules. Note that caveolae are involved both in endocytosis and transcytosis of plasma molecules. (Reprinted with permission from Simionescu et al. 2002)

sulin and insulin growth factors, transferrin and ceruloplasmin (for review, see Simionescu 2001). Lipoproteins are fully degraded in lysosomes, and the ensuing cholesterol, amino acids and phospholipids are used for cell metabolism. Native LDL is taken up by the arterial endothelium via coated pits and vesicles that perform receptor-mediated endocytosis and by caveolae that function in transcytosis. Both processes, especially transcytosis, are markedly enhanced in hyperlipaemia, leading to progressive accumulation of modified LDL in the subendothelium. Modified lipoproteins are removed from the circulation by EC scavenger receptors (CD36, SREC and LOX-1) that contribute to the preservation of plasma homeostasis (reviewed in Steinbrecher 1999). In ageing and diabetes, accumulated plasma advanced glycation end-product (AGE) proteins are endocytosed by AGE-receptors (R-AGE), thus ensuring the plasma clearance of these injurious proteins (Schmidt et al. 1994).

A special type of endocytosis regulates IgG plasma homeostasis. EC caveolae take up IgG by fluid phase endocytosis and transport the molecule to the endosomal compartment where, at low pH, IgG binds to the neonatal receptors (FcRn). Then, via carrier vesicles, IgG is delivered either to the apical or basal cell surface, whereas excess non-bound IgG is degraded within the lysosomal compartment; as a result, IgG plasma homeostasis is continuously maintained (Ghetie et al. 1996; Antohe et al. 2001). A new term, potocytosis, was proposed for all endocytic activities that use the caveolae endo-membrane system as vesicles for the sequestration and transport of small and large molecules (reviewed in Mineo and Anderson 2001).

7.2
Transcytosis

In simple terms, transcytosis signifies the bidirectional transport of macromolecules across the EC within a discrete compartmentalised organelle, the caveolae. As in endocytosis, the mechanisms of transcytosis entail either a non-specific fluid phase or adsorptive process or a specific receptor-mediated process.

Non-specific fluid phase transcytosis implies uptake of a fraction of plasma by caveolae that shuttle across the EC and discharge the content to the abluminal front. The rate of uptake depends on the size of the vesicle's opening, solute concentration and the steric competition. Adsorptive transcytosis entails an electrostatic interaction between the permeant molecule and the vesicle carrier in which the deciding factor is the distribution and density of the surface charge. Because most plasma molecules are anionic and the vesicle membrane is devoid of strong anionic sites, we consider the vesicles as devised to carry plasma proteins.

By receptor-independent transcytosis of cholesterol-carrying LDL, to supply the cells of the vessel wall or the underlying cells, the shuttling caveolae maintain the cholesterol homeostasis both in the plasma and in the surrounding cells and tissues.

Receptor-mediated transcytosis is a basic process shared by most epithelial cells including the ECs. Specific transcytosis was demonstrated for (1) LDL in arterial endothelium and lung capillaries, (2) transferrin in the microvessels of the brain and the bone marrow, (3) ceruloplasmin in liver sinusoidal capillaries, (4) insulin in aortic endothelium and (5) albumin in the lung, adipose tissue and skeletal muscle endothelia (reviewed in Simionescu et al. 2002; Tuma and Hubbard 2003).

As already stated, the caveolae are endowed with a large number of life and death receptors that distinguish the cargo molecule. It is reasonable to assume that the numerous receptors ascribed to caveolae are not functional in all vesicles, sustaining the argument that there are several classes of caveolae, each performing a specific task.

8
Changes of the Vascular Endothelium in Different Pathologies

Accumulated data have revealed that major pathological conditions such as atherosclerosis, diabetes, Alzheimer's disease (AD), inflammation, immune and autoimmune diseases, hypertension, respiratory distress syndrome and Fabry's disease have in common a dysfunctional endothelium. Whether the EC dysfunction is a primary (pathogenic) factor or a secondary (reactive) response to various insults, i.e. whether the EC dysfunction is a direct cause or a secondary consequence of a disease, remains to be established for each condition.

8.1
The Endothelial Cell Is the Key Player in All Stages of Atherosclerosis

Atherosclerosis, a continuous and progressive disorder of large and medium-size arteries, can be conventionally regarded as having two consecutive stages that differ in the type of cell and the altered mechanisms involved, and in the various active factors. The pre-lesional stage entails a lipid disorder and a critical inflammatory process, whereas the lesional stage results in the formation of a complex plaque; the ECs are implicated in both stages of atherogenesis, as follows.

8.1.1
Changes in the Endothelium During the Pre-lesional Stage of Atherosclerosis

Initially, hyperlipaemia induces modulation of two constitutive functions of the ECs, namely it increases the permeability (in particular for lipoproteins) and enhances the cell biosynthetic activity (initially of basal lamina components). The functional modifications that reflect the attempts of the cell to adapt to the modified environment are well mirrored in the structural changes of the EC (Simionescu et al. 1990; Simionescu 2004). The increase in permeability, as the initial event in atherogenesis, accompanied by the reduced

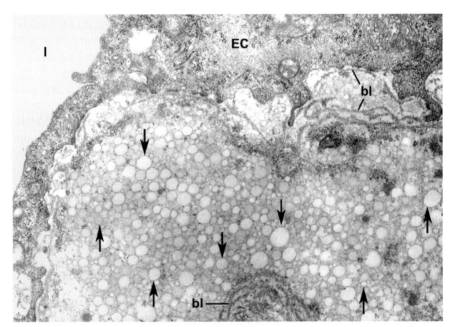

Fig. 9 Electron micrograph showing the accumulation under the endothelial cell (*EC*) of numerous modified lipoproteins that appear as vesicles of various sizes (*arrows*) within the meshes of the basal lamina (*bl*). *l*, lumen; × 22,000

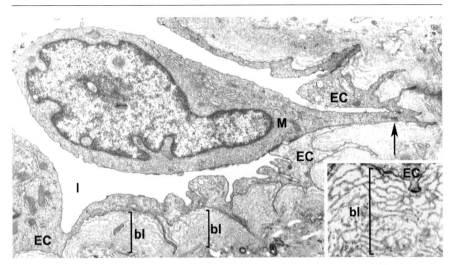

Fig. 10 Electron micrograph depicting diapedesis (*arrow*) of a monocyte (*M*) between two arterial endothelial cells (*EC*). Note the hyperplasic basal lamina (*bl*) made up of 10–20 layers (*inset*). *l*, lumen; × 10,000; *inset*, × 25,000. (Reprinted with permission from Simionescu et al. 1996)

efflux of lipoproteins from the vessel wall, leads to the accumulation of modified and reassembled lipoproteins (MRL) within the subendothelium (Fig. 9). Concurrently, the EC shifts to a secretory phenotype, characterised by an increased number of biosynthetic organelles that correlates with the appearance of a multilayer, hyperplasic basal lamina, sometimes consisting of 20–25 layers (Fig. 10), in the meshes of which MRL accumulate in large numbers.

The ECs are afflicted on both fronts—on the luminal side by hyperlipaemia and on the abluminal side by the accumulated MRL. These insults lead to a dysfunctional endothelium and a multipart inflammatory process in which the ECs express more or new adhesion molecules and synthesise factors [e.g. monocyte chemoattractant protein (MCP)-1] that attract and induce migration of plasma inflammatory cells such as T lymphocytes and monocytes to the subendothelium (Fig. 10). Within the subendothelium, the monocytes become activated macrophages that engulf MRL (via scavenger receptors) to become macrophage-derived foam cells, which release cytokines and factors that affect the ECs and induce migration of SMCs from the media to the intima.

8.1.2
Alterations in the Endothelium at the Lesional Stage of Atherosclerosis

Extensive hyperlipaemia amplifies the EC dysfunction, expressed by impairment of NO bioavailability, alteration of procoagulant and anticoagulant synthesis and secretion, increased secretion of matrix metalloproteinase (MMP)-1 and changes in the cross-talk with neighbouring SMCs. The result of this stage

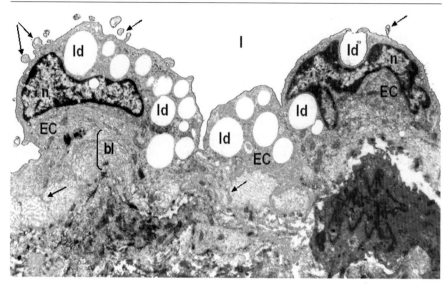

Fig. 11 Electron micrograph illustrating a late stage of hyperlipaemia in which the endothelial cell (*EC*) is loaded with lipid droplets (*ld*) and lies on a hyperplasic basal lamina (*bl*). The ECs have numerous apical and basal pseudopods (*arrows*). *n*, nucleus; *l*, lumen; × 14,000. (Reprinted with permission from Simionescu et al. 1997)

is the formation of the complicated plaque made up of macrophage-derived foam cells, SMC-derived foam cells, lymphocytes and mast cells, all embedded in a hyperplasic extracellular matrix. With time, the ECs become progressively loaded with lipid droplets, ultimately becoming EC-derived foam cells (Fig. 11) that are vulnerable and susceptible to physical disruption.

Severe dyslipidaemia leads to EC injury and death; desquamation of ECs exposes the subendothelial collagen and vWF that promote platelet adhesion, activation, thrombus formation and the occlusion of the vessel's lumen that triggers the acute coronary syndromes (reviewed in Simionescu 2004).

8.2
Diabetes Induces Modifications of Endothelial Cells

Diabetes, a complex disease characterised by abnormalities of glucose homeostasis, is now considered to be a vascular disease due to the macro- and microangiopathies that accompany this condition. Among the abnormalities induced by hyperglycaemia are increased oxidative stress, non-enzymatic glycosylation of proteins and increased accumulation of AGE-proteins and glycated LDL in the circulation and within the subendothelium. EC dysfunction is expressed by the increased plasma concentration of nitrites, nitrates, endothelin I, vWF, tissue-type plasminogen activator, PAI-I and endothelial adhesion molecules (i.e. E-selectin, ICAM I).

8.2.1
Alterations of Large Vessel Endothelial Cells

Diabetes accelerates the early development and progression of plaque formation that develop in arterial lesion-prone areas, leading to rapid calcification. In general, the sequence of events taking place in the aorta of diabetic animals is largely similar to those found in hyperlipaemia, except that the alterations occur at a faster rate. This is particularly true in simultaneous hyperglycaemia and hyperlipaemia. Briefly, the aortic endothelium changes to a secretory phenotype, and within the multilayer basal lamina numerous MRL are entrapped. Accumulation of subendothelial foam cells (derived from macrophages, smooth muscle cells and ultimately ECs) lead to a developed fibro-lipid plaque. By contrast, in coronary arteries under the continuous endothelium and proliferated basal lamina, numerous SMCs form a fibro-muscular plaque. Interestingly, the aortic endothelium maintains its integrity throughout the process of plaque formation and only at very late stages do the ECs undergo apoptosis or necrosis.

8.2.2
Modifications of Capillary Endothelial Cells

Myocardial Capillaries In diabetes, a fraction of heart capillaries are partially collapsed, the lumen narrows and a marked increase in the deposition of extracellular matrix impedes the diffusion of oxygen and transport of molecules

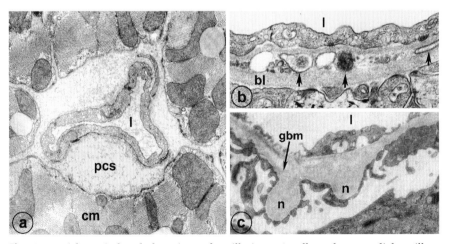

Fig. 12a–c Diabetes-induced alterations of capillaries. **a** A collapsed myocardial capillary is surrounded by pericapillary hyperplasic matrix (*pcs*) that separates the vessel from the neighbouring cardiomyocytes (*cm*). **b** Retinal capillaries with characteristic hyperplasic basal lamina (*bl*) that houses a fragmented pericyte (*arrows*). **c** In glomerular capillaries, the thickened basement membrane (gbm) contains numerous nodules (*n*). *l*, lumen. **a** × 14,000; **b** × 20,000; **c** × 8,000 (Reprinted with permission from Simionescu et al. 1996)

from the plasma to cardiomyocytes (Fig. 12a). In some locations, pericytes show signs of damage and death.

Retinal capillary endothelial cells display a prominent biosynthetic apparatus and a thickening of the basal lamina that entraps pericytes in various stages of degeneration and death: Cell fragments appear dispersed within the subendothelium (Fig. 12b).

Kidney, glomerular capillaries exhibit severe thickening of the glomerular basement membrane that displays marked irregularities and large protruding nodules (Fig. 12c).

8.3
Alterations of the Endothelium in Alzheimer's Disease

The ultrastructural alterations of capillary ECs in AD consist of atrophy, swelling, the presence of irregular nuclei, compromised morphology of tight junctions, degeneration of pericytes and changes in mitochondrial density. The capillary basal lamina exhibits consistent thickening and local disruption. All these modifications impair the function of the BBB (reviewed in Farkas and Luiten 2001). Accumulation of amyloid-β within the parenchyma or cerebral vasculature and its interaction with the putative receptor (R-AGE) stimulate molecular signalling that induces expression of EC adhesion molecules and ensuing migration of circulating monocytes across the BBB (Giri et al. 2002).

9
Conclusion: Lessons from a Brave Cell

The multitude of functions of the vascular endothelium in physiological conditions, along with its reactivity to aggressive factors in pathological conditions, highlights the attributes and the ability of these cells to tolerate and adapt to the surrounding milieu. It is remarkable that ECs have the innate mechanisms to adjust, confront and counterbalance the insults coming from the plasma or from neighbouring cells. The cells respond to aggressive factors by modulation of constitutive functions, followed by EC dysfunction, and only ultimately by cell damage and death. It is noteworthy that the ECs possess the complex mechanisms aimed at maintaining their structural–functional integrity and concurrently at protecting the cells of the tissues in which they reside and, on a larger scale, of the entire organism.

Acknowledgements We are grateful to all our collaborators, excellent scientists and superb technicians that over the years participated in the collection of data presented in this chapter. Also, we are greatly indebted to Mrs. Marilena Daju who assisted us with her expert skills and dedication throughout the preparation of this chapter.

References

Antohe F, Heltianu C, Simionescu N (1986) Further evidence for the distribution and nature of histamine receptors on microvascular endothelium. Microcirc Endothelium Lymphatics 3:163–185

Antohe F, Dobrila LN, Heltianu C (1989) Histamine receptors on mast cells. Microcirc Endothelium Lymphatics 4:469–488

Antohe F, Poznansky MJ, Simionescu M (1999) Low density lipoprotein binding induces asymmetric redistribution of the low density lipoprotein receptors in endothelial cells. Eur J Cell Biol 78:407–415

Antohe F, Radulescu L, Gafencu A, Ghetie V, Simionescu M (2001) Expression of functionally active FcRn and the differentiated bidirectional transport of IgG in human placental endothelial cells. Hum Immunol 62:93–105

Bearer EL, Orci L (1976) Endothelial fenestral diaphragms: a quick-freeze, deep-etch study. J Cell Biol 100:418–428

Betsholtz C, Lindblom P, Gerhardt H (2005) Role of pericytes in vascular morphogenesis. EXS 94:115–125

Dejana E (2004) Endothelial cell-cell junctions: happy together. Nat Rev Mol Cell Biol 5:261–270

Drenckhahn D, Ness DW (1997) The endothelial contractile cytoskeleton. In: Born GVR, Schwartz CJ (eds) Vascular endothelium: physiology, pathology and therapeutic opportunities. Schattauer, Stuttgart, pp 1–15

Farkas E, Luiten PG (2001) Cerebral microvascular pathology in aging and Alzheimer's disease. Prog Neurobiol 64:575–611

Gafencu A, Stanescu M, Toderici AM, Heltianu C, Simionescu M (1998) Protein and fatty acid composition of caveolae from apical plasmalemma of aortic endothelial cells. Cell Tissue Res 293:101–110

Ghetie V, Hubbard JG, Kim JK, Tsen MF, Lee Y, Ward ES (1996) Abnormally short serum half-lives of IgG in beta 2-microglobulin-deficient mice. Eur J Immunol 26:690–696

Ghitescu L, Fixman A, Simionescu M, Simionescu N (1986) Specific binding sites for albumin restricted to plasmalemmal vesicles of continuous capillary endothelium: receptor-mediated transcytosis. J Cell Biol 102:1304–1311

Giri R, Selvaraj S, Miller CA, Hofman F, Yan SD, Stern D, Zlokovic BV, Kalra VK (2002) Effect of endothelial cell polarity on β-amyloid-induced migration of monocytes across normal and AD endothelium. Am J Physiol Cell Physiol 283:C89–C904

Haselton FR, Alexander JS, Mueller SN, Fishman AP (1992) Modulation of endothelial paracellular permeability. A mechanistic approach. In: Simionescu N, Simionescu M (eds) Endothelial cell dysfunctions. Plenum Press, New York, pp 103–126

Heltianu C, Simionescu M, Simionescu N (1982) Histamine receptors of the microvascular endothelium revealed in situ with a histamine-ferritin conjugate: characteristic high-affinity binding sites in venules. J Cell Biol 93:357–364

Lisanti MP, Scherer PE, Vidugiriene J, Tang Z, Hermanowski-Vosatka A, Tu YH, Cook RF, Sargiacomo M (1994) Characterization of caveolin-rich membrane domains isolated from an endothelial-rich source: implications for human disease. J Cell Biol 126:111–126

Martin AR, Bailie JR, Robson T, McKeown SR, Al-Assar O, McFarland A, Hirst DG (2000) Retinal pericytes control expression of nitric oxide synthase and endothelin-1 in microvascular endothelial cells. Microvasc Res 59:131–139

McIntosh DP, Schnitzer JE (1999) Caveolae require intact VAMP for targeted transport in vascular endothelium. Am J Physiol 277:H2222–H2232

Mineo C, Anderson RG (2001) Potocytosis. Histochem Cell Biol 116:109–118

Minshall RD, Sessa WC, Stan RV, Anderson RG, Malik AB (2003) Caveolin regulation of endothelial function. Am J Physiol Lung Cell Mol Physiol 285:L1179–L1183

Muller WA, Gimbrone MA Jr (1986) Plasmalemmal proteins of cultured vascular endothelial cells exhibit apical-basal polarity: analysis by surface-selective iodination. J Cell Biol 103:2389–2402

Oh P, McIntosh DP, Schnitzer JE (1998) Dynamin at the neck of caveolae mediates their budding to form transport vesicles by GTP-driven fission from the plasma membrane of endothelium. J Cell Biol 141:101–114

Page C, Rose M, Yacoub M, Pigott R (1992) Antigenic heterogeneity of vascular endothelium. Am J Pathol 141:673–683

Palade GE (1953) An electron microscope study of the mitochondrial structure. J Histochem Cytochem 1:188–211

Pino RM (1986) The cell surface of a restrictive fenestrated endothelium. I. Distribution of lectin-receptor monosaccharides on the choriocapillaries. Cell Tissue Res 243:145–156

Razani B, Woodman SE, Lisanti MP (2002) Caveolae: from cell biology to animal physiology. Pharmacol Rev 54:431–467

Rothberg KG, Heuser JE, Donzell WC, Ying YS, Glenney JR, Anderson RG (1992) Caveolin, a protein component of caveolae membrane coats. Cell 68:673–682

Sage H, Trueb B, Bornstein P (1983) Biosynthetic and structural properties of endothelial cell type VIII collagen. J Biol Chem 258:13391–13401

Schmelz M, Franke WW (1993) Complexus adhaerentes, a new group of desmoplakin-containing junctions in endothelial cells: the syndesmos connecting retothelial cells of lymph nodes. Eur J Cell Biol 61:274–289

Schmidt AM, Hasu M, Popov D, Zhang JH, Chen J, Yan SD, Brett J, Cao R, Kuwabara K, Costache G, Simionescu N, Simionescu M, Stern D (1994) Receptor for advanced glycation end products (AGEs) has a central role in vessel wall interactions and gene activation in response to circulating AGE proteins. Proc Natl Acad Sci USA 91:8807–8811

Schnitzer JE, Liu J, Oh P (1995) Endothelial caveolae have the molecular transport machinery for vesicle budding, docking, and fusion including VAMP, NSF, SNAP, annexins, and GTPases. J Biol Chem 270:14399–14404

Shepro D, Dunham B (1986) Endothelial cell metabolism of biogenic amines. Annu Rev Physiol 48:335–345

Simionescu M (1991) Cellular organization of the alveolar capillary unit: structural–functional correlations. In: Said SI (ed) The pulmonary circulation and acute lung injury. Futura Publishing Company. Mount Kisko, New York, pp 13–42

Simionescu M (2001) Effect of hyperlipemia-hyperglycemia on the vascular endothelium. In: Catravas JD, Callow AD, Ryan US, Simionescu M (eds) Vascular endothelium source and target for inflammatory mediators. NATO ASI Series, New York, pp 87–99

Simionescu M (2004) Endothelial cell—a key player in all stages of atherosclerosis. In: Simionescu M, Sima A, Popov D (eds) Cellular dysfunction in atherosclerosis and diabetes—reports from bench to bedside. Romanian Academy Publishing House, Bucharest, pp 73–97

Simionescu M, Simionescu N (1991) Endothelial transport of macromolecules. Transcytosis and endocytosis. a look from cell biology. Cell Biol Rev 25:47–61

Simionescu M, Simionescu N, Palade GE (1975b) Segmental differentiations of cell junctions in the vascular endothelium. The microvasculature. J Cell Biol 67:863–885

Simionescu M, Simionescu N, Palade GE (1982a) Differentiated microdomains on the luminal surface of the capillary endothelium. I. Distribution of lectin receptors. J Cell Biol 94:406–413

Simionescu M, Simionescu N, Palade GE (1982b) Preferential distribution of anionic sites on the basement membrane and the abluminal aspect of the endothelium in fenestrated capillaries. J Cell Biol 95:425–434

Simionescu M, Popov D, Sima A, Hasu M, Costache G, Faitar S, Vulpanovici A, Stancu C, Stern D, Simionescu N (1996) Pathobiochemistry of combined diabetes and atherosclerosis studied on a novel animal model. The hyperlipemic-hyperglicemic hamster. Am J Pathol 148:997–1014

Simionescu M, Sima A, Popov D, Hasu M, Costache G (1997) Pathobiology of the wall in combined diabetes and atherosclerosis. In: Motta PM (ed) Recent advances in microscopy of cell, tissues and organs. Antonio Delfino Editore, Rome, pp 339–343

Simionescu M, Gafencu A, Antohe F (2002) Transcytosis of plasma macromolecules in endothelial cells: a cell biological survey. Microsc Res Tech 57:269–288

Simionescu N (1979) The microvascular endothelium: segmental differentiations, transcytosis, selective distribution of anionic sites. In: Weissman G, Samuelson B, Paoletti R (eds) Advances in inflammation research. Raven Press, New York, pp 61–70

Simionescu N, Simionescu M, Palade GE (1975a) Permeability of muscle capillaries to small hemepeptides. Evidence for the existence of patent transendothelial channels. J Cell Biol 64:586–607

Simionescu N, Simionescu M, Palade GE (1981) Differentiated microdomains on the luminal surface of the capillary endothelium. I. Preferential distribution of anionic sites. J Cell Biol 90:605–613

Simionescu N, Lupu F, Simionescu M (1983) Rings of membrane sterols surround the opening of vesicles and fenestrae in capillary endothelium. J Cell Biol 97:1592–1600

Simionescu N, Mora R, Vasile E, Lupu F, Filip DA, Simionescu M (1990) Prelesional modifications of the vessel wall in hyperlipidemic atherogenesis. Extracellular accumulation of modified and reassembled lipoproteins. Ann NY Acad Sci 598:1–16

Sims DE (2000) Diversity within pericytes. Clin Exp Pharmacol Physiol 27:842–846

Stan RV, Tkachenko E, Niesman IR (2004) PV1 is a key structural component for the formation of the stomatal and fenestral diaphragms. Mol Biol Cell 15:3615–3630

Steinbrecher UP (1999) Receptors for oxidized low density lipoprotein. Biochim Biophys Acta 1436:279–298

Tiruppathi C, Song W, Bergenfeldt M, Sass P, Malik AB (1997) Gp60 activation mediates albumin transcytosis in endothelial cells by tyrosine kinase-dependent pathway. J Biol Chem 272:25968–25975

Tuma PL, Hubbard AL (2003) Transcytosis: crossing cellular barriers. Physiol Rev 83:871–932

Weibel ER, Palade GE (1964) New cytoplasmic components in arterial endothelia. J Cell Biol 23:101–112

Development of the Endothelium

A. M. Suburo[1] · P. A. D'Amore[2] (✉)

[1]Facultad de Ciencias Biomédicas, Universidad Austral, B1629AHJ Buenos Aires, Argentina
[2]Schepens Eye Research Institute and Harvard Medical School, Boston MA, 02114, USA
pdamore@vision.eri.harvard.edu

1	**Introduction**	72
2	**Early Endothelial Precursors**	73
2.1	Haemangioblasts in the Yolk Sac	74
2.2	Development of Primitive Intraembryonic Vessels	75
2.3	The Haemogenic Endothelium	75
3	**Molecular Differentiation of EC**	77
3.1	The Yolk Sac and Extraembryonic Vasculogenesis	77
3.2	Endothelial Differentiation in Embryoid Bodies	78
3.3	Intraembryonic Differentiation of EC	79
3.4	VEGF-A Transcription and Signalling in Differentiation of EC	80
4	**Development of Mural Cells**	82
4.1	Regulation of Pericyte/SMC Phenotype	83
4.2	Differentiation of Pericytes and SMC	84
4.2.1	S1P Phosphate and S1P Receptors	84
4.2.2	Wnts	85
4.2.3	Platelet-Derived Growth Factors Family	86
4.2.4	Angiopoietins and the Tie Receptors	86
4.2.5	Transforming Growth Factor-β1	88
4.2.6	Interactions Between Signalling Cascades	89
5	**Endothelium Morphogenesis**	90
5.1	Angiogenic Sprouting	90
5.2	Attraction and Repulsion of Angiogenic Sprouts	91
5.2.1	Patterning of the Embryonic Midline	91
5.2.2	Semaphorins	91
5.2.3	Netrins and Their Receptors	92
5.2.4	Calcineurin/NFAT	92
6	**Development of Arteries and Veins**	93
6.1	Ephrins and Eph Receptors	93
6.2	Hedgehog in Arteriogenesis	94
6.3	VEGF-A in Arteriogenesis	94
6.4	Notch Pathways	95
6.5	TGF-β1 Receptors	96
7	**Concluding Remarks**	97
	References	97

Abstract Our understanding of the regulation of vascular development has exploded over the past decade. Prior to this time, our knowledge of vascular development was primarily based on classic descriptive studies. The identification of stem cells, lineage markers, specific growth factors and their receptors, and signalling pathways has facilitated a rapid expansion in information regarding details of the mechanisms that govern development of the vascular system.

Keywords Embryo · VEGF · Haemangioblasts · Endothelial determination and differentiation · Mural cells

1
Introduction

Endothelial cells (EC) derive from early precursors that proliferate and then coalesce to form complex vascular networks. During this developmental process, EC precursors receive appropriate developmental signals, inducing expression of specific genes and stimulating proliferation and migration. At the same time, EC are able to direct differentiation of neighbouring tissues, including cells that will form periendothelial vascular structures and the parenchyma served by the developing vessels. The result is a quiescent tissue, finely tuned to functional demands of nearby tissues. This review will describe fundamental steps of endothelial developmental processes as a pathway to the phenotypic diversity that is seen throughout the vascular system. In addition, we will review the anomalies of endothelial development and the possibility of reactivation of developmental processes under situations of stress and disease.

Differentiation of EC precursors is followed by formation of primitive endothelial tubes, and development and maturation of a vascular network. These processes involve changes in shape and adhesivity of EC and their precursors, sprouting and splitting of primitive vascular tubules, and remodelling of existing vessels plus their investment with mural cells-vascular smooth muscle cells (SMC) and pericytes.

Co-ordinated operation of numerous receptor-mediated signalling pathways and the activation of specific transcription factors are required for EC differentiation. Expression of receptors for vascular endothelial growth factor (VEGF)-A, which has been implicated in virtually all aspects of cardiovascular system formation, including heart development, haematopoiesis, vasculogenesis, angiogenesis and endothelial survival (Zachary 2003), is considered a hallmark of endothelial development. However, VEGF-A signals must be co-ordinated with many other intra- and extracellular messengers that contribute to the development of structurally and functionally mature blood vessels.

2
Early Endothelial Precursors

Vasculogenesis is the differentiation and coalescence of mesodermal precursor cells to form vessels, whereas angiogenesis involves the migration and division of EC from pre-existing vessels to form new vasculature. The existence of the haemangioblast, a common progenitor for endothelial and haematopoietic lineages, was first postulated at the beginning of the last century, and it was considered that separation of both lineages occurred in early stages of yolk sac development. Contemporary findings, however, indicate a more complicated differentiation pathway (summarised in Fig. 1).

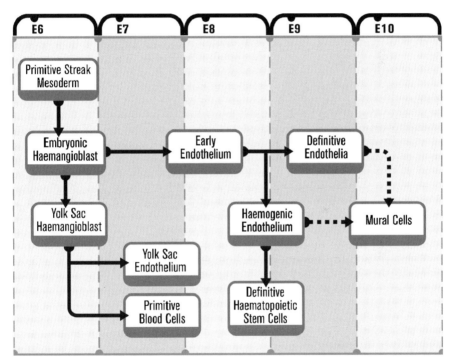

Fig. 1 Timetable of endothelial differentiation. In the mouse embryo, major steps of endothelial differentiation take place between embryonic day E6 and E10. Haemangioblasts differentiate within the mesoderm of the primitive streak and migrate to the yolk sac where they form blood islands that give rise to endothelium and primitive blood cells. Blood islands fuse to form the extraembryonic vessels. Within the embryo, endothelial precursors, presumably derived from similar haemangioblasts, differentiate to the endothelium of large intraembryonic vessels. Through angiogenesis, this early endothelium is the origin of the rest of the vasculature. Certain regions of the early endothelium are specialised into the haemogenic endothelium, which is the source of definitive haematopoietic cells. Some evidence suggests that endothelium and haematopoietic cells may be able to differentiate into mural cells

2.1
Haemangioblasts in the Yolk Sac

Haemangioblasts have recently been defined as a subpopulation of mesoderm cells that originate in the posterior region of the primitive streak. They co-express brachyury (also known as T) and VEGF-A receptor 2 (VEGFR-2; Flk1 in mouse and KDR in human) genes, and are first detected at the mid-streak stage of gastrulation (Huber et al. 2004). Thus, the earliest stages of haemangioblast differentiation probably occur before their migration to the extraembryonic mesoderm of the presumptive yolk sac (Fig. 2). Haemangioblasts aggregate in presumptive blood islands (also known as mesodermal cell masses or angioblastic cords) that appear in the extraembryonic mesoderm between mouse embryonic day (E)7 and E7.5. Cells at the outer aspect

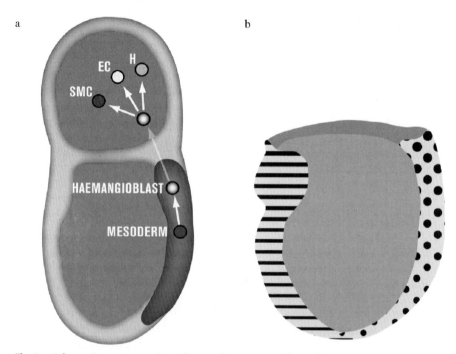

Fig. 2 a Schematic representation of a 7.0-day mouse embryo illustrating haemangioblast development and migration to the yolk sac. The haemangioblast is a Bry$^+$ and VEGFR-2$^+$ cell derived from mesodermal Bry$^+$ cells located in the region of the primitive streak (*black*). Haemangioblasts migrate onto the yolk sac where they differentiate into haematopoietic cells (H), EC and SMC. Adapted from Huber et al. (2004). **b** Representation of the spatial distribution of VEGF-A and VEGFR-2 transcripts in an E7.75 embryo transversely sectioned through the amnion. VEGF-A is present throughout the whole embryo, but is at higher levels in the cephalic region (*striped region*) where the neural plate is developing. Conversely, VEGFR-2 is also widely distributed but predominates caudally (*dots*) where EC precursors arise in the region of the primitive streak. (Adapted from Hiratsuka et al. 2005)

of the blood islands assume a spindle shape as they differentiate into EC, whereas inner cells progressively lose their intercellular attachments as they differentiate into primitive blood cells. Shortly thereafter, blood islands fuse to form the first endothelial tubes. A three-dimensional network, the primary vascular plexus, takes shape and then undergoes reorganisation, sprouting and remodelling to form the large vitelline vessels. Remodelling is accompanied by the recruitment and differentiation of vascular SMC (Drake and Fleming 2000).

At the three-somite stage, vascular development has spread throughout the yolk sac, but primitive red blood cells remain restricted to the blood islands of the proximal yolk sac, suggesting that there are haemangiogenic and angiogenic regions within the yolk sac (McGrath et al. 2003). On the other hand, the presence of cells giving rise to both endothelial and haematopoietic lineages in the allantois, placenta and somitic tissue (Alvarez-Silva et al. 2003; Finkelstein and Poole 2003), indicates that haemangioblasts could be far more extensively dispersed than previously thought.

2.2
Development of Primitive Intraembryonic Vessels

Vasculogenesis and angiogenesis are regulated by the capacity of EC and their precursors to adhere to each other and form new tubes. These cells can undergo dramatic changes in their shape, and their plasma membranes can engage in extensive protrusive activity with directionally oriented processes recognising and contacting neighbouring EC precursors to form cord-like cellular assemblies. At the same time, EC flatten and assume the spindle shape characteristic of differentiated EC. Tensional forces contribute to the creation of a single cell-layered vascular lumen. Continued vascular fusion can combine neighbouring small-calibre vessels into larger ones. The earliest intraembryonic endothelial populations appear in regions fated to give rise to the heart before vasculogenesis. The quantity of these cells increases dramatically before the aortic primordia first become discernible. Intraembryonic vasculogenesis is initiated in the cranial region of E7.3 embryos. Bilateral aortic primordia become discernible by E7.8 and their fusion is completed by E8.3. The lateral vascular networks are formed between E8.2 and E8.5. These early vascular channels develop before links with the vitelline vessels are established (Drake and Fleming 2000).

2.3
The Haemogenic Endothelium

Groups of 25–100 rounded cells, possessing the same ultrastructural features of primitive haematopoietic cells of the yolk sac blood islands (Tavian et al. 1996; Godin and Cumano 2002), are attached to the ventral luminal wall of

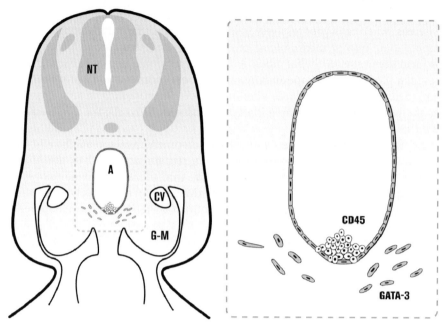

Fig. 3 Schematic representation of the embryo at the level of the truncal aorta-gonad-mesonephros (*AGM*). The area of haemogenic activity, including the aorta and subaortic patches, is outlined. *NT* is the neural tube and *CV* is the cardinal vein. The enlargement of the aortic region illustrates the intra-aortic clusters, which are restricted to the ventral part (floor) of the vessel and exhibit CD45. The subaortic patches are found bilaterally. (Based on Tavian et al. 1996)

the main arteries, aorta, omphalomesenteric and umbilical arteries (Fig. 3). These cells, which exhibit haematopoietic markers, are only observed during a brief stage in gestation (E9-11.5 in mice and ED30-40 in humans). This time period coincides with the one in which multipotent definitive haematopoietic stem cells can be isolated from the aorta-gonad-mesonephros (AGM) region, defined as the region of the murine embryonic splanchnopleuric mesoderm bounded by the dorsal aorta, gonadal ridge and pro/mesonephros. No intra-aortic clusters are visible outside the AGM in the post-umbilical caudal region of the embryo. Cytological features of the aortic floor, such as the presence of "bottled-shaped" cells and the absence of a basal membrane, suggest that cell migration can occur across this endothelium (Godin and Cumano 2002). A special group of mesenchymal cells, the subaortic patches, are located below the haematopoietic clusters, but their relationship with differentiation of the intra-aortic clusters has still to be clarified (Fraser et al. 2003).

3
Molecular Differentiation of EC

3.1
The Yolk Sac and Extraembryonic Vasculogenesis

Early haemangioblasts ($Bry^+/VEGFR\text{-}2^+$) apparently arise in the primitive streak region; however, the yolk sac probably provides them with a suitable environment inducing divergence of primitive EC and primitive blood cells. The yolk sac is composed of two cell layers, an extraembryonic mesodermal layer and a visceral endoderm layer. Members of the GATA family of transcription factors are important for mesodermal development. In mouse embryos, the loss of *GATA-1* leads to a qualitative defect in primitive erythroid cell differentiation, whereas the loss of *GATA-2* has a modest quantitative effect at the yolk sac (Fujiwara et al. 2004). In later stages, definitive haematopoietic stem cells are highly dependent on *GATA-2*, which is expressed in the aortic endothelium and neighbouring mesenchymal cells (Ling et al. 2004).

The haematopoietically expressed homeobox (Hex) gene is transiently expressed in the nascent blood islands of the visceral yolk sac and later in embryonic angioblasts and endocardium. Hex is required for the transition from the definitive haemangioblast to a definitive haematopoietic stem cell, and to a somewhat lesser extent, EC, since $Hex^{-/-}$ embryos can form some vessels before they die at day 12 (Guo et al. 2003). Other transcription factors, encoded by the stem cell leukaemia (SCL, also known as TAL-1) and LMO-2 genes, are essential for the development of both primitive erythropoiesis and definitive haematopoiesis. SCL is expressed in the presumptive yolk sac region in the mid/late streak stage of mouse embryos, coincident with VEGFR-2, and continues to be expressed in haemangioblasts, definitive haematopoietic stem cells, some haematopoietic lineages and, at lower levels, in EC precursors and some EC. Expression of SCL follows expression of VEGFR-2, and is not detected in $VEGFR\text{-}2^{-/-}$ embryos (Ema et al. 2003). $SCL^{-/-}$ mouse embryos contain no primitive or definitive haematopoietic cells in the yolk sac and die around E10.5 because of defective embryonic haematopoiesis. Although these embryos generate EC, suggesting that this transcription factor is only required for blood cell commitment, they also show defective remodelling of primary vascular networks (Gottgens et al. 2002).

Signalling from the endoderm is a critical early determinant of haematopoietic and vascular development. Indian hedgehog (Ihh) but not Sonic hedgehog (Shh) is expressed in the visceral endoderm of gastrulating mouse embryos and mature yolk sacs. Ihh alone is sufficient to activate embryonic haematopoiesis and vasculogenesis in epiblasts in the absence of visceral endoderm (Dyer et al. 2001), and $Ihh^{-/-}$ yolk sacs can form blood vessels, but they are fewer in number and smaller, perhaps owing to their inability to undergo vascular remodelling (Byrd et al. 2002).

VEGF-A signalling is pivotal for vascular differentiation because its inhibition prevents vascular development from its beginning and consistently inhibits tumour vascularisation. The VEGF ligand family includes VEGF-A, VEGF-B, placenta growth factor (PlGF), VEGF-C and VEGF-D. VEGF-A interacts with three tyrosine kinase receptors, VEGFR-1 (Flt1), VEGFR-2 and VEGFR-3 (Flt4). VEGF-A function is required for development of the yolk sac mesenchyme and recruitment of haematopoietic precursors to the yolk sac, expansion of the primitive erythroid compartment, survival of primitive erythrocytes, and angiogenic sprouting of blood vessels, but not for EC specification (Duan et al. 2003; Martin et al. 2004). The extraembryonic visceral endoderm and the yolk sac mesodermal sheet are the first tissues to express VEGF-A, and expression in the visceral endoderm seems to be necessary and sufficient for normal development of the yolk sac vasculature (Damert et al. 2002). In blood islands, outer EC are VEGFR-2^+, whereas "core" cells, representing the primitive haematopoietic lineage, are VEGFR-2^- (Drake and Fleming 2000) and CD41^+ (Ferkowicz et al. 2003). Embryos lacking VEGF-A or VEGFR-2 genes have few or no blood vessels (Shalaby et al. 1997). *VEGFR-$2^{-/-}$* mice do not develop yolk sac blood islands or blood vessels, and die between E8.5 and E9.5, whereas *VEGFR-$1^{-/-}$* die due to an overgrowth of vascular EC and disorganisation of blood vessels.

Transforming growth factor-β1 (TGF-β1)/bone morphogenetic protein (BMP) families of factors and their receptors are required for extraembryonic vasculogenesis. BMP4 is secreted by extraembryonic mesoderm at the posterior end of the primitive streak and, in *BMP4*-null mice that survive beyond gastrulation, both haematopoiesis and vasculogenesis are greatly reduced. BMP4 acts through activation of the Smad/5 downstream signalling molecules, and mice deficient in Smad1 or Smad5 also display defects in haematopoietic and vascular development (Tremblay et al. 2001). Deficiency of retinoic acid synthesis also generates embryos with multiple anomalies, including missing organised extraembryonic vessels in the yolk sac. Lack of retinoic acid leads to suppression of TGF-β1 and fibronectin production in EC and downregulation of VEGF-A, Ihh and fibroblast growth factor (FGF)-2 in visceral endoderm; these changes are correlated with enhanced EC growth, decreased visceral endoderm survival and lack of capillary plexus remodelling (Bohnsack et al. 2004).

3.2
Endothelial Differentiation in Embryoid Bodies

Under certain in vitro conditions, embryonic stem (ES) cells differentiate into embryoid bodies (EB) that contain precursors for multiple lineages. Differentiation of haematopoietic and endothelial lineages in this model parallels that of the normal mouse (Feraud et al. 2003). Thus, Bry$^+$ mesodermal progenitors can originate blast colony-forming cells (BL-CFCs) expressing VEGFR-2 and will grow blast colonies in response to VEGF-A (Faloon et al. 2000). Since blast colonies contain both haematopoietic and EC precursors, BL-CFCs are postulated to represent the haemangioblast (Chung et al. 2002). In serum-free conditions, ES cells develop only to the mesodermal stage. BMP4 is required for the transition of ES cells to mesoderm, from mesoderm to VEGFR-2^+ cells and from VEGFR-2^+ to SCL$^+$ cells. VEGF-A then acts through VEGFR-2 to expand SCL$^+$ cells. TGF-β1 and activin A further modulate the expansion of haematopoietic and EC lineages (Park et al. 2004). In addition, BMP-binding endothelial cell precursor-derived regulator (BMPER) is specifically expressed in VEGFR-2^+ cells and directly interacts with BMP2, BMP4 and BMP6, and antagonises Smad5 activation, possibly modulating local BMP activity during EC differentiation (Moser et al. 2003).

BL-CFCs have provided a suitable model system to analyse the divergence of haematopoietic and EC lineages in vitro. Initially, a subset of VEGFR-2^+/GATA-1^+ mesodermal cells, representing the primitive erythroid lineage, loses the capacity to give rise to EC (Fujimoto et al. 2001). The remaining VEGFR-2^+/GATA-1^- cells express vascular endothelium (VE)-cadherin, the major component of endothelial adherens junctions. A subset of VE-cadherin$^+$ cells, giving rise to definitive haematopoietic progenitors and to EC, probably represents the "haemogenic" EC (Fujimoto et al. 2001). Primitive endothelial-like cells derived from human ES cells also express platelet endothelial cell adhesion molecule-1 (PECAM-1; CD31), but not CD45, and give rise to endothelial and haematopoietic lineages (Wang et al. 2004a). Wild-type EB give rise to BL-CFCs differentiating into endothelial and haematopoietic cells, but *SCL$^{-/-}$* EB can only differentiate into EC (Faloon et al. 2000).

VEGF-A regulates cellular properties required for migration, including invasive activity, motility and adhesion/de-adhesion to matrix substrates. In cystic EB, VEGF-A expression is both temporally and spatially correlated with development of a vascular network. By contrast, EB derived from *VEGF-A*-null ES cells contain PECAM-1-positive EC that do not form tubes. Addition of VEGF-A partially rescues the formation of vascular networks in the VEGF-A-null EB, whereas addition of FGF-2 results in increased EC proliferation but does not rescue vascular morphogenesis (Ng et al. 2004).

3.3
Intraembryonic Differentiation of EC

Using mice embryos (E7.25-E7.75) in which the lacZ gene is driven under the control of the endogenous VEGFR-2 promoter, EC precursors can be traced as they migrate from the caudal to the cephalic region, where they are incorporated to the developing heart and aorta. EC precursors derived from wild-type or *VEGFR-$2^{+/-}$* mice rapidly move in a cephalic direction, whereas cells derived from *VEGFR-$2^{+/-}$* mice carrying a truncated VEGFR-1 migrate very little. Direction of migration is correlated with sites of VEGF-A synthesis, which is much higher in the cephalic than in the caudal region. VEGFR-1 and VEGFR-2 are mainly expressed caudally (Fig. 2b), where both receptors localise to the same cells. In vitro migration of embryo-derived VEGFR-2$^+$ cells is stimulated both by VEGF-A and PlGF, a specific ligand for VEGFR-1 (Hiratsuka et al. 2005).

PECAM-1 is expressed by early endothelial precursors, first within the yolk sac and then in aortic primordia at E7.8, whereas CD34, VE-cadherin, and Tie2 appear the next day. PECAM-1 expression is initially associated with the entire cell surface, but later becomes localised to sites of cell-cell contact (Drake and Fleming 2000). VE-cadherin promotes cell adhesion and is required for the assembly of the yolk sac primary plexus and remodelling of embryonic blood vessels (Bazzoni and Dejana 2004).

Cell clusters associated with the endothelial floor of the 5-week human embryonic aorta express, among other molecules, the transcription factors SCL, GATA-2, GATA-3 and Runx1 (Godin and Cumano 2002). The haemogenic endothelium expresses GATA-2, c-KIT, tenascin C, VWF, VEGFR-2, PECAM-1, CD34, endomucin, VEGFR-1, VEGFR-2, Flt3L, SCL, Tie2, VE-cadherin and VEGF-A (Godin and Cumano 2002). Embryonic cells selected by surface expression of CD34 or CD31 yield myelo-lymphoid cells in culture, thus supporting the haemogenic nature of intra-aortic clusters (Oberlin et al. 2002). A transient population of cells expressing both CD45 and VE-cadherin probably represents an intermediate stage between EC and blood cells (Fraser et al. 2003). VEGFR-2$^+$/CD34$^-$ cells persist in the para-aortic splanchnopleura or subaortic patches until the disappearance of aorta-associated haematopoietic cell clusters, and it is speculated that these cells represent the intraembryonic haemangioblastic precursor of haematopoietic and endothelial lineages (Cortes et al. 1999).

The transcription factor Runx1 (also known as AML1 or CBFA2), a frequent target of chromosome translocations in acute myeloid leukaemia, is first detected in mesenchymal cells of the yolk sac at E7.5. Clusters of Runx1$^+$ cells, also expressing the pan-leucocyte marker CD45, can be detected inside the aorta, vitelline and umbilical arteries (Fraser et al. 2003). Although Runx1-null embryos show no dramatic defects in primitive erythropoiesis, they fail to generate definitive haematopoietic lineage cells. Main EC and haematopoietic differentiation markers are summarised in Fig. 4.

3.4
VEGF-A Transcription and Signalling in Differentiation of EC

Molecular responses to oxygen gradients contribute to the differentiation and maintenance of the cardiovascular system. Hypoxia-sensitive genes include erythropoietin, transferrin and its receptor, VEGF-A and its receptors, platelet-derived growth factor (PDGF)-B, FGF-2, and multiple genes encoding glycolytic enzymes (Ramirez-Bergeron et al. 2004). Hypoxia-inducible factor (HIF), consisting of HIF-1α (or HIF-2α) and aryl hydrocarbon receptor nuclear

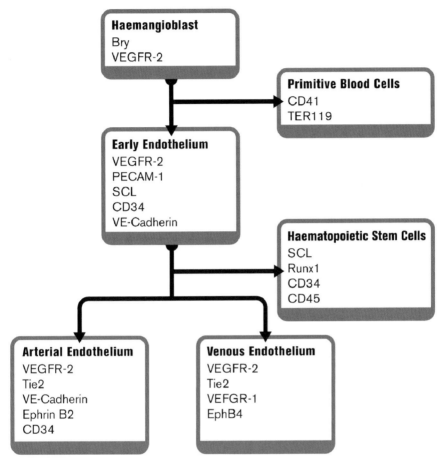

Fig. 4 Gene markers at different stages of endothelial and haematopoietic differentiation. Development of these lineages requires the concerted action of many genes, but those included in the chart have been shown to perform essential differentiation steps. Data were collected from several references included in the text

translocator (ARNT, also known as HIF-1β) subunits, activates multiple genes in response to oxygen deprivation. VEGF-A expression can be activated by HIF-1α or HIF-2α, but only the latter can activate expression of VEGFR-2 (Elvert et al. 2003). In differentiating ES cells, hypoxia accelerates the expression of Bry, BMP4 and VEGFR-2, and proliferation of BL-CFCs (Ramirez-Bergeron et al. 2004).

Other effectors, however, must be involved during early embryogenesis, since oxygen is distributed by diffusion and its levels seem to be almost the same throughout the embryo (Hiratsuka et al. 2005). Many transcriptional regulators have been associated with VEGF-A expression under pathological conditions, but few of them have been studied during embryonic development.

Ets transcription factors could be involved in the control of VEGF-A and other genes involved in angiogenesis, such as VEGFR-1, VEGFR-2, Tie1 and Tie2. Ets-1 is highly expressed in the lateral mesoderm when VEGFR-2 starts to be expressed in EC precursors, and HIF-2α co-operates with Ets-1 in activating transcription of this receptor (Elvert et al. 2003). ErbB2, one of the receptors for the family of epidermal growth factor (EGF) ligands, has also been implicated as a positive modulator of VEGF-A expression (Loureiro et al. 2005).

Most biologically relevant VEGF-A signalling in EC is mediated via VEGFR-2. Major pathways include survival signalling through phosphoinositide (PI)-3-kinase-dependent activation of the anti-apoptotic kinase Akt/protein kinase B (Zachary 2003). VEGFR-1 has a tenfold higher affinity for VEGF-A than VEGFR-2 but with a much weaker tyrosine kinase activity. VEGFR-1 is expressed as a full-length molecule in blood vessels and capillaries of developing organs, closely resembling the pattern of VEGFR-2 distribution, and as a soluble form that consists of the extracellular domain. Since VEGFR-1 lacking the tyrosine kinase domain is sufficient for normal development and angiogenesis in mice (Hiratsuka et al. 2005), it has been suggested that VEGFR-1 may function as a "decoy" receptor to negatively regulate VEGFR-2-mediated actions. Such a role is supported by increased VEGFR-2 tyrosine phosphorylation in differentiated ES cell cultures lacking VEGFR-1 (Roberts et al. 2004).

4
Development of Mural Cells

Pericytes are the mural cells of capillaries and post-capillary venules, whereas SMC are associated with arteries, arterioles and veins. Mural cells contribute to the developing vascular wall through cell proliferation and production of extracellular matrix components such as collagen, elastin and proteoglycans. Most mural cells are of mesodermal origin, but unlike other tissues, a discrete population of mural cell precursors cannot be distinguished in the developing organism. SMC in the proximal aorta, aortic arch and pulmonary trunk are derived from neural crest, whereas SMC in the coronary arteries are derived from epicardium, and those in the descending aorta originate from mesoderm and possibly from transdifferentiated endothelium (Mann et al. 2004). Various clonal lines of multipotent, self-renewing cells called mesoangioblasts have been isolated from embryonic dorsal aorta (Minasi et al. 2002).

In vitro experiments suggest that EC or EC precursors may give rise to mural cells. Thus, VEGFR-2^+ cells derived from ES cells can differentiate into both endothelial and mural cells and can form capillary-like structures in vitro. The same cells can also incorporate into blood vessels as either EC or pericytes when injected into chick embryos (Yamashita et al. 2000). SMC are also produced from ES-derived BL-CFCs, and VEGFR-2^+ cells retain the capacity to form this phenotype after the time of haematopoietic cell formation (Ema et al. 2003).

The absence of mural cells during vascular development results in endothelial hyperplasia, abnormal EC shape, alteration of junctional proteins, increased capillary diameter vessel dilation and microaneurysms, abnormal vascular remodelling and increase of permeability. Affected embryos frequently die from embryonic or perinatal haemorrhage (Hellstrom et al. 2001; Uemura et al. 2002).

4.1
Regulation of Pericyte/SMC Phenotype

Understanding phenotypic regulation of SMC during development is particularly important, since changes of SMC associated with diseased vascular tissue partially recapitulate normal fetal and neonatal development. Different molecular transitions occur during SMC differentiation, leading to the development of the cytoskeleton, acquisition of contractile function and differentiation of arterial and venous SMC. Transcripts for α-smooth muscle actin (α-SMA) and SMα22, a calponin-related protein, are expressed in the developing dorsal aorta at E9.5, in the umbilical vessels and other cephalic vessels at E10.5, and in most vessels at E14.5. These genes, however, are also expressed in the early tubular heart, myotome and skeletal muscles. A more specific marker, smooth muscle-myosin heavy chain (SM-MHC), does not appear in the aorta until E10.5 (Li et al. 1996). In the retina, mural cell precursors express NG2 proteoglycan (or its human homologue, high molecular weight-melanoma associated antigen) and α-SMA, whereas mature pericytes express NG2 and desmin. Calponin and caldesmon, required for the contractile response, are markers of highly differentiated SMC (Hughes and Chan-Ling 2004). Diversity of gene products generated by alternative splicing can be enormous and is especially relevant for development of different muscle phenotypes, e.g. the expression of different smoothelin isoforms in vascular and visceral SMC (Rensen et al. 2002). Tissue-specific alternative splicing characterises the differentiated vascular SMC phenotype and is rapidly lost during vascular disease.

Little is known about the maturation of vascular SMC, but Notch3 (see Sect. 6.4) and angiotensin receptor 2 (AT2) may be involved. In fetal blood vessels, the AT2 receptor is expressed at late gestation but decreases to very low levels in the adult. Levels of the regulatory proteins calponin and caldesmon are below normal in the aorta of $AT2^{-/-}$ mice. Since AT2 is re-expressed in vascular injury, it may have a role in late vascular remodelling; however, this remains controversial (Perlegas et al. 2005).

Most SMC genes are under the control of the serum response factor (SRF) that binds to a *cis* element known as a CArG box. The SM-MHC gene includes three positive-acting CArG elements that are selectively required for the different SMC phenotypes. Mutation of an intronic CArG results in an arterial phenotype, with complete silencing of SM-MHC expression in the aorta, common carotid arteries and the main trunks of subclavian arteries (Manabe and

Owens 2001). Three CArG sites also present in the SMα22 promoter region appear to be involved in vascular SMC differentiation (Ding et al. 2004). Myocardin and related molecules MRTF-A and MRTF-B are SRF co-activators that are expressed in a subset of vascular and visceral SMC, usually preceding expression of SMC-specific genes. Interfering with myocardin expression results in embryonic death at E11.5 from a lack of vascular SMC. It has been proposed that the reversible association of myocardin with SRF could be the basis of the switch between muscle-specific and growth-regulated genes during embryological and pathological SMC differentiation (Wang and Olson 2004).

4.2
Differentiation of Pericytes and SMC

Mural cells are expanded and recruited to angiogenic sprouts by proliferation and migration (Beck and D'Amore 1997). Association of mural cells with newly formed blood vessels appears to regulate EC proliferation, survival, migration, differentiation and stability (Antonelli-Orlidge et al. 1989; Hirschi et al. 1999). Differentiation of mesenchymal cell precursors (10T1/2 cells) into pericytes is not only accompanied by the expression of α-SMA and NG2, but also by the induction of VEGF-A (Hirschi et al. 1998; Darland and D'Amore 2001a). Vascular development is conveniently studied in the retinas of mice, which are vascularised postnatally. In this model, a subset of pericytes was shown to express VEGF-A, further supporting the observation that contact-induced pericyte differentiation leads to a localised source of VEGF-A (Darland et al. 2003) and other growth factors (see Sects. 4.2.3 and 4.2.4). Pericytes as a source of a local survival factor may explain the regression of pericyte-deficient vessels, and the prevention of regression by the administration of VEGF-A. Conversely, pericytes suppress EC proliferation and migration in vitro (Orlidge and D'Amore 1987; Sato and Rifkin 1989), possibly explaining lesions observed in diabetic retinopathy (Hammes et al. 2002) and various mouse mutants (Hellstrom et al. 2001), where the loss of pericytes precedes retinal EC proliferation. These interactions between EC and mural cells are critical to mural cell differentiation and vessel remodelling, and reflect the collective activity of several signalling molecules, including those described in the following sections.

4.2.1
S1P Phosphate and S1P Receptors

Sphingosine-1 (S1P) is a lipid mediator derived from sphingomyelin that can signal through S1P receptors (S1P1-S1P5), a family of G protein-coupled receptors also known as endothelial differentiation genes (EDG). These receptors and sphingosine kinase are expressed in pre-vascularised embryonic tissues and during vasculogenesis and angiogenesis (Allende et al. 2003). Exogenous S1P or sphingosine, but not VEGF-A or FGF-2, can replace the requirement for serum in promoting vasculogenesis in cultured allantois explants. In the absence of S1P, failure of the cells to move, coupled with the continued proliferation due to the mitogenic effects of VEGF-A, results in small vascular networks with abnormally high cell numbers (Argraves et al. 2004).

The receptor S1P1 is highly expressed in EC and developing SMC, whereas S1P2, is strongly expressed in adult SMC (Lockman et al. 2004). Mice lacking S1P1 die around E12.5-E14.5 from severe haemorrhage, and exhibit incomplete SMC ensheathment of dorsal aorta and large arteries. Endothelial-specific deletion of S1P1 leads to a similar phenotype, whereas deletion targeted to vascular SMC produces viable animals (Allende et al. 2003). Other receptors are probably involved, since S1P stimulates expression of multiple SMC differentiation markers in primary SMC cultures and in 10T1/2 cells, through the activation of an SRF co-factor (Lockman et al. 2004).

4.2.2
Wnts

Wnts are secreted glycoproteins that are likely to play an important role in normal and pathologic angiogenesis and in neointimal hyperplasia (Goodwin and D'Amore 2002). Three major Wnt signalling pathways have been identified: the canonical or β-catenin-dependent cascade, the Wnt/Ca^{++} pathway and the planar cell polarity (PCP) pathway that co-ordinates polarisation of cells within the plane of epithelial sheets (Huelsken and Behrens 2002).

EC and SMC in culture express components of the canonical pathway, including the Frizzled (Fzd) receptors Fzd-1, Fzd-2 and Fzd-3. The mouse gene Fzd5 is strongly expressed in the yolk sac after E9.5, and the placental blood vessels as late as E10.5. Fzd5 ligands, Wnt5a and Wnt10b, are also expressed in the early yolk sac. Homozygous Fzd5 knock-out mice are lethal, owing to defects in the yolk sac vasculogenesis. Wnt2 is also a Fzd5 ligand, and Wnt2-deficient embryos show placental defects suggesting its importance for vascular growth during later stages of development (Ishikawa et al. 2001).

Engagement of Fzd receptors results in recruitment of dishevelled (Dvl), which inhibits β-catenin phosphorylation. About 50% of Dvl2-deficient mice die perinatally due to severe cardiovascular outflow tract defects that have been related to alterations of neural crest (Hamblet et al. 2002). Dvl2, which mediates both the canonical and PCP pathways, has recently been detected in the cytoplasm of cultured EC (Wechezak and Coan 2005). Secreted Fzd-related proteins (FRP) compete with Fzd receptors for Wnt binding. The secreted Frizzled FrzA (or sFRP-1) promotes EC migration and organization into capillary-like structures (Ezan et al. 2004), probably explaining the reduction in size of experimental infarct in mice overexpressing this protein (Barandon et al. 2003). In vitro experiments suggest that Wnt-1 is also co-localised with β-catenin in adherens junctions, probably accounting for the enhanced adhesiveness of transfected EC (Wechezak and Coan 2003).

4.2.3
Platelet-Derived Growth Factors Family

The PDGF family of growth factors is composed of four different polypeptide chains: PDGF-A, PDGF-B, PDGF-C and PDGF-D, which form five dimeric ligands. PDGF-B is secreted by vascular endothelium, PDGF-C by vascular SMC and PDGF-D by adventitial fibroblasts, whereas the receptor PDGFR-β is present in vascular mural cells (Hoch and Soriano 2003). Endothelial expression of PDGF-B occurs during vascular development and is downregulated in quiescent EC. Thus, as development progresses, PDGF-B expression becomes restricted to short capillary segments probably representing angiogenic sprouts. PDGFR-β is expressed by developing pericytes and SMC of arteries/arterioles (Hellstrom et al. 2001).

The ability of EC from different sources to recruit presumptive mural cell precursors is blocked by a neutralising antibody to PDGF-B (Hirschi et al. 1998), indicating that this ligand is a chemotactic, and perhaps survival, signal for PDGFR-β-expressing pericyte/SMC progenitors. Mice lacking PDGF-B or PDGFR-β die perinatally with extensive haemorrhaging, as a result of absence of microvascular pericytes and subsequent microaneurysm formation and capillary rupture (Hoch and Soriano 2003).

Deletion of the extracellular retention motif of PDGF-B by gene targeting in mice results in defective pericyte investment in the microvasculature and delayed formation of the renal glomerulus mesangium. In these mutants, pericytes appear partially detached and with processes directed away from the vessels, suggesting that extracellular retention of PDGF-B may act to restrict pericyte migration to the abluminal surface of microvessels (Lindblom et al. 2003).

4.2.4
Angiopoietins and the Tie Receptors

The two endothelial-specific receptors, Tie1 and Tie2 (tyrosine kinase receptors with immunoglobulin and EGF homology domains), are expressed in the vascular system from the earliest embryonic stages and remain endothelial-specific throughout adult life (Thurston 2003). Angiopoietins (Ang-1 to -4) are the ligands for the Tie2 receptor, but the identity of the Tie1 ligand(s) remains unknown. Ang-1 is expressed by perivascular cells during development and in adult tissues. Ang-1 and -4 stimulate Tie2, whereas Ang-2 and -3 block Ang-1-induced tyrosine phosphorylation of Tie2.

4.2.4.1
Angiopoietin-1 and Tie2

Ang-1 consists of four alternatively spliced isoforms. The 1.5-kb isoform is the activating ligand of Tie-2, whereas the smaller isoforms probably represent dominant-negative regulatory molecules. Both *Ang-1* and *Tie2* knock-out mice exhibit reduced embryonic pericyte/SMC formation and die with cardiac failure and haemorrhage. Initial phases of blood vessel formation occur normally, but there is no remodelling, and vascular networks exhibit no hierarchical organisation (Thurston 2003). Intravitreal Ang-1 injections to newborn mice slightly accelerate the rate of vascular development and partially restore defects induced in neonatal retinal vasculature by depletion of mural cells (Uemura et al. 2002).

Endothelial loss of Tie2 expression correlates with EC apoptosis in haemorrhagic regions of the embryo (Jones et al. 2001), probably reflecting the inactivation of the Akt survival pathway. Akt effects are mediated through members of the FOXO subclass of forkhead transcription factors. Deletion of FOXO1 (but not that of FOXO3a or 4) causes embryonic death on E10.5 because of incomplete vascular development (Hosaka et al. 2004). Since FOXO1 regulates EC apoptosis as well as many genes associated with vascular destabilisation and remodelling (including Ang-2), Ang-1 blockade of the FOXO1 cascade promotes vessel stability (Daly et al. 2004).

Some familial forms of venous malformations, characterised by the formation of low-resistance vessels with insufficient SMC investment, have been associated with point mutations in the kinase domain of Tie2. The means by which Tie2 mutation leads to these abnormal vessels is unclear (Morris et al. 2005).

4.2.4.2
Angiopoietin-2

Ang-2, produced by EC and stored in Weibel-Palade granules, binds Tie2 but does not transduce a signal (Fiedler et al. 2004). Ang-2 controls EC quiescence and responsiveness, probably by inhibition of Ang-1-mediated Tie2 activation. Ang-2 is not essential for embryonic vascular development, but it is required for subsequent postnatal vascular remodelling. Newborn pups lacking Ang-2 have the beginnings of a normal eye vasculature, with well-formed hyaloid vessels. However, the hyaloid vasculature does not regress and the peripheral retina remains avascular; this defect is not rescued by expression of Ang-1 (Gale et al. 2002). Ang-2-null mice also exhibit defects in their lymphatic vasculature, which can be rescued by Ang-1. Mice overexpressing Ang-2 display vascular anomalies similar to mice lacking Ang-1 (Thurston 2003). Availability of VEGF-A appears to switch Ang-2 functions from anti- to pro-angiogenic. In the pupillary membrane, Ang-2, in the presence of VEGF-A, promotes a rapid increase in capillary diameter, remodelling of the basal lamina and sprouting of new blood vessels. By contrast, Ang-2, in the absence of VEGF-A, promotes EC death and vessel regression (Lobov et al. 2002).

4.2.4.3
Tie1

Mice deficient in Tie1 die between E13.5 and E18.5, depending on the genetic background. These embryos show signs of oedema, local haemorrhage and microvessel rupture, but the major blood vessels appear intact (Thurston 2003). Tie1 and Tie2 are also expressed in haematopoietic cells and they are specifically required during postnatal bone marrow haematopoiesis (Puri and Bernstein 2003).

4.2.5
Transforming Growth Factor-β1

Signalling by TGF-β1 family members occurs through a receptor complex formed by two type I (also termed activin-receptor-like kinases, ALKs) and two type II transmembrane serine/threonine kinases. In most cells, TGF-β1 signals via a type II receptor and ALK5 to induce Smad2 and Smad3 phosphorylation, whereas in EC, TGF-β1 also activates an ALK1-promoting Smad1/5 phosphorylation. Smad3 can be proangiogenic through stimulation of VEGF-A expression, whereas Smad2 can be antiangiogenic via thrombospondin-1 (TSP-1) expression (Nakagawa et al. 2004). Thus, EC regulation of the various TGF-β1 intracellular cascades remains to be elucidated. Effects of members of the TGF-β1 superfamily are mediated through a consensus TGF-β1-controlling element (TCE), which is common to regulatory regions of SMC-marker genes. TCE-binding factors act as potent repressors of SMC differentiation marker genes (Ding et al. 2004).

Mice lacking TGF-β1 show defects in the yolk sac vasculature, including decreased vessel wall integrity, reduced contact between EC and mesenchymal cells, and incomplete maturation of SMC. The yolk sac vessels are large and leaky with abnormal endothelial adhesion. Mice lacking the TGF-β1 type II receptor exhibit a similar vascular phenotype, with additional abnormalities in other organ systems (Oshima et al. 1996). Conversely, diverse cell types, including 10T1/2, a line of multipotent mesenchymal cells, murine ES cells and rat neural crest stem cells, differentiate into SMC upon TGF-β1 treatment (Mann et al. 2004). TGF-β1 is also involved in the inhibition of EC growth induced by pericytes and SMC (Antonelli-Orlidge et al. 1989) and cord formation in EC and 10T1/2 co-cultures (Darland and D'Amore 2001b). EC, SMC and 10T1/2 secrete latent TGF-β1 that is locally activated upon contact between the EC and either SMC or 10T1/2 cells (Antonelli-Orlidge et al. 1989; Hirschi et al. 1998). 10T1/2 cells engineered to form defective gap junctions cannot activate endogenous TGF-β1 but can respond to exogenous TGF-β1 (Hirschi et al. 2003). Other members of the TGF-β1 family might also be involved in the control of the SMC phenotype; however, their role during embryonic vascular development has yet to be studied.

4.2.6
Interactions Between Signalling Cascades

Complex interactions exist between PDGF-B, Ang-1 and TGF-β1 (Fig. 5). In mural cell precursors, PDGF-B upregulates Ang-1 and TGF-β1 expression, via the PI3-kinase and PKC pathways for Ang-1 and the MAPK/ERK pathway for TGF-β1. In addition, TGF-β1 partially inhibits endogenous Ang-1 expression and completely blocks expression induced by PDGF-B. In EC, either Ang-1 or TGF-β1 alone marginally downregulates PDGF-B expression, but a combination of these factors produces a much stronger downregulation (Nishishita and Lin 2004).

S1P and PDGF-B seem to co-ordinate EC-mural cell interactions required for development and stability of the vessel wall. In vitro, S1P potently stimulates PDGF-A and -B chain messenger RNA (mRNA) and protein expression in vascular SMC (Usui et al. 2004). On the other hand, PDGF-B acts on SMC to stimulate S1P release, resulting in stimulation of cell migration via activation of muscular S1P receptors in an autocrine/paracrine fashion (Hobson et al. 2001). More recent evidence suggests that PDGFR-β integrates a pre-formed complex with the S1P1 receptor that, upon PDGF stimulation, is internalised through endocytic vesicles and activates a MAPK cascade (Waters et al. 2005).

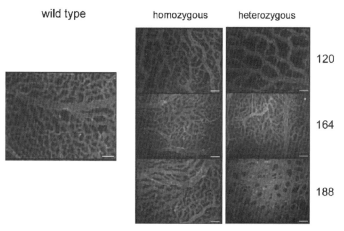

Fig. 5 Yolk sac vasculature of E10.5 mice that express single VEGF-A isoforms. Shown are yolk sacs isolated from embryos of wild-type mice that express all three VEGF-A isoforms and mice that express VEGF120 alone, VEGF164 alone or VEGF188 alone. Yolk sacs were stained with anti-PECAM antisera to visualise the vasculature

Pericyte growth and differentiation are differentially regulated by antagonistic signalling cascades involving FGF-2 and TGF-β1. FGF-2 markedly stimulates pericyte growth, whereas its removal and/or the addition of TGF-β1 causes the withdrawal of pericytes from the growth cycle and the induction of a contractile phenotype (Papetti et al. 2003).

5
Endothelium Morphogenesis

5.1
Angiogenic Sprouting

Angiogenic sprouting involves specialised endothelial tip cells that respond to chemoattractant and repellent guidance cues. Tip cells display long filopodia that sense extracellular VEGF-A gradients through VEGFR-2. Whereas tip cells do not proliferate, activation of VEGFR-2 is interpreted differently by sprout stalk cells, which are induced to proliferate (Gerhardt et al. 2003).

Different VEGF-A protein isoforms, VEGF120, VEGF164 and VEGF188, have a different affinity for heparan sulphate proteoglycans (HSPG) and heparin (Ng et al. 2001). This is the basis for the selective spatial distribution of VEGF-A, a primary mechanism controlling directed EC migration and the vas-

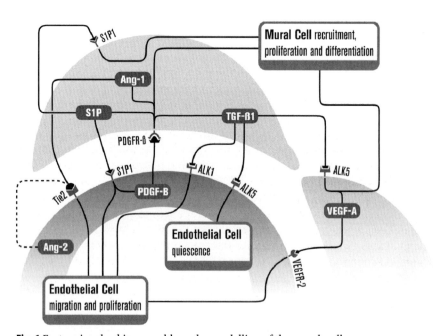

Fig. 6 Factors involved in assembly and remodelling of the vessel wall

cular pattern (Fig. 6). HSPG-binding properties have also been demonstrated for a wide range of growth factors, including members of the FGF, TGF-β1, EGF, insulin-like growth factor (IGF), PDGF-B, Wnt families and many other chemokines and cytokines (Iozzo and San Antonio 2001).

5.2
Attraction and Repulsion of Angiogenic Sprouts

5.2.1
Patterning of the Embryonic Midline

Vessel formation takes place throughout the embryonic disc, with the exception of the midline region surrounding the notochord, where no vessels grow during the early stages of development. This vascular exclusion zone is not determined by a lack of endothelial growth factors, but by notochordal production of the BMP antagonists Chordin and Noggin, which provide strong inhibitory cues (Reese et al. 2004). The neural tube, a localised source of VEGF-A, plays a role in patterning the midline vasculature, since it recruits somite precursors that develop into the perineural vascular plexus surrounding the developing brain and spinal cord. Sprouts from this plexus do not invade the neural tissue until later in development, suggesting that negative or repulsive cues also originate from the neural tube (Hogan et al. 2004).

5.2.2
Semaphorins

Neuropilin 1 (NRP-1) and NRP-2 are related transmembrane receptors that respond to two different extracellular ligands, class 3 semaphorins (SEMA3) and VEGF164, which are competitive inhibitors of one another in binding and in EC motility assays. Transgenic mice lacking both NRP-1 and NRP-2 die in utero at E8.5 with avascular yolk sacs. *NRP-1*-null mice die between E11 and E14 with cardiovascular and neuronal defects, whereas many *NRP-2*-deficient mice survive to adulthood but show lymphatic and neurologic defects. Cardiovascular defects in *NRP-1*-null mice include transposition of great vessels and persistent aorticopulmonary truncus (Takashima et al. 2002). *NRP-1*-deficient mice exhibit a defect in tip cell guidance that leads to paucity of sprouting, which in the presence of EC proliferation results in development of aneurysmatic malformations (Gerhardt et al. 2004). A knock-in mouse expressing the variant NRP-1^{Sema-}, unreactive to semaphorin but retaining VEGF-A 165 responses, survives until birth and has normal cardiac outflow tracts, indicating that semaphorin-NRP-1 signalling is not critical for embryonic viability (Gu et al. 2003).

Semaphorins induce the association of NRPs with transmembrane proteins of the plexin family such as plexinD1, which is expressed by most embryonic and adult vascular EC. PlexinD1-null embryos show severe defects of the cardiac outflow tract and a deficiency of differentiated SMC in the developing 4th and 6th aortic arch arteries (Gitler et al. 2004). SEMA3E can bind directly to plexinD1 without intervention of a neuropilin. This property is not shared by any of the other known SEMA3. In E10.5-E11.5 mouse embryos, SEMA3E expression is localised to the somites, where it acts as a repulsive cue for plexinD1-expressing EC of adjacent intersomitic vessels (Gu et al. 2005). SEMA3A signalling inhibits integrin-mediated adhesion to the ECM, and no vascular remodelling is found in $SEMA3A^{-/-}$ embryos (Serini et al. 2003).

5.2.3
Netrins and Their Receptors

Netrins are guidance molecules related to laminin. Two families of netrin receptors are known, the deleted in colorectal cancer (DCC18) and UNC-5 families. DCC18 receptors mediate attraction, while UNC-5 mediates repulsion (Mehlen and Mazelin 2003). Receptor UNC-5B, selectively localised to arterial EC and endothelial tips, controls filopodial activity. UNC-5B mutant embryos develop a normal vascular plexus, but remodelling produces 40% more branching points than in wild-type embryos. Mutants die around E12.5 with heart failure probably resulting from increased peripheral resistance. Increased branching is associated with a larger number of tip filopodial extensions, and reflects the lack of UNC-5B negative regulation by netrin-1 stimulation. Intravitreal injection of netrin-1 during retinal angiogenesis leads to a marked decrease in filopodial extension (Lu et al. 2004).

5.2.4
Calcineurin/NFAT

Calcineurin, a protein phosphatase that is downstream of VEGFR-2, activates the nuclear factor of activated T cells (NFATc1-c4). This pathway leads to the transcriptional activation of various proangiogenic genes and can be counterbalanced by upregulation of the Down syndrome critical region 1 (DSCR-1) gene, a calcineurin inhibitor with antiangiogenic properties (Yao and Duh 2004). Signals transduced by Ca^{2+}, calcineurin, and NFATc3/c4 promote the proper anatomical patterning of the developing vascular system, as shown by disorganised vascular growth in mice doubly mutant for the NFATc3 and c4 genes. In these mutants, intersomitic vessels ignore somitic or neural boundaries, suggesting that NFAT signalling normally prevents abnormal growth of vessels into these tissues (Graef et al. 2001). EC show a low degree of NFATc4 expression, but perivascular mesenchyme typically expresses high levels of NFATc4, reflecting its importance for recruitment of pericytes and SMC. Calcineurin and NFATc1 direct neural crest stem cells to a SMC fate, whereas

DSCR-1 decreases SMC differentiation. DSCR-1 and NFATc1 are upregulated in response to TGF-β1, and expression of either calcineurin or NFATc1 mimics the effects of TGF-β1 on neural crest stem cells, suggesting that TGF-β1-dependent differentiation of SMC is mediated by calcineurin signalling (Mann et al. 2004).

6
Development of Arteries and Veins

Developmental remodelling includes structural and functional differentiation of arteries and veins, and establishment of an organ-specific microvascular network. Circulatory dynamics were thought to play a major role in establishing these differences; however, it has been demonstrated recently that the identities of arterial and venous endothelium are defined early in development, even before the start of circulation (Wang et al. 1998). Ephrins and their receptors, Eph, seem to be the earliest markers of arteriovenous differences, except for the recent description of the apelin (APJ) receptor as an even earlier marker for developing retinal veins (Saint-Geniez et al. 2003).

6.1
Ephrins and Eph Receptors

Eph, receptor tyrosine kinases that are typically activated by ligands anchored to the membrane of adjacent cells, regulate cellular adhesion, migration or chemorepulsion, and tissue/cell boundary formation. Reverse signalling, downstream of membrane-anchored ephrin ligands, can also occur. In all vertebrates, ephrin-B2 is expressed in arterial EC, while its receptor, EphB4, is expressed predominantly in venous EC. Ephrin-B2 also appears in perivascular mesenchyme and developing mural cells (Wang et al. 1998). Ephrin-B1 is co-expressed with ephrin-B2 in EC, whereas EphB3 and ephrin-B3 are co-expressed with EphB4 in venous EC. In the adult vasculature, expression of ephrin-B2 and EphB4 extends into the smallest-diameter capillaries, suggesting that they can also have arterial and venous identity (Shin et al. 2001). Eph-ephrin signalling is the basis for endothelial propulsive and repulsive activities that mediate EC guidance signals during angiogenesis, as well as the positional control of EphB receptor- and ephrin-B ligand-expressing cells towards each other. Forward EphB4 signals may direct EC in a repulsive manner avoiding areas where ephrin-B2 is expressed, whereas promotion of EC migration may occur if ephrin-B2-expressing EC are activated by EphB4. These propulsive and repulsive activities may also segregate EC from each other to limit cellular intermingling and control arterio-venous positioning of cells (Fuller et al. 2003).

Ephrin-B2 and EphB4 are also involved in mural cell development. Stromal cells expressing ephrin-B2 support the proliferation of ephrin-B2$^+$ EC, suppress the proliferation of ephrin-B2 EC, promote vascular network formation and induce the recruitment and proliferation of α-SMA$^+$ cells. Conversely, stromal cells expressing EphB4 inhibit vascular network formation, ephrin-B2$^+$ EC proliferation and α-SMA cell recruitment and proliferation (Zhang et al. 2001).

Targeted disruption of either ephrin-B2 or EphB4 results in embryonic lethality at E11 and E9.5-10, respectively, due to defects in angiogenic remodelling of arteries and veins, and alterations of myocardial trabeculation. Early vasculogenesis is also abnormal, since EphB4-deficient EB display delayed expression of VEGFR-2 (Wang et al. 2004b). The initial commitment of ephrin-B2$^+$ or EphB4$^+$ EC could be the trigger for determining the arterial or venous fate of developing vessels. However, determination of arterial or venous fates probably requires the action of other upstream signals (see Sects. 6.2 and 6.5).

6.2
Hedgehog in Arteriogenesis

Hh proteins act as morphogens in many tissues during embryonic development. Signalling requires the interaction of Hh protein with its receptor, Patched-1 (Ptc1), leading to activation of a transcription factor, Gli, that induces expression of downstream target genes including Ptc and Gli themselves. Zebrafish embryos lacking Shh activity fail to express ephrin-B2a within their blood vessels, and a similar failure occurs in embryos lacking VEGF-A or Notch. In these embryos, ectopically expressed Shh induces ectopic formation of ephrin-B2-expressing vessels (Lawson and Weinstein 2002). A determinant role for Hh proteins in arteriogenesis of higher vertebrates has not been as clearly demonstrated as in Zebrafish. However, in the murine corneal angiogenesis assay, Shh produces large, branching vessels, whereas VEGF-A results in capillaries of lesser lumenal calibre. Moreover, Shh is involved in arteriogenesis during revascularisation of adult ischaemic tissues (Pola et al. 2003).

6.3
VEGF-A in Arteriogenesis

The association of peripheral nerves and expression of arterial markers during development has led to the suggestion that neurally derived VEGF-A directs arteriogenesis. Nerves express VEGF-A at a higher level than surrounding mesenchymal tissue. Moreover, expression of ephrin-B2 can be induced in embryonic EC by incubation with VEGF-A or co-culture with neurons or Schwann cells. In these experiments, only 50% of EC cultures express the arterial ephrin, suggesting that VEGF-A could represent a permissive inducing signal rather than an instructive determinant of arterial identity (Mukouyama et al. 2002). Since major receptors for VEGF-A are expressed on all EC, the arteriogenic effect of this factor has been ascribed to the co-receptor NRP-1 that is preferentially expressed on arteries, whereas NRP-2 tends to be expressed in veins/lymphatic vessels (Yuan et al. 2002).

Defective vascular development in mice expressing single VEGF-A isoforms illustrates the complexity of VEGF-A signalling in arterial specification. In the early developing retina, prior to mural cell differentiation, the arterial marker ephrin-B2 is detected in about 50% of the retinal vessels, and NRP-1 shows a similar distribution, being localised in retinal arterioles with very low expression in retinal venules. Arteries and veins develop normally in $VEGF^{164/164}$ mice, but severe arterial defects accompanied by relatively normal veins and capillaries appear in $VEGF^{188/188}$ mice. $VEGF^{120/120}$ mice show severe retinal vascular defects, but 50% of early retinal vessels express ephrin-B2, suggesting unimpaired arterial specification. After remodelling, however, arterial development appears to lag behind venous development, suggesting that expression of NRP-1 is not the only mechanism driving the arterial specificity of the VEGF-A-response (Stalmans et al. 2002).

6.4
Notch Pathways

Notch receptor-ligand interaction results in proteolytic cleavage of the Notch receptor, producing a C-terminal intracellular fragment (NotchIC) that translocates to the nucleus. NotchIC binds to a transcriptional repressor, derepressing or co-activating the expression of various lineage-specific genes. Since the Notch cascade has a role in determining cell identities, it is probably involved in the events distinguishing EC from mural cells, artery from vein, pulmonary from systemic vessels, and large vessels from capillaries (Iso et al. 2003).

Several Notch pathway ligands and receptors are selectively localised in EC and their supporting cells. $Notch1^{-/-}$ and $Notch1^{+/-}/Notch4^{-/-}$ embryos arrest early in development with severe defects in the yolk sac and embryonic vessels. The primary vascular plexus develops normally, but both small capillaries and large vitelline collecting vessels fail to form, and embryonic large blood vessels are severely malformed (Krebs et al. 2000). Constitutive activation of Notch4 causes defects in vascular remodelling, whereas mice deficient in Jagged1, one of the Notch ligands, die from haemorrhage early during embryogenesis (Uyttendaele et al. 2001; Leong et al. 2002). Notch4 activation in EC promotes mesenchymal transformation, evidenced by downregulation of EC-specific proteins such as VE-cadherin, and upregulation of mesenchymal proteins, such as α-SMA, fibronectin and PDGFR-β (Noseda et al. 2004).

The Notch ligands, Jagged1, Jagged 2 and Dll4, as well as the receptors Notch1, Notch3 and Notch4, are selectively expressed in arteries. Notch1 and Notch4 are expressed in EC, whereas Notch3 is localised specifically to SMC (Villa et al. 2001). Heterozygous deletion of Dll4 results in absence of well-defined arterial vessels, including the internal carotid artery, although a relatively normal venous plexus is present. SMC coverage of large arterial vessels is often lacking or markedly deficient (Gale et al. 2004). In $Dll4^{-/-}$ mice embryos, EC do not express the arterial markers ephrin-B2, connexin37 and connexin40 (Duarte et al. 2004).

Effectors of the Notch cascade are also involved in arterial differentiation. Loss of RBP-J (mammalian suppressor of hairless), one of the primary transcriptional mediators, results in the production of arteriovenous malformations (AVM), including fusion of the dorsal aorta with the common cardinal vein (Krebs et al. 2004). *Hey1* and *Hey2*, two other targets of Notch signalling, are preferentially expressed in embryonic arteries. $Hey1^{-/-}/Hey2^{-/-}$ mice display a phenotype resembling that produced by Notch1 deficiency, including defects in yolk sac vascular remodelling and lack of the arterial markers CD44, neuropilin1 and ephrin-B2 (Fischer et al. 2004). In Zebrafish, Notch-induced arterial differentiation is downstream of VEGF-A signalling (Lawson and Weinstein 2002). This is likely to be the case in mammals, since in vitro VEGF-A stimulation upregulates Notch1 and Dll4 transcription (Liu et al. 2003).

In humans, mutations of the ligand Jagged 1 are associated with Alagille syndrome, a developmental disorder that includes vascular defects (Gridley 2003). Cerebral cavernous malformation (CCM), a vascular malformation characterised by thin-walled vascular cavities that haemorrhage, has been linked to loss-of-function mutations in a locus termed CCM1. $CCM1^{-/-}$ mouse embryos exhibit progressive dilatation of cephalic vessels, with marked enlargement of the aorta and branchial arch arteries, downregulation of Dll4 and Notch4, and lack of ephrin-B2 expression and SMC recruitment in arteries. Consistent with the murine data, Notch4 is not detected in human cavernous lesions, and is markedly reduced in brain arteries adjacent to the vascular malformations (Whitehead et al. 2004).

Missense mutations in Notch3 have been implicated in a neurovascular disorder known as cerebral autosomal dominant arteriopathy with subcortical infarcts and leukoencephalopathy (CADASIL), an arteriopathy that involves regression of arterial vascular SMC. In mice, the absence of Notch3 function is compatible with normal angiogenesis and remodelling, but arterial SMC is severely affected and resembles venous SMC, both by its orientation and by the lack of smoothelin (Domenga et al. 2004).

6.5
TGF-β1 Receptors

Hereditary haemorrhagic telangiectasia (HHT) is a vascular dysplasia characterised by localised vascular malformations. Mutations in endoglinCD34(ENG, CD105) have been linked to HHT type 1, whereas mutations in the gene coding for ALK1 are associated with HHT type 2. ALK1, a receptor for TGF-B1 and activins, is predominantly expressed in arterial capillary EC. In ALK1-null mice, there is downregulation of ephrin-B2, loss of arterial-specific haematopoiesis, defects in development of mural cells, and arteriovenous malformations between major arteries and veins (Seki et al. 2003).

ENG is a component of the TGF-β1 receptor complex that is uniformly expressed in all vessels, including liver sinusoids (Jonker and Arthur 2002). The most recent evidence indicates that ENG stimulates TGF-β1/ALK1-induced Smad1/5 responses and indirectly inhibits the TGF-β1/ALK5 signalling pathway, thereby promoting endothelial activation (Lebrin et al. 2004). The loss of ALK1 or ENG does not disrupt de novo assembly of large vessels, but impairs the ability to maintain the arterial and venous beds as distinct circuits during remodelling (Sorensen et al. 2003). CD34 is a cell-surface glycoprotein that is expressed on the surface of haematopoietic, as well as EC, but is normally expressed at a much higher level on arterial endothelium. In early $ALK1^{-/-}$ and $ENG^{-/-}$ embryos, CD34 is strongly expressed in venous vessels, suggesting a progressive conversion of venous endothelium to arterial haemogenic endothelium. The appearance of venous endothelial haematopoiesis could reflect an intrinsic defect in definitive haematopoietic stem cells, which also express ENG (Chen et al. 2002).

7
Concluding Remarks

The identification of a large number of growth factors and their signalling pathways, in conjunction with observations of mice in which these molecules have been genetically deleted, has provided an enormous body of information regarding their roles in vascular development. These data have made it clear that the formation of the vasculature is a highly complex process that involves a large number of growth factors and cell-cell interactions. Although use of knock-out mice has indicated a role for many factors, the precise role that each molecule plays is not known. In particular, the contextual role of such factors has not been elucidated concerning how the actions of a specific factor are modified by the environment and/or by the presence of other factors. Further, the tissue specificity of the various developmental pathways has not been systematically studied. Thus, though there has been a virtual explosion of knowledge regarding the development of the vascular system, many important questions remain to be answered.

References

Allende ML, Yamashita T, Proia RL (2003) G-protein-coupled receptor S1P1 acts within endothelial cells to regulate vascular maturation. Blood 102:3665–3667

Alvarez-Silva M, Belo-Diabangouaya P, Salaun J, Dieterlen-Lievre F (2003) Mouse placenta is a major hematopoietic organ. Development 130:5437–5444

Antonelli-Orlidge A, Smith SR, D'Amore PA (1989) Influence of pericytes on capillary endothelial cell growth. Am Rev Respir Dis 140:1129–1131

Argraves KM, Wilkerson BA, Argraves WS, Fleming PA, Obeid LM, Drake CJ (2004) Sphingosine-1-phosphate signaling promotes critical migratory events in vasculogenesis. J Biol Chem 279:50580–50590

Barandon L, Couffinhal T, Ezan J, Dufourcq P, Costet P, Alzieu P, Leroux L, Moreau C, Dare D, Duplaa C (2003) Reduction of infarct size and prevention of cardiac rupture in transgenic mice overexpressing FrzA. Circulation 108:2282–2289

Bazzoni G, Dejana E (2004) Endothelial cell-to-cell junctions: molecular organization and role in vascular homeostasis. Physiol Rev 84:869–901

Beck L Jr, D'Amore PA (1997) Vascular development: cellular and molecular regulation. FASEB J 11:365–373

Bohnsack BL, Lai L, Dolle P, Hirschi KK (2004) Signaling hierarchy downstream of retinoic acid that independently regulates vascular remodeling and endothelial cell proliferation. Genes Dev 18:1345–1358

Byrd N, Becker S, Maye P, Narasimhaiah R, St-Jacques B, Zhang X, McMahon J, McMahon A, Grabel L (2002) Hedgehog is required for murine yolk sac angiogenesis. Development 129:361–372

Chen CZ, Li M, de Graaf D, Monti S, Gottgens B, Sanchez MJ, Lander ES, Golub TR, Green AR, Lodish HF (2002) Identification of endoglin as a functional marker that defines long-term repopulating hematopoietic stem cells. Proc Natl Acad Sci U S A 99:15468–15473

Chung YS, Zhang WJ, Arentson E, Kingsley PD, Palis J, Choi K (2002) Lineage analysis of the hemangioblast as defined by FLK1 and SCL expression. Development 129:5511–5520

Cortes F, Debacker C, Peault B, Labastie MC (1999) Differential expression of KDR/VEGFR-2 and CD34 during mesoderm development of the early human embryo. Mech Dev 83:161–164

Daly C, Wong V, Burova E, Wei Y, Zabski S, Griffiths J, Lai KM, Lin HC, Ioffe E, Yancopoulos GD, Rudge JS (2004) Angiopoietin-1 modulates endothelial cell function and gene expression via the transcription factor FKHR (FOXO1). Genes Dev 18:1060–1071

Damert A, Miquerol L, Gertsenstein M, Risau W, Nagy A (2002) Insufficient VEGFA activity in yolk sac endoderm compromises haematopoietic and endothelial differentiation. Development 129:1881–1892

Darland DC, D'Amore PA (2001a) Cell-cell interactions in vascular development. Curr Top Dev Biol 52:107–149

Darland DC, D'Amore PA (2001b) TGF beta is required for the formation of capillary-like structures in three-dimensional cocultures of 10T1/2 and endothelial cells. Angiogenesis 4:11–20

Darland DC, Massingham LJ, Smith SR, Piek E, Saint-Geniez M, D'Amore PA (2003) Pericyte production of cell-associated VEGF is differentiation-dependent and is associated with endothelial survival. Dev Biol 264:275–288

Ding R, Darland DC, Parmacek MS, D'Amore PA (2004) Endothelial-mesenchymal interactions in vitro reveal molecular mechanisms of smooth muscle/pericyte differentiation. Stem Cells Dev 13:509–520

Domenga V, Fardoux P, Lacombe P, Monet M, Maciazek J, Krebs LT, Klonjkowski B, Berrou E, Mericskay M, Li Z, Tournier-Lasserve E, Gridley T, Joutel A (2004) Notch3 is required for arterial identity and maturation of vascular smooth muscle cells. Genes Dev 18:2730–2735

Drake CJ, Fleming PA (2000) Vasculogenesis in the day 6.5 to 9.5 mouse embryo. Blood 95:1671–1679

Duan LJ, Nagy A, Fong GH (2003) Gastrulation and angiogenesis, not endothelial specification, is sensitive to partial deficiency in vascular endothelial growth factor-A in Mice. Biol Reprod 69:1852–1858

Duarte A, Hirashima M, Benedito R, Trindade A, Diniz P, Bekman E, Costa L, Henrique D, Rossant J (2004) Dosage-sensitive requirement for mouse Dll4 in artery development. Genes Dev 18:2474–2478

Dyer MA, Farrington SM, Mohn D, Munday JR, Baron MH (2001) Indian hedgehog activates hematopoiesis and vasculogenesis and can respecify prospective neurectodermal cell fate in the mouse embryo. Development 128:1717–1730

Elvert G, Kappel A, Heidenreich R, Englmeier U, Lanz S, Acker T, Rauter M, Plate K, Sieweke M, Breier G, Flamme I (2003) Cooperative interaction of hypoxia-inducible factor-2alpha (HIF-2alpha) and Ets-1 in the transcriptional activation of vascular endothelial growth factor receptor-2 (Flk-1). J Biol Chem 278:7520–7530

Ema M, Faloon P, Zhang WJ, Hirashima M, Reid T, Stanford WL, Orkin S, Choi K, Rossant J (2003) Combinatorial effects of Flk1 and Tal1 on vascular and hematopoietic development in the mouse. Genes Dev 17:380–393

Ezan J, Leroux L, Barandon L, Dufourcq P, Jaspard B, Moreau C, Allieres C, Daret D, Couffinhal T, Duplaa C (2004) FrzA/sFRP-1, a secreted antagonist of the Wnt-Frizzled pathway, controls vascular cell proliferation in vitro and in vivo. Cardiovasc Res 63:731–738

Faloon P, Arentson E, Kazarov A, Deng CX, Porcher C, Orkin S, Choi K (2000) Basic fibroblast growth factor positively regulates hematopoietic development. Development 127:1931–1941

Feraud O, Prandini MH, Vittet D (2003) Vasculogenesis and angiogenesis from in vitro differentiation of mouse embryonic stem cells. Methods Enzymol 365:214–228

Ferkowicz MJ, Starr M, Xie X, Li W, Johnson SA, Shelley WC, Morrison PR, Yoder MC (2003) CD41 expression defines the onset of primitive and definitive hematopoiesis in the murine embryo. Development 130:4393–4403

Fiedler U, Scharpfenecker M, Koidl S, Hegen A, Grunow V, Schmidt JM, Kriz W, Thurston G, Augustin HG (2004) The Tie-2 ligand angiopoietin-2 is stored in and rapidly released upon stimulation from endothelial cell Weibel-Palade bodies. Blood 103:4150–4156

Finkelstein EB, Poole TJ (2003) Vascular endothelial growth factor: a regulator of vascular morphogenesis in the Japanese quail embryo. Anat Rec 272A:403–414

Fischer A, Schumacher N, Maier M, Sendtner M, Gessler M (2004) The Notch target genes Hey1 and Hey2 are required for embryonic vascular development. Genes Dev 18:901–911

Fraser ST, Ogawa M, Yokomizo T, Ito Y, Nishikawa S, Nishikawa S (2003) Putative intermediate precursor between hematogenic endothelial cells and blood cells in the developing embryo. Dev Growth Differ 45:63–75

Fujimoto T, Ogawa M, Minegishi N, Yoshida H, Yokomizo T, Yamamoto M, Nishikawa S (2001) Step-wise divergence of primitive and definitive haematopoietic and endothelial cell lineages during embryonic stem cell differentiation. Genes Cells 6:1113–1127

Fujiwara Y, Chang AN, Williams AM, Orkin SH (2004) Functional overlap of GATA-1 and GATA-2 in primitive hematopoietic development. Blood 103:583–585

Fuller T, Korff T, Kilian A, Dandekar G, Augustin HG (2003) Forward EphB4 signaling in endothelial cells controls cellular repulsion and segregation from ephrinB2 positive cells. J Cell Sci 116:2461–2470

Gale NW, Thurston G, Hackett SF, Renard R, Wang Q, McClain J, Martin C, Witte C, Witte MH, Jackson D, Suri C, Campochiaro PA, Wiegand SJ, Yancopoulos GD (2002) Angiopoietin-2 is required for postnatal angiogenesis and lymphatic patterning, and only the latter role is rescued by angiopoietin-1. Dev Cell 3:411–423

Gale NW, Dominguez MG, Noguera I, Pan L, Hughes V, Valenzuela DM, Murphy AJ, Adams NC, Lin HC, Holash J, Thurston G, Yancopoulos GD (2004) Haploinsufficiency of delta-like 4 ligand results in embryonic lethality due to major defects in arterial and vascular development. Proc Natl Acad Sci U S A 101:15949–15954

Gerhardt H, Golding M, Fruttiger M, Ruhrberg C, Lundkvist A, Abramsson A, Jeltsch M, Mitchell C, Alitalo K, Shima D, Betsholtz C (2003) VEGF guides angiogenic sprouting utilizing endothelial tip cell filopodia. J Cell Biol 161:1163–1177

Gerhardt H, Ruhrberg C, Abramsson A, Fujisawa H, Shima D, Betsholtz C (2004) Neuropilin-1 is required for endothelial tip cell guidance in the developing central nervous system. Dev Dyn 231:503–509

Gitler AD, Lu MM, Epstein JA (2004) PlexinD1 and semaphorin signaling are required in endothelial cells for cardiovascular development. Dev Cell 7:107–116

Godin I, Cumano A (2002) The hare and the tortoise: an embryonic haematopoietic race. Nat Rev Immunol 2:593–604

Goodwin AM, D'Amore PA (2002) Wnt signaling in the vasculature. Angiogenesis 5:1–9

Gottgens B, Nastos A, Kinston S, Piltz S, Delabesse EC, Stanley M, Sanchez MJ, Ciau-Uitz A, Patient R, Green AR (2002) Establishing the transcriptional programme for blood: the SCL stem cell enhancer is regulated by a multiprotein complex containing Ets and GATA factors. EMBO J 21:3039–3050

Graef IA, Chen F, Chen L, Kuo A, Crabtree GR (2001) Signals transduced by Ca(2+)/calcineurin and NFATc3/c4 pattern the developing vasculature. Cell 105:863–875

Gridley T (2003) Notch signaling and inherited disease syndromes. Hum Mol Genet 12 Spec No 1:R9–R13

Gu C, Rodriguez ER, Reimert DV, Shu T, Fritzsch B, Richards LJ, Kolodkin AL, Ginty DD (2003) Neuropilin-1 conveys semaphorin and VEGF signaling during neural and cardiovascular development. Dev Cell 5:45–57

Gu C, Yoshida Y, Livet J, Reimert DV, Mann F, Merte J, Henderson CE, Jessell TM, Kolodkin AL, Ginty DD (2005) Semaphorin 3E and plexin-D1 control vascular pattern independently of neuropilins. Science 307:265–268

Guo Y, Chan R, Ramsey H, Li W, Xie X, Shelley WC, Martinez-Barbera JP, Bort B, Zaret K, Yoder M, Hromas R (2003) The homeoprotein Hex is required for hemangioblast differentiation. Blood 102:2428–2435

Hamblet NS, Lijam N, Ruiz-Lozano P, Wang J, Yang Y, Luo Z, Mei L, Chien KR, Sussman DJ, Wynshaw-Boris A (2002) Dishevelled 2 is essential for cardiac outflow tract development, somite segmentation and neural tube closure. Development 129:5827–5838

Hammes HP, Lin J, Renner O, Shani M, Lundqvist A, Betsholtz C, Brownlee M, Deutsch U (2002) Pericytes and the pathogenesis of diabetic retinopathy. Diabetes 51:3107–3112

Hellstrom M, Gerhardt H, Kalen M, Li X, Eriksson U, Wolburg H, Betsholtz C (2001) Lack of pericytes leads to endothelial hyperplasia and abnormal vascular morphogenesis. J Cell Biol 153:543–553

Hiratsuka S, Kataoka Y, Nakao K, Nakamura K, Morikawa S, Tanaka S, Katsuki M, Maru Y, Shibuya M (2005) Vascular endothelial growth factor A (VEGF-A) is involved in guidance of VEGF receptor-positive cells to the anterior portion of early embryos. Mol Cell Biol 25:355–363

Hirschi KK, Rohovsky SA, D'Amore PA (1998) PDGF, TGF-beta, and heterotypic cell-cell interactions mediate endothelial cell-induced recruitment of 10T1/2 cells and their differentiation to a smooth muscle fate. J Cell Biol 141:805–814

Hirschi KK, Rohovsky SA, Beck LH, Smith SR, D'Amore PA (1999) Endothelial cells modulate the proliferation of mural cell precursors via platelet-derived growth factor-BB and heterotypic cell contact. Circ Res 84:298–305

Hirschi KK, Burt JM, Hirschi KD, Dai C (2003) Gap junction communication mediates transforming growth factor-beta activation and endothelial-induced mural cell differentiation. Circ Res 93:429–437

Hobson JP, Rosenfeldt HM, Barak LS, Olivera A, Poulton S, Caron MG, Milstien S, Spiegel S (2001) Role of the sphingosine-1-phosphate receptor EDG-1 in PDGF-induced cell motility. Science 291:1800–1803

Hoch RV, Soriano P (2003) Roles of PDGF in animal development. Development 130:4769–4784

Hogan KA, Ambler CA, Chapman DL, Bautch VL (2004) The neural tube patterns vessels developmentally using the VEGF signaling pathway. Development 131:1503–1513

Hosaka T, Biggs WH 3rd, Tieu D, Boyer AD, Varki NM, Cavenee WK, Arden KC (2004) Disruption of forkhead transcription factor (FOXO) family members in mice reveals their functional diversification. Proc Natl Acad Sci U S A 101:2975–2980

Huber TL, Kouskoff V, Fehling HJ, Palis J, Keller G (2004) Haemangioblast commitment is initiated in the primitive streak of the mouse embryo. Nature 432:625–630

Huelsken J, Behrens J (2002) The Wnt signalling pathway. J Cell Sci 115:3977–3978

Hughes S, Chan-Ling T (2004) Characterization of smooth muscle cell and pericyte differentiation in the rat retina in vivo. Invest Ophthalmol Vis Sci 45:2795–2806

Iozzo RV, San Antonio JD (2001) Heparan sulfate proteoglycans: heavy hitters in the angiogenesis arena. J Clin Invest 108:349–355

Ishikawa T, Tamai Y, Zorn AM, Yoshida H, Seldin MF, Nishikawa S, Taketo MM (2001) Mouse Wnt receptor gene Fzd5 is essential for yolk sac and placental angiogenesis. Development 128:25–33

Iso T, Hamamori Y, Kedes L (2003) Notch signaling in vascular development. Arterioscler Thromb Vasc Biol 23:543–553

Jones N, Voskas D, Master Z, Sarao R, Jones J, Dumont DJ (2001) Rescue of the early vascular defects in Tek/Tie2 null mice reveals an essential survival function. EMBO Rep 2:438–445

Jonker L, Arthur HM (2002) Endoglin expression in early development is associated with vasculogenesis and angiogenesis. Mech Dev 110:193–196

Krebs LT, Xue Y, Norton CR, Shutter JR, Maguire M, Sundberg JP, Gallahan D, Closson V, Kitajewski J, Callahan R, Smith GH, Stark KL, Gridley T (2000) Notch signaling is essential for vascular morphogenesis in mice. Genes Dev 14:1343–1352

Krebs LT, Shutter JR, Tanigaki K, Honjo T, Stark KL, Gridley T (2004) Haploinsufficient lethality and formation of arteriovenous malformations in Notch pathway mutants. Genes Dev 18:2469–2473

Lawson ND, Weinstein BM (2002) Arteries and veins: making a difference with Zebrafish. Nat Rev Genet 3:674–682

Lebrin F, Goumans MJ, Jonker L, Carvalho RL, Valdimarsdottir G, Thorikay M, Mummery C, Arthur HM, ten Dijke P (2004) Endoglin promotes endothelial cell proliferation and TGF-beta/ALK1 signal transduction. EMBO J 23:4018–4028

Leong KG, Hu X, Li L, Noseda M, Larrivee B, Hull C, Hood L, Wong F, Karsan A (2002) Activated Notch4 inhibits angiogenesis: role of beta 1-integrin activation. Mol Cell Biol 22:2830–2841

Li L, Miano JM, Cserjesi P, Olson EN (1996) SM22 alpha, a marker of adult smooth muscle, is expressed in multiple myogenic lineages during embryogenesis. Circ Res 78:188–195

Lindblom P, Gerhardt H, Liebner S, Abramsson A, Enge M, Hellstrom M, Backstrom G, Fredriksson S, Landegren U, Nystrom HC, Bergstrom G, Dejana E, Ostman A, Lindahl P, Betsholtz C (2003) Endothelial PDGF-B retention is required for proper investment of pericytes in the microvessel wall. Genes Dev 17:1835–1840

Ling KW, Ottersbach K, van Hamburg JP, Oziemlak A, Tsai FY, Orkin SH, Ploemacher R, Hendriks RW, Dzierzak E (2004) GATA-2 plays two functionally distinct roles during the ontogeny of hematopoietic stem cells. J Exp Med 200:871–882

Liu ZJ, Shirakawa T, Li Y, Soma A, Oka M, Dotto GP, Fairman RM, Velazquez OC, Herlyn M (2003) Regulation of Notch1 and Dll4 by vascular endothelial growth factor in arterial endothelial cells: implications for modulating arteriogenesis and angiogenesis. Mol Cell Biol 23:14–25

Lobov IB, Brooks PC, Lang RA (2002) Angiopoietin-2 displays VEGF-dependent modulation of capillary structure and endothelial cell survival in vivo. Proc Natl Acad Sci U S A 99:11205–11210

Lockman K, Hinson JS, Medlin MD, Morris D, Taylor JM, Mack CP (2004) Sphingosine 1-phosphate stimulates smooth muscle cell differentiation and proliferation by activating separate serum response factor co-factors. J Biol Chem 279:42422–42430

Loureiro RM, Maharaj AS, Dankort D, Muller WJ, D'Amore PA (2005) ErbB2 overexpression in mammary cells upregulates VEGF through the core promoter. Biochem Biophys Res Commun 326:455–465

Lu X, Le Noble F, Yuan L, Jiang Q, De Lafarge B, Sugiyama D, Breant C, Claes F, De Smet F, Thomas JL, Autiero M, Carmeliet P, Tessier-Lavigne M, Eichmann A (2004) The netrin receptor UNC5B mediates guidance events controlling morphogenesis of the vascular system. Nature 432:179–186

Manabe I, Owens GK (2001) CArG elements control smooth muscle subtype-specific expression of smooth muscle myosin in vivo. J Clin Invest 107:823–834

Mann KM, Ray JL, Moon ES, Sass KM, Benson MR (2004) Calcineurin initiates smooth muscle differentiation in neural crest stem cells. J Cell Biol 165:483–491

Martin R, Lahlil R, Damert A, Miquerol L, Nagy A, Keller G, Hoang T (2004) SCL interacts with VEGF to suppress apoptosis at the onset of hematopoiesis. Development 131:693–702

McGrath KE, Koniski AD, Malik J, Palis J (2003) Circulation is established in a stepwise pattern in the mammalian embryo. Blood 101:1669–1675

Mehlen P, Mazelin L (2003) The dependence receptors DCC and UNC5H as a link between neuronal guidance and survival. Biol Cell 95:425–436

Minasi MG, Riminucci M, De Angelis L, Borello U, Berarducci B, Innocenzi A, Caprioli A, Sirabella D, Baiocchi M, De Maria R, Boratto R, Jaffredo T, Broccoli V, Bianco P, Cossu G (2002) The meso-angioblast: a multipotent, self-renewing cell that originates from the dorsal aorta and differentiates into most mesodermal tissues. Development 129:2773–2783

Morris PN, Dunmore BJ, Tadros A, Marchuk DA, Darland DC, D'Amore PA, Brindle NP (2005) Functional analysis of a mutant form of the receptor tyrosine kinase Tie2 causing venous malformations. J Mol Med 83:58–63

Moser M, Binder O, Wu Y, Aitsebaomo J, Ren R, Bode C, Bautch VL, Conlon FL, Patterson C (2003) BMPER, a novel endothelial cell precursor-derived protein, antagonizes bone morphogenetic protein signaling and endothelial cell differentiation. Mol Cell Biol 23:5664–5679

Mukouyama YS, Shin D, Britsch S, Taniguchi M, Anderson DJ (2002) Sensory nerves determine the pattern of arterial differentiation and blood vessel branching in the skin. Cell 109:693–705

Nakagawa T, Li JH, Garcia G, Mu W, Piek E, Bottinger EP, Chen Y, Zhu HJ, Kang DH, Schreiner GF, Lan HY, Johnson RJ (2004) TGF-beta induces proangiogenic and antiangiogenic factors via parallel but distinct Smad pathways. Kidney Int 66:605–613

Ng YS, Rohan R, Sunday ME, Demello DE, D'Amore PA (2001) Differential expression of VEGF isoforms in mouse during development and in the adult. Dev Dyn 220:112–121

Ng YS, Ramsauer M, Loureiro RM, D'Amore PA (2004) Identification of genes involved in VEGF-mediated vascular morphogenesis using embryonic stem cell-derived cystic embryoid bodies. Lab Invest 84:1209–1218

Nishishita T, Lin PC (2004) Angiopoietin 1, PDGF-B, and TGF-beta gene regulation in endothelial cell and smooth muscle cell interaction. J Cell Biochem 91:584–593

Noseda M, Chang L, McLean G, Grim JE, Clurman BE, Smith LL, Karsan A (2004) Notch activation induces endothelial cell cycle arrest and participates in contact inhibition: role of p21Cip1 repression. Mol Cell Biol 24:8813–8822

Oberlin E, Tavian M, Blazsek I, Peault B (2002) Blood-forming potential of vascular endothelium in the human embryo. Development 129:4147–4157

Orlidge A, D'Amore PA (1987) Inhibition of capillary endothelial cell growth by pericytes and smooth muscle cells. J Cell Biol 105:1455–1462

Oshima M, Oshima H, Taketo MM (1996) TGF-beta receptor type II deficiency results in defects of yolk sac hematopoiesis and vasculogenesis. Dev Biol 179:297–302

Papetti M, Shujath J, Riley KN, Herman IM (2003) FGF-2 antagonizes the TGF-beta1-mediated induction of pericyte alpha-smooth muscle actin expression: a role for myf-5 and Smad-mediated signaling pathways. Invest Ophthalmol Vis Sci 44:4994–5005

Park C, Afrikanova I, Chung YS, Zhang WJ, Arentson E, Fong Gh, Rosendahl A, Choi K (2004) A hierarchical order of factors in the generation of FLK1- and SCL-expressing hematopoietic and endothelial progenitors from embryonic stem cells. Development 131:2749–2762

Perlegas D, Xie H, Sinha S, Somlyo AV, Owens GK (2005) ANG II type 2 receptor regulates smooth muscle growth and force generation in late fetal mouse development. Am J Physiol Heart Circ Physiol 288:H96–H102

Pola R, Ling LE, Aprahamian TR, Barban E, Bosch-Marce M, Curry C, Corbley M, Kearney M, Isner JM, Losordo DW (2003) Postnatal recapitulation of embryonic hedgehog pathway in response to skeletal muscle ischemia. Circulation 108:479–485

Puri MC, Bernstein A (2003) Requirement for the TIE family of receptor tyrosine kinases in adult but not fetal hematopoiesis. Proc Natl Acad Sci U S A 100:12753–12758

Ramirez-Bergeron DL, Runge A, Dahl KDC, Fehling HJ, Keller G, Simon MC (2004) Hypoxia affects mesoderm and enhances hemangioblast specification during early development. Development 131:4623–4634

Reese DE, Hall CE, Mikawa T (2004) Negative regulation of midline vascular development by the notochord. Dev Cell 6:699–708

Rensen SS, Thijssen VL, De Vries CJ, Doevendans PA, Detera-Wadleigh SD, Van Eys GJ (2002) Expression of the smoothelin gene is mediated by alternative promoters. Cardiovasc Res 55:850–863

Roberts DM, Kearney JB, Johnson JH, Rosenberg MP, Kumar R, Bautch VL (2004) The vascular endothelial growth factor (VEGF) receptor Flt-1 (VEGFR-1) modulates Flk-1 (VEGFR-2) signaling during blood vessel formation. Am J Pathol 164:1531–1535

Saint-Geniez M, Argence CB, Knibiehler B, Audigier Y (2003) The msr/apj gene encoding the apelin receptor is an early and specific marker of the venous phenotype in the retinal vasculature. Gene Expr Patterns 3:467–472

Sato Y, Rifkin DB (1989) Inhibition of endothelial cell movement by pericytes and smooth muscle cells: activation of a latent transforming growth factor-beta 1-like molecule by plasmin during co-culture. J Cell Biol 109:309–315

Seki T, Yun J, Oh SP (2003) Arterial endothelium-specific activin receptor-like kinase 1 expression suggests its role in arterialization and vascular remodeling. Circ Res 93:682–689

Serini G, Valdembri D, Zanivan S, Morterra G, Burkhardt C, Caccavari F, Zammataro L, Primo L, Tamagnone L, Logan M, Tessier-Lavigne M, Taniguchi M, Puschel AW, Bussolino F (2003) Class 3 semaphorins control vascular morphogenesis by inhibiting integrin function. Nature 424:391–397

Shalaby F, Ho J, Stanford WL, Fischer KD, Schuh AC, Schwartz L, Bernstein A, Rossant J (1997) A requirement for Flk1 in primitive and definitive hematopoiesis and vasculogenesis. Cell 89:981–990

Shin D, Garcia-Cardena G, Hayashi S, Gerety S, Asahara T, Stavrakis G, Isner J, Folkman J, Gimbrone MA Jr, Anderson DJ (2001) Expression of ephrinB2 identifies a stable genetic difference between arterial and venous vascular smooth muscle as well as endothelial cells, and marks subsets of microvessels at sites of adult neovascularization. Dev Biol 230:139–150

Sorensen LK, Brooke BS, Li DY, Urness LD (2003) Loss of distinct arterial and venous boundaries in mice lacking endoglin, a vascular-specific TGFbeta coreceptor. Dev Biol 261:235–250

Stalmans I, Ng YS, Rohan R, Fruttiger M, Bouche A, Yuce A, Fujisawa H, Hermans B, Shani M, Jansen S, Hicklin D, Anderson DJ, Gardiner T, Hammes HP, Moons L, Dewerchin M, Collen D, Carmeliet P, D'Amore PA (2002) Arteriolar and venular patterning in retinas of mice selectively expressing VEGF isoforms. J Clin Invest 109:327–336

Takashima S, Kitakaze M, Asakura M, Asanuma H, Sanada S, Tashiro F, Niwa H, Miyazaki Ji J, Hirota S, Kitamura Y, Kitsukawa T, Fujisawa H, Klagsbrun M, Hori M (2002) Targeting of both mouse neuropilin-1 and neuropilin-2 genes severely impairs developmental yolk sac and embryonic angiogenesis. Proc Natl Acad Sci U S A 99:3657–3662

Tavian M, Coulombel L, Luton D, Clemente HS, Dieterlen-Lievre F, Peault B (1996) Aorta-associated CD34+ hematopoietic cells in the early human embryo. Blood 87:67–72

Thurston G (2003) Role of Angiopoietins and Tie receptor tyrosine kinases in angiogenesis and lymphangiogenesis. Cell Tissue Res 314:61–68

Tremblay KD, Dunn NR, Robertson EJ (2001) Mouse embryos lacking Smad1 signals display defects in extra-embryonic tissues and germ cell formation. Development 128:3609–3621

Uemura A, Ogawa M, Hirashima M, Fujiwara T, Koyama S, Takagi H, Honda Y, Wiegand SJ, Yancopoulos GD, Nishikawa S (2002) Recombinant angiopoietin-1 restores higher-order architecture of growing blood vessels in mice in the absence of mural cells. J Clin Invest 110:1619–1628

Usui S, Sugimoto N, Takuwa N, Sakagami S, Takata S, Kaneko S, Takuwa Y (2004) Blood lipid mediator sphingosine 1-phosphate potently stimulates platelet-derived growth factor-A and -B chain expression through S1P1-Gi-Ras-MAPK-dependent induction of Kruppel-like factor 5. J Biol Chem 279:12300–12311

Uyttendaele H, Ho J, Rossant J, Kitajewski J (2001) Vascular patterning defects associated with expression of activated Notch4 in embryonic endothelium. Proc Natl Acad Sci U S A 98:5643–5648

Villa N, Walker L, Lindsell CE, Gasson J, Iruela-Arispe ML, Weinmaster G (2001) Vascular expression of Notch pathway receptors and ligands is restricted to arterial vessels. Mech Dev 108:161–164

Wang DZ, Olson EN (2004) Control of smooth muscle development by the myocardin family of transcriptional coactivators. Curr Opin Genet Dev 14:558–566

Wang HU, Chen ZF, Anderson DJ (1998) Molecular distinction and angiogenic interaction between embryonic arteries and veins revealed by ephrin-B2 and its receptor Eph-B4. Cell 93:741–753

Wang L, Li L, Shojaei F, Levac K, Cerdan C, Menendez P, Martin T, Rouleau A, Bhatia M (2004a) Endothelial and hematopoietic cell fate of human embryonic stem cells originates from primitive endothelium with hemangioblastic properties. Immunity 21:31–41

Wang Z, Cohen K, Shao Y, Mole P, Dombkowski D, Scadden DT (2004b) Ephrin receptor, EphB4, regulates ES cell differentiation of primitive mammalian hemangioblasts, blood, cardiomyocytes, and blood vessels. Blood 103:100–109

Waters CM, Connell MC, Pyne S, Pyne NJ (2005) c-Src is involved in regulating signal transmission from PDGFbeta receptor-GPCR(s) complexes in mammalian cells. Cell Signal 17:263–277

Wechezak AR, Coan DE (2003) Subcellular distribution of Wnt-1 at adherens junctions and actin-rich densities in endothelial cells. Exp Cell Res 288:335–343

Wechezak AR, Coan DE (2005) Dvl2 silencing in postdevelopmental cells results in aberrant cell membrane activity and actin disorganization. J Cell Physiol 202:867–873

Whitehead KJ, Plummer NW, Adams JA, Marchuk DA, Li DY (2004) Ccm1 is required for arterial morphogenesis: implications for the etiology of human cavernous malformations. Development 131:1437–1448

Yamashita J, Itoh H, Hirashima M, Ogawa M, Nishikawa S, Yurugi T, Naito M, Nakao K (2000) Flk1-positive cells derived from embryonic stem cells serve as vascular progenitors. Nature 408:92–96

Yao YG, Duh EJ (2004) VEGF selectively induces Down syndrome critical region 1 gene expression in endothelial cells: a mechanism for feedback regulation of angiogenesis? Biochem Biophys Res Commun 321:648–656

Yuan L, Moyon D, Pardanaud L, Breant C, Karkkainen MJ, Alitalo K, Eichmann A (2002) Abnormal lymphatic vessel development in neuropilin 2 mutant mice. Development 129:4797–4806

Zachary I (2003) VEGF signalling: integration and multi-tasking in endothelial cell biology. Biochem Soc Trans 31:1171–1177

Zhang XQ, Takakura N, Oike Y, Inada T, Gale NW, Yancopoulos GD, Suda T (2001) Stromal cells expressing ephrin-B2 promote the growth and sprouting of ephrin-B2(+) endothelial cells. Blood 98:1028–1037

Transport Across the Endothelium: Regulation of Endothelial Permeability

R. D. Minshall · A. B. Malik (✉)

Department of Pharmacology (m/c 868), University of Illinois, 835 S. Wolcott Avenue, Chicago IL, 60612, USA
abmalik@uic.edu

1	Permeability Pathways: Paracellular and Transcytosis	108
1.1	Transcellular Permeability	109
1.1.1	Albumin-Binding Proteins and Their Role in Transcytosis	110
1.1.2	Role of Caveolae in Mediating Endothelial Permeability via Transcytosis	112
1.1.3	Signalling Regulation of Caveolae-Mediated Transcytosis	114
1.1.4	Endothelial Barrier Function: Adjustments in Caveolin-1 Knockout Mice	117
1.2	Paracellular Permeability	118
1.2.1	Endothelial Retraction and Disruption of Cell-Cell Junctions	118
1.2.2	Role of Ca^{2+} Signalling in Mechanism of Increased Endothelial Permeability	119
1.2.3	Caveolin-1 Regulation of Ca^{2+} Signalling: Implication for Increased Paracellular Permeability	121
1.2.4	Regulation of Integrity of Inter-endothelial Junctions	122
1.2.5	Role of RhoA in Endothelial Barrier Regulation	124
2	Mechanisms of Dysregulation of Endothelial Permeability in Inflammation	125
2.1	Starling Forces Underlying Oedema Formation	125
2.1.1	Formation of Oedema	125
2.1.2	Role of Starling Forces in Oedema Formation	126
2.1.3	Causes of Increased Endothelial Permeability	126
2.1.4	Mediators of Increased Endothelial Permeability	127
2.1.5	Role of Lymphatics in Tissue Fluid Homeostasis	129
2.2	Role of "Safety Factors" in Tissue Fluid Homeostasis	130
3	Restoration of Endothelial Permeability	130
3.1	Cyclic Adenosine Monophosphate	131
3.2	Sphingosine-1-Phosphate	131
3.3	Angiopoietin-1	132
3.4	Nitric Oxide	132
3.5	RhoGTPase Cdc42	133
4	Concluding Remarks	133
	References	134

Abstract An important function of the endothelium is to regulate the transport of liquid and solutes across the semi-permeable vascular endothelial barrier. Two cellular pathways controlling endothelial barrier function have been identified. The transcellular pathway transports plasma proteins of the size of albumin or greater via the process of transcytosis in vesicle carriers originating from cell surface caveolae. Specific signalling cues are able to induce the internalisation of caveolae and their movement to the basal side of the endothelium. Caveolin-1, the primary structural protein required for the formation of caveolae, is also important in regulating vesicle trafficking through the cell by controlling the activity and localisation of signalling molecules that mediate vesicle fission, endocytosis, fusion and finally exocytosis. An important function of the transcytotic pathways is to regulate the delivery of albumin and immunoglobulins, thereby controlling tissue oncotic pressure and host-defence. The paracellular pathway induced during inflammation is formed by gaps between endothelial cells at the level of adherens and tight junctional complexes. Paracellular permeability is increased by second messenger signalling pathways involving Ca^{2+} influx via activation of store-operated channels, protein kinase $C\alpha$ (PKCα), and Rho kinase that together participate in the stimulation of myosin light chain phosphorylation, actin-myosin contraction, and disruption of the junctions. In this review of the field, we discuss the current understanding of the signalling pathways regulating paracellular and transcellular endothelial permeability.

Keywords Caveolae · Transcytosis · Interendothelial junctions · Actin-myosin contraction

1
Permeability Pathways: Paracellular and Transcytosis

Transvascular exchange of molecules and fluid between the blood and interstitial space is controlled by a monolayer of endothelial cells which lines blood vessels, essentially forming a semi-permeable vascular barrier (Michel and Curry 1999). Transport of plasma proteins (such as albumin) across the vascular endothelial barrier can occur via two discrete structural features of the endothelium: the paracellular pathway, consisting of the restrictive interendothelial cell junctions (IEJ), and the transcellular pathway, consisting of a highly mobile set of vesicles that shuttle across the endothelial barrier from luminal-to-abluminal side (Fig. 1). Junctional permeability is regulated by complexes present in IEJs, adherens junctions (AJs) and tight junctions (TJs), and interactions of these complexes with the actin cytoskeleton (Lum and Malik 1994). Junctional transport is increased in response to inflammatory mediators-such as thrombin, bradykinin, vascular endothelial growth factor (VEGF), platelet activating factor (PAF) and histamine-that "dilate" the intercellular space, resulting in increased endothelial permeability to plasma proteins and liquid (Lum and Malik 1994; Dvorak et al. 1995; Garcia et al. 1996; Moy et al. 1996; Rabiet et al. 1996). However, in the absence of a pathological insult, these junctions are normally impermeable to albumin and other plasma proteins. Electron micrographic studies have shown that this pathway is closed (restricted) and excludes macromolecule tracers (Milici et al. 1987; Predescu and Palade 1993; Predescu et al. 1994, 1997, 2004). The transport of albumin and other macromolecules across the endothelium under normal circumstances can be fully explained by transcytosis involving the plasma membrane vesicular structures or caveolae (Predescu and Palade 1993; Schnitzer et al. 1994).

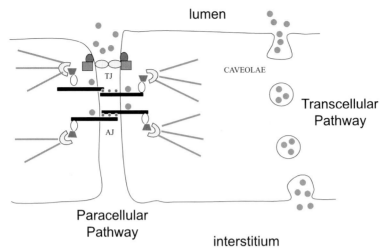

Fig. 1 Endothelial cell transport pathways. The exchange of molecules and fluid occurs through two distinct pathways. The paracellular pathway, comprising tight junctions (*TJ*) and adherens junctions (*AJ*) between neighbouring cells, is normally a restrictive barrier to macromolecular transport. VE-cadherin molecules form Ca^{2+}-dependent homotypic adhesions with VE-cadherin molecules in adjacent cells and are connected to the actin cytoskeleton via the catenins. The transcellular pathway comprises membrane-attached and cytosolic caveolae that transmigrate across the endothelium, delivering macromolecules from the blood to the interstitium. Caveolae-mediated endocytosis of albumin, the primary plasma macromolecule, is initiated by albumin-binding protein gp60 activation of *Src*-family kinases

The following sections will discuss in detail these pathways and their modes of regulation.

1.1
Transcellular Permeability

Transcellular transport, or transcytosis, is the primary mechanism by which albumin, lipids, steroid hormones, fat-soluble vitamins and other substances that bind avidly to albumin cross the normally restrictive microvessel barrier lined with continuous endothelia. Studies in microvascular endothelial cells have identified specific interactions between the 60-kDa endothelial cell surface albumin-binding glycoprotein, termed gp60, and caveolin-1, the primary structural protein of caveolae (Tiruppathi et al. 1997; Minshall et al. 2000). These interactions are required for albumin transport (Minshall et al. 2000, 2002; John et al. 2003). Signalling pathways activated by the association of gp60 with caveolin-1 are crucial in regulating albumin permeability in endothelial cells via transcytosis (Tiruppathi et al. 1997, 2003; Minshall et al. 2000; Shajahan et al. 2004a, b). Endothelial cells also transport insulin and transferrin via a transcellular mechanism; however, in contrast to albumin transport, transferrin uptake relies on clathrin-coated pits (King and Johnson 1984; Goldberg et al. 1987; Anderson 1991). Thus, studies during the last 20 years have established that endothelial albumin transport is mediated primarily via caveolae (Ghitescu et al. 1986; Milici et al. 1987; Predescu et al. 1994, 2004; Schnitzer et al. 1994; Minshall et al. 2000; Vogel et al. 2001a; John et al. 2003; Tiruppathi et al. 2003).

1.1.1
Albumin-Binding Proteins and Their Role in Transcytosis

A key event initiating the release of caveolae from the plasma membrane is the binding of albumin to a set of defined albumin-binding proteins (ABPs) (Tiruppathi et al. 1996, 1997; Schnitzer 1992; Schnitzer et al. 1988). These proteins, as identified by ligand blotting and crosslinking studies, have molecular weights of 18, 31, 60 and 75 kDa (Ghitescu et al. 1986; Ghinea et al. 1988, 1989; Schnitzer et al. 1988, Siflinger-Birnboim et al. 1991; Schnitzer 1992; Antohe et al. 1993; Tiruppathi et al. 1996; Predescu et al. 2002). Despite their potential importance in albumin transport, their identity and function remain poorly characterised. Some ABPs, specifically the 60-kDa (gp60) and 18-kDa forms, are particularly abundant in lung microvascular endothelial cell membranes (Tiruppathi et al. 1996; Schnitzer et al. 1992; Schnitzer and Bravo 1993). Functional studies to date have primarily concentrated on gp60 because it has been shown to bind native albumin and regulate transcellular albumin transport (Schnitzer et al. 1988; Schnitzer 1992; Schnitzer and Oh 1994; Tiruppathi et al. 1996, 1997; Minshall et al. 2000; Vogel et al. 2001a, b; John et al. 2003). Albumin binding to cell surface gp60 appears to be a crucial event signalling caveolae-mediated endocytosis of albumin (Minshall et al. 2000). Vesicles containing gp60-bound albumin as well as albumin in the fluid phase of vesicles were shown to internalise and translocate to the basolateral membrane, where they released their contents into the subendothelial space (Ghitescu et al. 1986; Milici et al. 1987; Simionescu and Simionescu 1991; Vogel et al. 2001a, b).

In contrast to gp60, the 18- and 31-kDa polypeptides bind to conformationally modified or denatured albumin forms (e.g. albumin-gold complex and formaldehyde- or maleic anhydride-treated albumin) with a 1,000-fold greater affinity than monomeric albumin (Schnitzer et al. 1992; Schnitzer and Bravo 1993; Schnitzer and Oh 1994). These proteins appear to be similar in their function to scavenger receptors on macrophages (Brown and Goldstein 1983) and may transfer albumin to the acidic lysosomal compartment for degradation. They are not likely to be important in transcytosis of albumin. Gp60 binding to albumin avoids lysosomal degradation of albumin (Vogel et al. 2001a); however, the mechanism by which albumin bypasses lysosomes and degradation is unclear.

Gp60 was initially characterised by its affinity to galactose-binding lectins, *Limax flavus* agglutinin and *Ricinus communis* agglutinin, which in competition assays inhibited albumin binding to rat fat tissue microvessel endothelial cells (Schnitzer et al. 1988). Siflinger-Birnboim et al. (1991) showed that R. *communis* agglutinin precipitated gp60 from bovine lung endothelial cell membranes and, importantly, that it inhibited transendothelial albumin transport. With the availability of anti-gp60 antibodies (Abs), studies have shown that the Ab blocked albumin binding and albumin permeability in rat lung microvascular bed (Schnitzer and Oh 1994). These results collectively point to an important role of gp60 in the transendothelial transport of albumin.

Other studies addressing the role of gp60 in albumin transport have shown that anti-gp60 Ab inhibited the specific binding of albumin to the endothelial cell surface at 4°C and that activation of gp60 by Ab-induced crosslinking stimulated albumin uptake and migration of vesicles to the basolateral membrane (Fig. 2; Tiruppathi et al. 1996, 1997; Minshall et al. 2000; Vogel et al. 2001a; John et al. 2003). These studies provide prima facie evidence of a potentially important functional role of gp60 in activating endothelial permeability of albumin by means of increasing transendothelial vesicle trafficking.

The Ab-induced crosslinking of gp60 shows that gp60 exhibits some interesting features of an "albumin receptor". Incubation of endothelial cells at 22°C with fluorescently tagged anti-gp60 Ab, followed by addition of a secondary Ab, resulted in formation of punctate structures resembling clusters of gp60 in vesicles beneath the plasma membrane (Tiruppathi et al. 1997). This

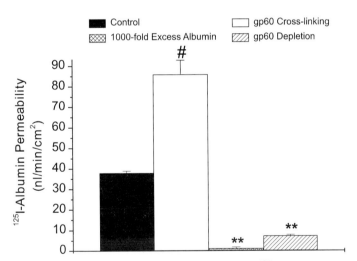

Fig. 2 Gp60-mediated transendothelial transport is shown. ^{125}I-Albumin transport was measured in cultured microvascular endothelial cell monolayers grown to confluence on Transwell filter inserts. The data show that albumin permeability is stimulated by gp60-crosslinking (anti-gp60 Ab plus secondary Ab) and blocked by excess unlabelled albumin or pretreatment of monolayers with anti-gp60 Ab to deplete cell surface gp60. These studies indicate that albumin transport requires gp60 activation and that the mechanism is inconsistent with a diffusion model due to leakage through junctions. (Modified from John et al. 2003)

membrane-receptor clustering phenomenon suggested that gp60, upon binding albumin, signals endocytosis in a receptor-dependent manner (Tiruppathi et al. 1997; Minshall et al. 2000; John et al. 2003). Membrane gp60 clustering increased the endocytosis of albumin as well as transendothelial albumin flux (John et al. 2003). As predicted by the model in which budding of plasmalemmal vesicles should also carry with it fluid phase solutes (Simionescu and Simionescu 1991), it was shown that gp60 clustering induced (1) endocytosis and transport of horseradish peroxidase (Tiruppathi et al. 1996), a tracer without any identified cell surface binding proteins, and (2) myeloperoxidase which binds specifically to albumin (Tiruppathi et al. 2004). As proof of such a mechanism operating in the intact microcirculation, Vogel et al. (2001a, b) showed that gp60 activation is also capable of inducing active transport of albumin across the continuous endothelial cell barrier of skeletal muscle and pulmonary microvessels. These studies demonstrated that gp60 activation increased transendothelial albumin transport, but did so without increasing liquid permeability (as measured by vessel wall hydraulic conductivity) (Vogel et al. 2001b). Thus, gp60 activation uncoupled hydraulic conductivity (which occurs via the diffusive paracellular pathway) from the transcellular pathway involving the back-and-forth shuttling of vesicles (Minshall et al. 2000).

Confocal imaging studies have further delineated the nature of this transcellular pathway. Studies using fluorescent-tagged albumin and Cy3 fluorophore-labelled anti-gp60 Ab showed that both probes were co-localised in vesicles near the luminal plasma membrane (Minshall et al. 2000). Gp60 activation increased transendothelial migration of water-soluble and lipophilic styryl pyridinium dye-labelled vesicles [used as a marker of vesicle trafficking since the dye fluorescence increased significantly when present at lipid-liquid interfaces; see Niles and Malik (1999)] (Minshall et al. 2000; Vogel et al. 2001b). Thus, gp60 activates membrane trafficking and increases transendothelial albumin permeability via the transcellular pathway.

1.1.2
Role of Caveolae in Mediating Endothelial Permeability via Transcytosis

Caveolae are cholesterol-rich and glycosphingolipid-rich membrane microdomains that function as mobile signalling platforms in the plasma membrane. They are, in fact, a ubiquitous feature of endothelial cells, comprising 95% of cell surface vesicles and approximately 15% of total endothelial cell volume (Predescu and Palade 1993). Caveolae released from the plasma membrane by dynamin-dependent membrane fission (Schnitzer et al. 1996; Oh et al. 1998; Shajahan et al. 2004a) can have several fates (Nabi and Le 2003). What determines whether a vesicle is destined to the basolateral membrane (the definition of transcytosis), endosomal compartment or some other intracellular organelle remains unknown.

Caveolin-1, the 22-kDa protein that coats the cytoplasmic surface of caveolae, is the defining protein constituent of caveolae (Rothberg et al. 1992; Kurzchalia et al. 1992). These characteristic flask-shaped caveolae structures are absent in endothelial cells from caveolin-1 knockout mice (Drab et al. 2001; Razani et al. 2001; Zhao et al. 2002; Predescu et al. 2004), indicating the importance of caveolin-1 in formation of the caveolar structure. Besides its presumptive function as a transcytotic vesicle carrier, caveolin-1 regulates the cholesterol content of caveolae (Smart et al. 1996). Caveolin-1 binds to cholesterol and shuttles it from the endoplasmic reticulum to the plasma membrane (Murata et al. 1995; Smart et al. 1996). However, the function of cholesterol in these "cholesterol-rich membrane microdomains" is not clear. Endocytosis of fluorescently tagged albumin or cholera toxin subunit B (CTB) (another marker of caveolae that binds to ganglioside GM1 enriched in caveolae; see Gilbert et al. 1999) in endothelial cells was blocked by filipin and methyl-β-cyclodextrin (Minshall et al. 2000; John et al. 2003; Shajahan et al. 2004a), sterol-binding agents that disassemble cholesterol-rich caveolae (Rothberg et al. 1990, 1992; Schnitzer et al. 1994; Keller and Simons 1998).

Numerous signalling molecules [such as heterotrimeric and monomeric G proteins, kinases, and endothelial nitric oxide (NO) synthase] are associated with caveolin-1 (Li et al. 1996a; Okamoto et al. 1998; Anderson 1998; Murthy and Makhlouf 2000; Minshall et al. 2000; Predescu et al. 2001). This association may maintain these enzymes in a quiescent or inhibited state (Li et al. 1996a), although this has not been specified for all of the binding partners. One function of protein-protein interactions in caveolae may be that caveolin-1 concentrates signalling molecules, allowing their rapid activation upon demand by post-translational protein modification, such as through phosphorylation or dephosphorylation (Li et al. 1996b; Minshall et al. 2002, 2003). Thus, caveolin-1 through its regulation of protein-protein interactions functions as an organising protein in caveolae, enabling "fine-tuning" of endothelial signalling (Minshall et al. 2003; Gratton et al. 2004). The signalling responses controlled by caveolin-1 include Ca^{2+} entry via specific plasma membrane channels (Lockwich et al. 2000) and activation of endothelial nitric oxide synthase (eNOS) (Isshiki et al. 2002), *Src* family tyrosine kinases, and dynamin-2 (Minshall et al. 2000; Shajahan et al. 2004a, b).

Caveolin-1 self-assembles into oligomers that associate with the cytoplasmic face of cholesterol-rich plasma membrane microdomains (Anderson 1998). Oligomerisation of caveolin-1 is required for formation of the characteristic flask-shaped caveolar structure (Fernandez et al. 2002) and regulates caveolae-mediated endocytosis, since caveolin oligomers stabilise caveolae at the plasma membrane and engage the signalling machinery required for endocytosis of caveolae (Nabi and Le 2003). Because caveolin-1 is the essential scaffolding protein in caveolae, it has also been hypothesised to function as a "master-regulator" of signalling molecules in caveolae (Okamoto et al. 1998; Anderson 1998; Minshall et al. 2002, 2003; Conner and Schmid 2003; Gratton et al. 2004).

Transcytosis, the primary means of albumin transport across continuous endothelia in the basal state (see Sect. 1.1), is the result of endocytosis (vesicle budding and fission) at the apical membrane and exocytosis (membrane fusion and release of vesicular contents) at the basolateral membrane (Tuma and Hubbard 2003). The key signalling event regulating transcytosis is *Src*-induced tyrosine phosphorylation of caveolin-1 and the GTPase dynamin-2, which are required for the induction of endocytosis (Minshall et al. 2000; Shajahan et al. 2004a, b, c). Caveolae release induced in this manner from the plasma membrane is the first step in migration of vesicles to the basal membrane (Schnitzer et al. 1996; Oh et al. 1998; Niles and Malik 1999; Minshall et al. 2000; John et al. 2003). Caveolae that detach from the plasmalemma shuttle to the basal membrane, where they fuse and release their contents (Ghitescu et al. 1986; Milici et al. 1987; Predescu et al. 1994, 1997; Minshall et al. 2000; Vogel et al. 2001a).

1.1.3
Signalling Regulation of Caveolae-Mediated Transcytosis

The details of the signalling pathways that mediate release of caveolae from the plasma membrane are incompletely understood, although it is clear that *Src* phosphorylation of caveolin-1 and dynamin-2 are crucial initial steps in the process (Minshall et al. 2003). These phosphorylation events are important, as shown by the findings that phosphatase inhibition increased caveolar fission while kinase inhibition decreased such fission (Parton et al. 1994; Mineo and Anderson 2001; Shajahan et al. 2004a). Caveolin-1 is phosphorylated by *Src* family kinases (Glenney 1989) on tyrosine residue 14 (Li et al. 1996b; Tiruppathi et al. 1997; Shajahan et al. 2004a, b, c). Studies have demonstrated a causal relationship between *Src* tyrosine kinase activity and release of caveolae from the membrane (Parton et al. 1994; Tiruppathi et al. 1997; Minshall et al. 2000; Conner and Schmid 2003; Shajahan et al. 2004a, b, c).

The heterotrimeric GTP-binding protein G_i, which binds to caveolin-1 (Li et al. 1996a; Song et al. 1997; Okamoto et al. 1998; Minshall et al. 2000), appears to play a fundamental role in the mechanism of caveolae-mediated transcytosis. We showed that caveolae-mediated endocytosis was pertussis toxin-sensitive and $G_{\alpha i}$-minigene peptide-sensitive (Minshall et al. 2000). Shajahan et al. (2004b) demonstrated that $G\beta\gamma$ signalling of *Src* activation induced caveolae-mediated transcytosis. ct-βARK expression, known to sequester $G\beta\gamma$ dimers and block signalling (Drazner et al. 1997), prevented the gp60-induced activation of *Src* kinase and subsequent phosphorylation of caveolin-1 and dynamin-2 (Shajahan et al. 2004b). In addition, ct-βARK expression blocked the *Src* phosphorylation-dependent association between dynamin-2 and caveolin-1 at the plasma membrane (Shajahan et al. 2004b), suggesting that $G\beta\gamma$-dependent *Src* activation helps to organise the endocytic machinery at the plasma membrane. The cell-permeant peptide myristoylated (m)SIRK, which promotes

Gβγ-dependent signalling in the absence of receptor stimulation or nucleotide exchange (Goubaeva et al. 2003; Ghosh et al. 2003), also activated *Src*, resulting in phosphorylation of dynamin-2 and caveolin-1, and internalisation of fluorescent CTB (Alexa 488-CTB) by 75% in endothelial cells. mSIRK-induced *Src* activation, phosphorylation of caveolin-1 and dynamin-2, and CTB endocytosis were prevented by ct-βARK and the *Src* kinase inhibitor PP2 (Shajahan et al. 2004b). These results together describe a model (Fig. 3) in which Gβγ is essential for the activation of *Src*, and hence caveolae-mediated endocytosis.

As indicated, *Src* family tyrosine kinases stimulate caveolae-mediated endocytosis of albumin by phosphorylating caveolin-1 and dynamin-2 in endothelial cells (Tiruppathi et al. 1997; Minshall et al. 2000; Shajahan et al. 2004a, b, c). Other studies have also reported that endocytosis via caveolae is critically dependent on stimulation of tyrosine kinase signalling (Parton et al. 1994; Aoki et al. 1999; Chen and Norkin 1999; Liu and Anderson 1999; Pelkmans et al. 2002; Singh et al. 2003; Sharma et al. 2004). *Src* phosphorylation of caveolin-1 at Tyr^{14} (Shajahan et al. 2004c; Aoki et al. 1999; Mastick et al. 1995; Li et al. 1996b; Tiruppathi et al. 1996; Rizzo et al. 2003) is believed to signal caveolae-mediated endocytosis (Minshall et al. 2000; Shajahan et al. 2004a, c; Aoki et al. 1999; Sharma et al. 2004). Thus, coincident with endocytosis of albumin occurring

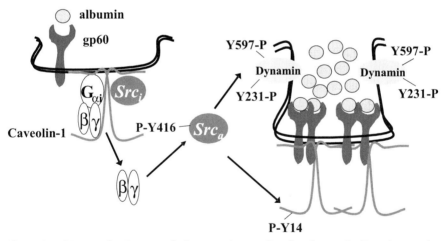

Fig. 3 Signalling mechanisms regulating caveolae-mediated endocytosis. Caveolae are the primary vesicular transporters or "carriers" of albumin in endothelial cells. Gp60, an albumin-binding protein, initiates the endocytosis of albumin by first associating with caveolin-1 and subsequently through activation of *Src*-family tyrosine kinase signalling. Caveolin-1 plays a central role, as it serves a scaffolding function for components of the "caveolar release complex"-G_i/βγ, *Src* and dynamin-2-the signalling machinery responsible for endocytosis. *Src*-family kinases, activated by Gβγ subunits upon stimulation of G_i (autophosphorylation of Y416), phosphorylate tyrosine residues on gp60, caveolin-1 Y14, and dynamin Y231 and Y597. The caveolar release complex engaged by *Src* activation and phosphorylation of caveolin-1 and dynamin-2 induces caveolar fission

within 1 min after gp60 activation, caveolin-1 and dynamin-2 were tyrosine phosphorylated at residues 14 and 231/597, respectively (Shajahan et al. 2004a, b, c). In both cases, pretreatment of cells with *Src* kinase inhibitor PP1 or PP2 abolished phosphorylation. Dephosphorylation may also be involved in the control of caveolae-mediated endocytosis. It is possible that protein tyrosine phosphatases may dephosphorylate caveolin-1 and dynamin-2 in endothelial cells. Csk, a negative regulator of *Src*, was shown to bind specifically to phosphorylated caveolin-1 (Cao et al. 2002), suggesting a mechanism of negative feedback regulation.

The GTPase dynamin-2 mediates fission of caveolae from the plasma membrane (Shajahan et al. 2004a, b; Oh et al. 1998; Henley et al. 1999; Conner and Schmid 2003). *Src* phosphorylation of dynamin increases its GTPase activity, assembly into oligomers (Ahn et al. 1999, 2002), and association with caveolin-1 at the plasma membrane (Kim and Bertics 2002; Shajahan et al. 2004a). Interestingly, SV40-induced internalisation of caveolae was also shown to be dependent on tyrosine kinase activity (Chen and Norkin 1999) and recruitment of dynamin to the membrane (Pelkmans et al. 2002).

The functional importance of these events in caveolae-mediated endocytosis was investigated in pulmonary microvessel endothelial cells stably expressing non-*Src* phosphorylatable caveolin-1 or dynamin-2 mutants (Minshall et al. 2003; Shajahan et al. 2004a, c). Expression of either Y14F caveolin-1 or Y597F dynamin-2 abolished albumin and CTB endocytosis (Shajahan et al. 2004a, c), indicating *Src* phosphorylation of these residues is required for signalling caveolae-mediated endocytosis. Association of caveolin and dynamin was also increased when dynamin was phosphorylated at Y597 and reduced by the non-phosphorylatable dynamin mutant (Shajahan et al. 2004a). This finding suggests that *Src* phosphorylation of dynamin may enable its localisation to caveolae, specifically the neck region, thereby "pinching" caveolae from the membrane (Conner and Schmid 2003).

Sequestration of the Gβγ heterodimer has also been shown to inhibit endocytosis via clathrin-coated vesicles (Lin et al. 1998; Kim et al. 2003), in part by interfering with actin polymerisation (Lin et al. 1998). Although the role of actin in caveolae-mediated endocytosis remains unclear, both *Src* and dynamin are known to participate in actin cytoskeletal remodelling by regulating cortactin (McNiven et al. 2000; Cao et al. 2003; Krueger et al. 2003). It is therefore possible that *Src* controls the function of actin or associated binding proteins and thereby regulates caveolar movement along the actin filaments or microtubule "tracks" (Krueger et al. 2003; Mundy et al. 2002; van Deurs et al. 2003). This would be an additional control exerted by *Src* beyond Gβγ-dependent *Src* activation and the subsequent phosphorylation of caveolin-1 and dynamin-2 (Shajahan et al. 2004a, b, c) described above.

1.1.4
Endothelial Barrier Function: Adjustments in Caveolin-1 Knockout Mice

Caveolae-mediated endocytosis sets into motion the transport of plasma proteins across the vascular endothelial barrier. It stands to reason from the above-described role of caveolin-1 that deletion of the caveolin-1 gene (*CAV1*) would result in the absence of plasmalemmal vesicles and inability to transport albumin across the endothelium. Caveolin-1 knockout mice show uncontrolled endothelial cell proliferation and lung fibrosis, increased NO production, impaired Ca^{2+} signalling and defective endocytosis of albumin (Zhao et al. 2002; Drab et al. 2001; Razani et al. 2001; Schubert et al. 2002; Predescu et al. 2004). These changes could be reversed by expression of caveolin-1 complementary DNA (cDNA). Deletion of the *CAV1* gene curiously was not lethal, suggesting that compensatory mechanisms, such as increased junctional permeability (Zhao et al. 2002; Schubert et al. 2002; Predescu et al. 2004), are responsible for survival of these mice.

Ultrastructural analysis of microvessels in the caveolin-1 knockout mouse model showed the absence of caveolae (Zhao et al. 2002; Drab et al. 2001; Razani et al. 2001; Predescu et al. 2004) but the presence of fenestrae and larger vesicular structures resembling vesicular-vacuolar organelles (VVOs) (Minshall et al. 2003). The assembly of these cellular structures apparently did not require the presence of caveolin-1. Interestingly, somewhat larger (100–120 nm diameter) uncoated vesicles resembling caveolae were present in endothelia of certain vascular beds, albeit fewer in number than caveolae in wild-type mice (Zhao et al. 2002; Drab et al. 2001; Razani et al. 2001; Predescu et al. 2004). This finding indicates that there may be an additional pool of vesicles in endothelial cells that are neither clathrin nor caveolin-1 coated. Recently, Kirkham and co-workers (2005) described the presence of uncoated caveolin-independent early endocytic vesicles in $CAV1^{-/-}$ mouse fibroblasts. These vesicle structures contained GPI-linked proteins and internalised fluid phase markers, which appeared to be the primary structures mediating CTB uptake in these cells. However, their role in transendothelial transport remains to be elucidated.

Perhaps the most striking observation regarding the phenotype of $CAV1^{-/-}$ mice was the fivefold increase in plasma NO level (Zhao et al. 2002). This finding supports the hypothesis that caveolin-1 has a function in regulating eNOS (Garcia-Cardena et al. 1997; Bucci et al. 2000; see below). The mechanism of caveolin-1 regulation of caveolae-associated proteins such as eNOS is not entirely clear, but it could be secondary to maintaining the correct lipid composition and interactions with the kinase PKB/Akt (Liu et al. 2002).

Together, these observations are consistent with data showing an important role of caveolin-1 in albumin transport, regulation of eNOS activity and cell proliferation (Minshall et al. 2000; Bucci et al. 2000; Zhao et al. 2002). However, the picture is far from complete. Additional studies are needed to determine (1) precisely how caveolin-1 regulates endothelial barrier function, (2) the role of elevated NO levels as determined by eNOS-caveolin-1 interactions in controlling endothelial barrier function and (3) the basis by which caveolin-1 keeps endothelial cells in a contact-inhibited state in order to maintain vessel wall integrity.

1.2
Paracellular Permeability

Capillary endothelial cells form the primary barrier between the plasma and interstitial fluid. Intercellular contacts between endothelial cells and cellular adhesion to the underlying subendothelial matrix are responsible for the junctional barrier properties of endothelium. The manner in which certain pathological conditions such acute lung injury (ALI) and other types of inflammatory diseases induce barrier dysfunction is not fully understood. Increased endothelial permeability is the result of loss of contact between microvascular endothelial cells and weakening of their adhesion to the basement membrane. Mediators elaborated during inflammation, such as thrombin, VEGF, histamine, PAF and bradykinin, are key to the disruption of endothelial barrier function by a direct action on the endothelium, which increases permeability by opening intercellular junctions. We discuss below the current understanding of the signalling mechanisms mediating increased endothelial permeability via the paracellular or IEJ pathway.

1.2.1
Endothelial Retraction and Disruption of Cell-Cell Junctions

The endothelium is the target of pro-inflammatory and thrombogenic mediators and growth factors, many of which have receptors in endothelial cells and thus can directly affect endothelial permeability. These mediators are capable of disrupting IEJs and increasing endothelial permeability, thus allowing the passage of plasma proteins through IEJs. The inflammatory mediator thrombin results in increased endothelial permeability by causing endothelial cell retraction and shape change (Vogel et al. 2000; Tiruppathi et al. 2003). The signal transduction pathways that promote loss of barrier function involve a complex series of signalling events leading ultimately to rapid and sustained phosphorylation of myosin light chain (MLC) and simultaneous inhibition of MLC-associated phosphatase, which functions to prevent dephosphorylation of MLC and prolong the contractile response (Dudek and Garcia 2001; Tiruppathi et al. 2003; Birukova et al. 2004). Endothelial cell retraction is likely to be precipitated by disruption of endothelial AJs secondary to the traction imposed by actomyosin-mediated endothelial contractility (Sandoval et al. 2001; Tiruppathi et al. 2003). Phosphorylation of MLC by Ca^{2+}/calmodulin-dependent myosin light chain kinase (MLCK) is required for actomyosin interaction and engagement of endothelial contractile apparatus. Filamentous actin within endothelial cells also associates with the cytoplasmic tail of the major AJ protein vascular endothelial (VE)-cadherin (Dejana et al. 1999). Contractile force may "unhinge" AJs, resulting in formation of IEJ gaps (Sandoval et al. 2001). These gaps, induced by thrombin within minutes (Sandoval et al. 2001), provide a plausible structural basis for increased paracellular permeability.

1.2.2
Role of Ca^{2+} Signalling in Mechanism of Increased Endothelial Permeability

The central role of Ca^{2+} signalling in mediating increased endothelial permeability is described in Fig. 4. Thrombin activates the GTP-binding protein coupled receptor PAR-1 [thrombin-ligated proteinase activated receptor (PAR) present in endothelial cells] (Tiruppathi et al. 2003; Vogel et al. 2000). The heterotrimeric GTP-binding protein G_q signals Ca^{2+} release from intracellular stores, and the subsequent Ca^{2+}-store depletion in turn signals Ca^{2+} entry via specific plasma membrane channels (Tiruppathi et al. 2003; see below). Upon PAR-1 activation, free Ca^{2+} in the cytosol binds to the Ca^{2+}-binding protein calmodulin; the Ca^{2+}/calmodulin complex activates MLCK, which in-

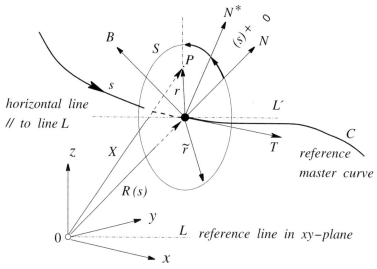

Fig. 4 Signalling functions of Ca^{2+}, PKCα and Rho in the mechanism of increased endothelial permeability. Activation of endothelial cell surface PAR-1 by thrombin results in inflammation/vascular leakage. G_q- and $G_{12/13}$-coupled signalling mechanisms activated by thrombin induce an elevation in intracellular Ca^{2+} and activation of PKCα and Rho GTPase. Crosstalk between G_q and $G_{12/13}$ signalling via PKCα is also an essential requirement for Rho and Rho kinase activation. Phosphorylation of myosin light chain (*MLC*) by Ca^{2+}/calmodulin (*CaM*)-dependent myosin light chain kinase (*MLCK*) and inhibition of MLC phosphatase via Rho kinase promote actin-myosin cross-bridge cycling, cell retraction and endothelial barrier dysfunction

duces phosphorylation of MLC (Dudek and Garcia 2001). In parallel with G_q, the $G_{12/13}$ G protein pathway-acting through cytoplasmic Rho GTPase and its effector Rho kinase-inhibits MLC dephosphorylation (Birukova et al. 2004). The combined effect of MLCK plus Rho kinase activity is to induce and to maintain MLC phosphorylation, resulting in formation of actomyosin contractile units (and actin stress fibres) that exert force on the IEJs, both the AJ and TJ complexes (Sandoval et al. 2001; Dudek and Garcia 2001; Birukova et al. 2004). In addition, there is the likelihood that not only actin but microtubule polymerisation is important in the mechanism of endothelial cell contractility (Dudek and Garcia 2001); thus, for endothelial cells to increase their permeability, there is likely to be crosstalk between multiple components of the cells' contractile machinery.

The role of Ca^{2+} signalling in mediating increased endothelial permeability is well established. Lum et al. (1989) showed-using bovine pulmonary arterial endothelial cells (BPAEC)-that a thrombin-induced increase in ^{125}I-albumin permeability can be reduced by 50% by chelating intracellular Ca^{2+} ($[Ca^{2+}]_i$) with quin-2. They showed that the increase in transendothelial albumin permeability was also dependent on both intracellular Ca^{2+} release and extracellular Ca^{2+} entry (Lum et al. 1989). Additionally, these authors have shown a temporal relationship between the inositol (1,4,5)-trisphosphate (IP_3)-induced increase in $[Ca^{2+}]_i$ and the increase in transendothelial albumin permeability using BPAEC monolayers (Lum et al. 1992). The thrombin-induced increase in $[Ca^{2+}]_i$ in endothelial cells is primarily the result of activation of PAR-1 (Ellis et al. 1999). Other studies showed abrogation of the thrombin-induced increase in lung microvascular permeability in PAR-1 knockout mice (Vogel et al. 2000). IP_3 formation induced by thrombin activation of PAR-1 is known to cause release of sequestered Ca^{2+} and elicit Ca^{2+} entry via store-operated channels (SOC) (Tiruppathi et al. 2002, 2003). Tiruppathi and associates have identified TRPC4 (transient receptor potential channel 4) as an essential constituent of the SOC in the mouse lung (Tiruppathi et al. 2002). Their data obtained using TRPC4 knockout mice support a causal relationship between Ca^{2+} entry via TRPC4 and elevated pulmonary microvascular permeability (Tiruppathi et al. 2002). Thus, increased Ca^{2+} influx leading to activation of endothelial retraction via actomyosin coupling may be a fundamental underlying basis of increased vascular permeability in vivo.

The increase in $[Ca^{2+}]_i$ is also coupled to activation of protein kinase C (PKC) isoforms, specifically PKCα, which leads to activation of Ca^{2+}/calmodulin-dependent MLCK (Garcia et al. 1995; Wysolmerski and Lagunoff 1990), another factor promoting actin-myosin interaction secondary to phosphorylation of the 20-kDa MLC_{20} (Garcia et al. 1995). This process can facilitate cytoskeletal reorganisation and induce endothelial cell shape change (Lum and Malik 1994; Dudek and Garcia 2001). Studies have shown that in endothelial cells the monomeric GTPase Rho can also contribute to mediating MLC_{20} phosphorylation, thus leading to increased permeability (van Nieuw Amerongen et al. 1998, 2000; Vouret-Craviari et al. 1998; Holinstat et al. 2003).

Van Nieuw Amerongen et al. (1998) investigated mechanisms in endothelial cells signalling responsible for increased endothelial permeability induced by thrombin and histamine. Chelation of $[Ca^{2+}]_i$ with BAPTA-AM (1,2-bis(2-aminophenoxy) ethane-N,N,N',N'-tetra-acetic acid) prevented the transient histamine-induced increase in endothelial monolayer permeability, but not the more prolonged thrombin-induced permeability increase, which depends on extracellular Ca^{2+} influx. By contrast, the tyrosine kinase inhibitor genistein and the RhoA inhibitor C3 transferase toxin given together prevented the thrombin-induced increase in permeability (Dudek and Garcia 2001). These studies have not implicated a role for PKCα activation in the mechanism of increased endothelial permeability. This observation contradicts earlier studies by Lynch et al. (1990) and others (Lum et al. 1993, Tiruppathi et al. 1992). Interestingly, recent findings in endothelial cells demonstrate that thrombin induces rapid PKCα-dependent phosphorylation of Rho-GDP guanine nucleotide dissociation inhibitor (GDI), and thereby facilitates Rho activation (Mehta et al. 2001). Prevention of PKCα activation abolished thrombin-induced Rho activation-indicating the requirement for PKCα in the mechanism of Rho activation in endothelial cells-and thereby increased endothelial permeability (Holinstat et al. 2003). This crosstalk between the Rho and PKCα signalling pathways appears to be mediated by the Rho exchange factor, p115RhoGEF (Holinstat et al. 2003).

1.2.3
Caveolin-1 Regulation of Ca^{2+} Signalling: Implication for Increased Paracellular Permeability

Several molecules involved in Ca^{2+} influx have been localised to caveolae, including an IP_3 receptor-like protein (Fujimoto et al. 1992), dihydropyridine-sensitive Ca^{2+} channels (Jorgensen et al. 1989), a Ca^{2+} ATPase (Fujimoto 1993), and TRP1 channels involved in capacitive Ca^{2+} entry (Lockwich et al. 2000). Electron microscopy studies showed a population of caveolae in close proximity with the endoplasmic reticulum (ER) (Sugi et al. 1982). The functional importance of caveolae with regards to Ca^{2+} release and re-uptake was assessed using the live cell Ca^{2+} sensor yellow cameleon (Isshiki et al. 2002). Using fusion proteins of yellow cameleon and caveolin-1, which target the Ca^{2+} indicator to the caveolae, it was demonstrated that caveolae are the preferred sites of Ca^{2+} entry upon ER Ca^{2+} store depletion, that is, SOC-dependent Ca^{2+} entry. The capacitive or SOC Ca^{2+} entry model described by Anderson and co-workers (Isshiki et al. 2002) suggests that caveolae function as organisers of Ca^{2+} signalling, providing a mechanism for regulating the "on-off" state of the Ca^{2+} signalling circuit that mediates increased endothelial permeability. Caveolae are thus the key compartments involved in regulating store-operated

Ca^{2+} entry. Ca^{2+} entering endothelial cells via caveolae may be a crucial factor regulating endothelial permeability via the junctional pathway; however, this question has not been extensively examined.

1.2.4
Regulation of Integrity of Inter-endothelial Junctions

Maintenance of cell shape, and thus integrity of the endothelial barrier, is the result of integrated actions of the contractile and adhesive forces that couple endothelial cells with each other and to the extracellular matrix (Dudek and Garcia 2001). Actin-myosin motor activation regulates the contractile force function of endothelial cells (Lum and Malik 1994). Endothelial AJs associate with the actin cytoskeleton and link neighbouring cells through transmembrane VE-cadherin molecules, and thereby contribute to the intercellular adhesive force (Fig. 5). VE-cadherins are located in intercellular AJs where they are linked in the cytoplasm to β-, γ- and p120-catenins, which in turn link them to α-catenin and the actin cytoskeleton (Lampugnani et al. 1995; Dejana 1996). The five extracellular cadherin repeats are involved in mediating adhesion via specific Ca^{2+}-binding sites (Sivasankar et al. 2001). Cadherins in a single cell oligomerise to form *cis*-oligomers and in adjacent cells to form *trans*-oligomers (Dejana et al. 1999). The cadherin cytoplasmic domain contains two functional sub-domains: juxtamembrane domain (JMD), a binding site for p120 catenin, and the C-terminal domain (CTD), a binding site for β-catenin and plakoglobin (or γ-catenin) that bind in a mutually exclusive manner (Dejana et al. 1999). Plakoglobin or γ-catenin associates with α-catenin, an actin-binding protein that links VE-cadherin to the actin cytoskeleton (Aberle et al. 1996).

Several lines of evidence now point to the essential role of VE-cadherin junctions in regulating IEJ permeability (Corada et al. 1999; Dejana 1996; Gao et al. 2000). Thrombin induced VE-cadherin disassembly, and the resulting loss of functional AJs, has been proposed as the basis of increased endothelial permeability (Rabiet et al. 1996; Corada et al. 1999; Sandoval et al. 2001). Calphostin C, a PKC inhibitor, prevented the thrombin-induced disorganisation of the VE-cadherin complex (Rabiet et al. 1996; Sandoval et al. 2001), supporting the role of PKC in mediating the permeability increase by a cadherin-dependent mechanism. Recent studies have also shown that histamine-induced loss of endothelial barrier function was associated with disassembly of VE-cadherin junctions (i.e. cell-cell tethering) (Winter et al. 1999). The role of Ca^{2+} and PKCα signalling in junctional disassembly received detailed attention by Sandoval et al. (2001), who studied the relationship between the level of cytosolic Ca^{2+} and increase in endothelial permeability. In this study, endothelial cells were exposed to thapsigargin or thrombin at concentrations that resulted in similar increases in [Ca^{2+}]$_i$. The rise in [Ca^{2+}]$_i$ in both cases was secondary to release of Ca^{2+} from intracellular stores and influx of extracellular Ca^{2+}. To the same degree, both agents decreased endothelial cell monolayer electrical resistance

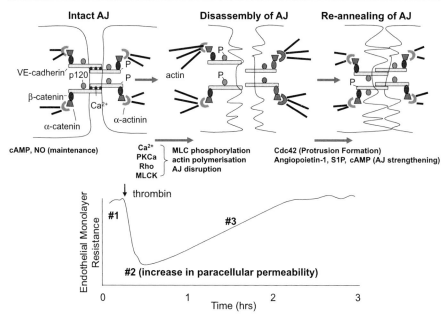

Fig. 5 Signalling of recovery of endothelial barrier function. Under normal physiological conditions, adherens junctions (*AJ*) are intact and restrict leakage of macromolecules and fluid into the tissue. AJ consist of VE-cadherin linked to β-, p120- and α-catenin, and thereby connect to the actin cytoskeleton. The tight barrier property of the endothelium is maintained by intracellular NO and cyclic adenosine monophosphate (*cAMP*). In response to an inflammatory mediator (e.g. thrombin), signalling molecules such as Ca^{2+}, PKCα and Rho facilitate myosin phosphorylation, actin polymerisation, disruption of AJs and gap formation between cells. Within 2 h of thrombin-stimulated disruption of endothelial monolayers, barrier integrity is restored by the formation of Cdc42-dependent membrane protrusions that fill the gaps between cells and re-establish interendothelial junctions. In addition, sphingosine-1-phosphate and angiopoietin-1 facilitate barrier recovery following injury and/or prevent barrier dysfunction (see text for details)

(a measure of endothelial cell shape change) and increased transendothelial ^{125}I-albumin permeability. Interestingly, thapsigargin induced activation of PKCα and discontinuities in VE-cadherin junctions without formation of actin stress fibres, whereas thrombin induced PKCα activation and similar alterations in VE-cadherin junctions, but in association with actin stress fibre formation. Both agents induced phosphorylation of VE-cadherin-associated proteins, which was prevented by the PKC inhibitor calphostin C. Further, thapsigargin failed to promote phosphorylation of MLC_{20}, whereas thrombin induced MLC_{20} phosphorylation consistent with formation of actin stress fibres. Calphostin C pretreatment also prevented disruption of VE-cadherin junctions and decrease in transendothelial electrical resistance caused by both agents. Thus, these findings collectively demonstrate that Ca^{2+} signalling is critical for activation of PKCα and disruption of VE-cadherin junctions, and the Ca^{2+} signalling thereby mediates increased endothelial permeability.

In addition, the IP_3-receptor antagonist 2-aminoethoxydiphenyl borate (2-APB) was shown to prevent thrombin-induced ER-stored Ca^{2+} release, Ca^{2+} influx and thrombin-induced decrease in transendothelial resistance (i.e. IEJ gap formation) in endothelial cells (Tiruppathi et al. 2003). This finding supports the concept that Ca^{2+} signalling is an essential determinant of increased endothelial permeability via IEJ disassembly. Additionally, Tiruppathi et al. (2001) showed that short-term exposure of human endothelial cells to tumour necrosis factor (TNF)-α augmented the thrombin-induced increase in endothelial permeability. This effect was not associated with increased IP_3 generation in response to thrombin, but it was ascribed to increased SOC-induced Ca^{2+} influx (Tiruppathi et al. 2001). These studies establish a causal relationship between SOC Ca^{2+} influx and increased endothelial permeability secondary to disassembly of AJs.

1.2.5
Role of RhoA in Endothelial Barrier Regulation

Thrombin stimulation of PAR-1 activates the heterotrimeric G proteins $G_{12/13}$, G_q and G_i (Tiruppathi et al. 2003). G_q in turn activates PLC_β, thus triggering Ca^{2+} mobilisation from the endoplasmic reticulum, PKC activation and PKC's activation of downstream effectors such as Ca^{2+}-regulated kinases and phosphatases. Both direct activation of $G_{12/13}$ and its cross-activation by G_q may activate the monomeric Rho GTPase, RhoA (Mehta et al. 2001; Holinstat et al. 2003). RhoA activity is controlled by its cycling between inactive GDP- and active GTP-bound states (Birukova et al. 2004). Three different classes of proteins are required for this cycling: (1) guanine nucleotide exchange factors (GEFs) that stimulate GDP to GTP exchange, (2) GTPase-activating proteins (GAPs) that stimulate GTP-hydrolysis, and (3) GDIs that bind and stabilise Rho-GDP (Hall 1998). Others groups in addition to our own have shown that thrombin, by activating RhoA, induces minute IEJ gaps that are responsible for the observed increase in vascular permeability (Mehta et al. 2001; Holinstat et al. 2003; Birukova et al. 2004). Formation of these gaps and loss of endothelial barrier function occurred as the result of change in cell shape. RhoA, by activating its downstream effector Rho kinase, prolonged actin-myosin driven contractile forces that are transmitted to the endothelial AJ complex. This series of events leads to the disruption of cell-cell adhesive forces, thus inducing IEJ gap formation (Rabiet et al. 1996; Sandoval et al. 2001). Another mechanism of RhoA regulation of endothelial cell shape and disruption of AJs is through its ability to regulate $[Ca^{2+}{}_i]$. RhoA may induce the translocation of the IP_3R in order to promote its association with TRP1 in the plasma membrane (Mehta et al. 2003). Additionally, RhoA has been implicated in inducing SOC activation (Bird and Putney 1993; Fasolato et al. 1993; Rosado and Sage 2000). This may occur by spatial reorganisation of actin filaments (a Rho-regulated phenomenon), thus favouring Ca^{2+} entry via activation of TRPC channels (Mehta et al. 2003). Thus, RhoA, by associating with IP_3R and the TRP1 channel, facilitates Ca^{2+} entry leading to disruption of AJs and an increase in endothelial permeability.

2
Mechanisms of Dysregulation of Endothelial Permeability in Inflammation

Tissue oedema, defined as fluid accumulation in the extravascular space, can interfere with functions of organs, for example resulting in impaired gas exchange and hypoxaemia in lung oedema (as in acute respiratory distress syndrome or ALI) and cardiac impairment in myocardial oedema. When an organ such as the lung fails to maintain fluid balance, liquid accumulates in the interstitium and ultimately invades the alveolar airspaces. We focus below on the role of the transcellular and paracellular permeability pathways in tissue oedema formation.

2.1
Starling Forces Underlying Oedema Formation

2.1.1
Formation of Oedema

Oedema formation is defined by the Starling forces generated across the microvessel wall. The mathematical relationship between fluid filtration rate (J_v), and transmural hydrostatic and oncotic pressures is

$$J_v = L_p S \left[(P_c - P_i) - \Phi(\pi_c - \pi_i) \right]$$

where Φ is albumin reflection coefficient, S vessel surface area, L_p hydraulic conductivity, P capillary hydrostatic pressure, π oncotic pressure, and subscripts c and I refer to capillary and interstitial compartments. The term $(P_c - P_i)$ gives the transmural hydrostatic pressure gradient and the product $\Phi(\pi_c - \pi_i)$ is the effective transmural oncotic pressure gradient; the difference between these four forces defines the "driving pressure" for net fluid filtration or reabsorption across the microvessel wall. A higher $(P_c - P_i)$ favours fluid filtration (increased J_v), whereas a higher $\Phi(\pi_c - \pi_i)$ favours fluid reabsorption. Normally, an equilibrium is achieved between fluid filtration and fluid reabsorption at proximal and distal segments of capillaries, and little or no net filtration occurs through capillary walls. The lymphatic system in most organs maintains a negative interstitial pressure by continuously withdrawing fluid from interstitium. Oedema develops when fluid filtration substantially exceeds its reabsorption and the capacity of the lymphatic system to remove fluid from the pulmonary interstitium.

2.1.2
Role of Starling Forces in Oedema Formation

An elevated net driving pressure without a marked increase in permeability underlies "pressure oedema" (i.e. oedema resulting from an increase in the capillary hydrostatic pressure). Hydrostatic oedema, when it is not associated with frank barrier breakdown, is generally protein-poor, at least in early stages of the syndrome, because the barrier properties tending to exclude large molecules are preserved. In hydrostatic oedema, the ratio of plasma to alveolar fluid protein concentration is usually less than 0.6 (Taylor and Parker 1985). The critical capillary pressure for formation of oedema due strictly to elevated hydrostatic pressure is a P_c above 25 mm Hg (Taylor and Parker 1985). Fluid accumulation in tissues is minimised by "safety factors" that are activated below this critical capillary pressure (see Sect. 3). The extravascular water content increases progressively as a result of the inability of these safety factors to reduce the fluid filtration rate when capillary hydrostatic pressure increases above the critical value. Most clinical manifestations of tissue oedema can be understood in terms of changes in Starling forces across the microvessel wall. A decrease in the plasma protein concentration, such as in hypoalbuminaemia, reduces the transmural oncotic pressure difference, thus favouring increased fluid filtration. In this case, the critical capillary pressure at which tissue begins to gain water decreases in direct proportion to the reduction in plasma oncotic pressure.

2.1.3
Causes of Increased Endothelial Permeability

IEJ rupture or breakdown underlies protein-rich oedema formation due to the loss of the normal restrictive properties of the capillary endothelial barrier (Mehta et al. 2004). Some evidence suggests that stimulation of protein transport via a transcellular pathway could also contribute to formation of protein-rich oedema fluid (van Nieuw Amerongen et al. 1998; Dvorak and Feng 2001). An active transcellular albumin transport process involving vesicular carriers is well established in pulmonary microvascular endothelial cells (John et al. 2003). These carriers, which are caveolae, occupy a remarkably high percentage (15%) of the endothelial cell volume (Predescu et al. 1997). An important area of investigation is whether pathologic conditions can stimulate transcytosis leading to protein-rich oedema fluid. For example, vesicular transport was suggested to be involved in the hyperpermeability response of the endothelium to VEGF by increasing the density of the otherwise rarely seen endothelial channel-like structures called VVOs (vesicular-vacuolar organelles; Dvorak and Feng 2001). Additional studies are required to determine if increased albumin permeability mediated by caveolae and the transcellular transport pathway participate in the formation of protein-rich oedema.

Vascular permeability can increase as a result of direct injury to endothelial cells (Goodman et al. 2003), alterations in the dimensions of IEJs (the paracellular pathway) (Rabiet et al. 1996) or a combination of these factors (Lum and Malik 1994). Figure 5 lists the primary intracellular signalling molecules thought to mediate the increase in endothelial monolayer permeability induced by thrombin, as well as the reversibility of the response. The figure also shows the time course and magnitude of the change in normalised transendothelial electrical resistance across a monolayer, indicating that recovery and re-annealing mechanisms re-establish an intact and restrictive barrier (see Sect. 1.2). The increase in vascular permeability is operationally defined in the Starling equation by an increased capillary filtration coefficient ($K_{f,c}$), which is equivalent to the $L_p S$ term in the equation. An increase in the $K_{f,c}$ corresponds to decreased barrier resistance to the movement of liquid across the capillary wall barrier. The albumin reflection coefficient (Φ_{Alb}) describes the albumin permeability of the vascular endothelial barrier and provides a quantitative measure of protein permeability (Malik et al. 2000). In high-permeability pulmonary oedema, the alveolar fluid protein concentration approximates the plasma protein concentration (Flick and Matthay 2000). The increase in vascular permeability shifts the relationship between P_c and extravascular water content towards a lower P_c, indicating that oedema occurs at a reduced driving pressure in the face of increased vascular permeability.

2.1.4
Mediators of Increased Endothelial Permeability

Vasoactive mediators such as thrombin, histamine, PAF, bradykinin and VEGF, which are released during thrombosis and inflammation, increase endothelial permeability by increasing $[Ca^{2+}]_i$, reactive oxygen species (ROS), and/or NO levels (Tiruppathi et al. 2003; Lo et al. 1992; Kubes 1995). The endothelial cell signalling mechanisms activated by vasoactive mediators are discussed in detail in Sect. 2.2. Increased vascular pressure also stimulates an increase in intracellular Ca^{2+} and ROS, thereby increasing permeability (Kuebler et al. 1999, 2002). Pro-inflammatory cytokines (interleukin-1β, TNF-α) released from macrophages and polymorphonuclear leucocytes (PMN) as part of the host defence against bacterial infection, and the bacterial product lipopolysaccharide (LPS; endotoxin), produce severe vascular endothelial injury, increased endothelial permeability and tissue oedema (Albelda et al. 1994; Horgan et al. 1991).

2.1.4.1
LPS, PMNs and Oxidants

The generation of oxidants by LPS has an important signalling function in up-regulating the pro-inflammatory gene *intercellular adhesion molecule 1* (*ICAM1*) in endothelial cells (Malik 1993; Rahman et al. 1999; Fan et al. 2002) whose protein product mediates stable ICAM-1-dependent endothelial adhesivity and firm adhesion of PMN to the endothelium (Issekutz et al. 1999). ICAM-1-dependent PMN binding to endothelial cells, and the subsequent PMN activation (characterised by release of ROS and intracellular proteases), are critical factors in the development of acute lung vascular injury (Albelda et al. 1994; Horgan et al. 1991) and tissue oedema (Horgan et al. 1991; Lo et al. 1992), the hallmarks of ALI (Abraham 2003). Studies have focussed on the cellular responses of the individual cell populations (i.e. PMN or endothelial cells) and have emphasised the role of cytokines, chemokines and oxidants in the pathogenesis of ALI (Abraham 2003). Although these studies have implicated PMN activation in the mechanism of ALI (Abraham 2003; Azoulay et al. 2002), little is known about the pathogenic role played by the PMN-endothelial cell interaction in mediating endothelial injury, beyond the generally accepted concept that PMN adhesion to the endothelium is a requirement for the induction of vessel wall injury.

LPS-induced activation of the PMN NADPH (nicotine adenine diphosphonucleotide, reduced) oxidase complex and generation of PMN oxidants play a critical role in promoting the activation of endothelial cells, a process which includes the induction of endothelial hyperadhesivity by ICAM-1 expression and induction of the LPS receptor Toll-like receptor 4 (TLR4) (Lo et al. 1993; Fan et al. 2002). Furthermore, endothelial cell activation-as defined by the activation of transcription factor nuclear factor (NF)-κB and resultant expression of ICAM-1, TLR4 and iNOS-is an essential requirement for the onset of endothelial injury. Firm adhesion of PMN to endothelial cells involves both ICAM-1 and CD11b/CD18, the ICAM-1 counter-receptor (Lo et al. 1992; Malik 1993). Thus, PMN NADPH oxidase-derived oxidant signalling induces not only ICAM-1 expression in endothelial cells but also CD11b/CD18 expression in PMN, which act in concert to promote the firm and stable adhesion of PMN to endothelial cells (Lo et al. 1993). In addition, oxidant signalling, generated by the PMN NADPH oxidase complex, up-regulates cell surface expression of TLR2 on endothelial cells (Fan et al. 2003). This raises the interesting possibility that oxidants, released by PMN, can activate the expression of TLR4 in endothelial cells, and thereby increase the responsiveness of endothelial cells to LPS.

2.1.4.2
Nitric Oxide

It is generally thought that a basal level of NO, generated by eNOS, is required to maintain endothelial integrity, while high levels of NO produced by inducible NOS (iNOS) during inflammation, result in endothelial injury and loss of barrier function (Kubes 1995; Cirino et al. 2003). Elevated NO levels stimulate the expression of macrophage inflammatory protein 2 (Skidgel et al. 2002) and react with superoxide to form peroxynitrite anion, a potent oxidant that nitrates proteins and lipids, inducing cellular damage (Beckman 1996). NOS inhibitors, for example N^G-monomethyl-L-arginine (L-NMMA), block LPS-induced increase in lung injury (increase in lung wet/dry weight ratio) and transcriptional activation of iNOS and interleukin (IL)-1β expression (Wang et al. 1998). To assess the effect of NO-mediated nitration of albumin, Predescu and colleagues (2002) examined the permeability properties of native albumin vs nitrated albumin in the mouse lung and heart microcirculation. These electron microscopy studies showed that nitrated-albumin extravasation was two- to four-fold greater than that of native albumin. While both compounds were found in plasmalemma vesicles (i.e. in the transcellular pathway), nitrated albumin was also present in open interendothelial junctions (Predescu et al. 2002). Thus, high-output NO-induced vascular injury may be due to disruption of AJs and increased paracellular permeability.

2.1.5
Role of Lymphatics in Tissue Fluid Homeostasis

Lymphatics are capable of removing excess extravascular fluid because of their effectiveness as a pump. Lymphatic propulsion is determined by the intrinsic contractility of lymphatic vessels and by unidirectional lymphatic valves (Malik et al. 2000). The extent to which lymphatic insufficiency is a factor in the mechanism of fluid accumulation is not clear. For example, in transplanted organs such as the lung, there is no longer a functioning lymphatic drainage system and this predisposes the lung to oedema; however, the increase in water content is usually transient, and therefore transcellular protein permeability in the reverse direction (tissue to blood) may help to maintain fluid balance (Greitz 2002).

Newly accumulated oedema fluid initially distends the interstitial compartment and then disrupts the interstitial protein lattice; proteolysis of interstitial structural proteins may occur, leading to increased interstitial compliance (Taylor and Parker 1985). Fluid that cannot be cleared by lymphatics accumulates in the connective tissue; in the lungs, this occurs specifically in tissue surrounding smaller vessels and bronchioles (Taylor 1981). The fluid then migrates down the interstitial fluid pressure gradient to interstitial spaces. If lymphatics in the connective tissue sheaths are unable to remove the excess fluid, undrained fluid becomes compartmentalised and forms perivascular cuffs. Normally the interstitial hydrostatic pressure in the lung is negative value (i.e. a value in lungs of −9 mm Hg). Because of the low interstitial compliance, excess fluid accumulation within the interstitium rapidly increases tissue pressure to slightly positive values (Taylor and Parker 1985). In lungs, the alveolar barrier breaks down at a pressure of 2 mm Hg, corresponding to an increase in the interstitial fluid volume of 35%–50%; tissue pressure values above this threshold will cause a precipitous alveolar oedema during which alveoli are flooded in an "all-or-nothing" manner. Initially, the distribution of alveolar flooding is patchy, but rapid severe flooding follows. The exact route by which fluid moves into the alveoli is not known. Fluid movement may involve bulk flow through large epithelial pores or channels, or may be the result of increased transport through intercellular pathways in respiratory epithelium of terminal bronchioles (Flick and Matthay 2000). There is also the possibility of epithelial injury involving detachment of epithelial cells from the underlying matrix, resulting in movement of fluid directly into the alveoli (Flick and Matthay 2000).

2.2
Role of "Safety Factors" in Tissue Fluid Homeostasis

Several safety factors protect against tissue oedema formation; these are (1) a decrease in albumin exclusion volume, (2) the lymphatic system as a whole, and (3) an increase in P_i. In high-pressure oedema, protein-poor fluid begins to accumulate in the interstitial space by ultrafiltration. A decrease in the exclusion volume for albumin (defined as volume of distribution for albumin) becomes important in decreasing the interstitial protein concentration and thereby decreasing π_i. Such a decrease in π_i, according to the Starling equation, reduces net fluid filtration and augments fluid reabsorption across the microvessel wall. The lymph flow is capable of increasing by a large factor in response to increased interstitial fluid volume. Lymph flow is actually dependent on P_i, which in turn is a function of interstitial volume and compliance. Beyond a critical fluid volume, lymph flow can no longer increase in proportion to the increase in P_i. Until this maximal value is attained, lymphatic drainage tracks the rate of oedema fluid formation and thereby limits fluid accumulation. An increase in P_i also represents the short-term protective mechanism to limit oedema formation. The low interstitial compliance in some organs such as the lung reflects an unusually low interstitial volume (Malik et al. 2000). This means that P_i undergoes a large rise for a relatively small increase in interstitial volume; such an increase in P_i favours fluid reabsorption, and in this sense it qualifies as an important safety factor.

3
Restoration of Endothelial Permeability

Mechanisms that strengthen the endothelial barrier (decrease permeability) and facilitate barrier recovery following microvascular injury are poorly understood. As depicted in Fig. 5, pro-inflammatory mediators disrupt AJs within 5–10 min, preceding an increase in endothelial permeability (Sandoval et al. 2001). AJs typically reform within 2 h, restoring AJ integrity and decreasing endothelial permeability. An important, unresolved question is how endothelial AJ integrity is re-established. One mechanism of AJ re-annealing may be the formation of actin-driven membrane protrusions that mediate initial cell-cell contact (Vasioukhin et al. 2000). Some understanding of the signalling mechanisms responsible for AJ re-annealing can be derived from the few examples of mediators which decrease endothelial permeability. The barrier-enhancing effects of cyclic adenosine monophosphate (cAMP), NO, sphingosine-1 phosphate (S1P) and angiopoietin-1 (Ang-1) are described in the following sections. These barrier-restoring agents may in part activate Rho GTPases (Rho, Rac and cdc42) which catalyse the reorganisation of the actin cytoskeleton, initiating membrane ruffling and protrusion formation (lamellipodia and filopodia) that eventually re-establish a connection to neighbouring cells via tight and adherens junctions. Of the Rho family members, Cdc42 is activated 1 h after thrombin-induced disruption of IEJs (Kouklis et al. 2004), suggesting that it plays a role in barrier recovery.

3.1
Cyclic Adenosine Monophosphate

An increase in the intracellular concentration of cAMP induced by agents such as cholera toxin, forskolin and isoprenaline decreases pulmonary vascular endothelial permeability (Stelzner et al. 1989) and inhibits the permeability-increasing effects of thrombin (Minnear et al. 1989) and histamine (Carson et al. 1989). This effect of cAMP is associated with an increase in the peripheral F-actin band, which enhances monolayer integrity (Stelzner et al. 1989) and inhibition of F-actin reorganisation caused by thrombin and histamine (Minnear et al. 1989; Carson et al. 1989; Patterson et al. 1994, 2000). cAMP decreased endothelial permeability to small molecules (sucrose and inulin) to a greater extent than to large molecules (ovalbumin and albumin), indicating it primarily reduced transport through the paracellular pathway. The protective effect of cAMP may be through the activation of protein kinase A (PKA) and subsequent actin reorganisation that strengthens cell-cell and cell-matrix contacts (Lum et al. 1999; Liu et al. 2001). Vasodilator-stimulated phosphoprotein (VASP), a substrate of PKA, was recently shown to induce endothelial barrier recovery (Comerford et al. 2002) by negatively regulating actin nucleation and polymerisation (Harbeck et al. 2000).

3.2
Sphingosine-1-Phosphate

S1P, a bioactive lipid which is stored and released by activated platelets as well as other cells types (Spiegel and Merrill 1996), is involved in angiogenesis, wound healing and tissue injury repair. S1P reversed a thrombin-mediated increase in permeability by stimulating endothelial differentiation gene (Edg) receptor activation (Edg-1 and Edg-3) followed by $G\alpha_i$ protein signalling, Rho kinase and tyrosine kinase activation and actin reorganisation (Garcia et al. 2001). S1P per se increased endothelial barrier integrity (transendothelial electrical resistance) by activating Rac and p21-associated kinase that increased cortical actin assembly and recruitment of cofilin, an actin regulatory protein (Garcia et al. 2001).

3.3
Angiopoietin-1

Activation of the endothelial-specific receptor tyrosine kinase Tie-2 with Ang-1 stabilises endothelial cell interactions with the extracellular matrix and enhances the integrity and restrictiveness of the endothelial barrier (Suri et al. 1996). Transgenic mice over-expressing Ang-1 or mice transduced with adenoviral vector containing Ang-1 were protected against the pro-inflammatory mediators VEGF and PAF (Thurston et al. 1999, 2000). In addition, Ang-1 pretreatment of human vascular endothelial cells (HUVEC) blocked the increase in endothelial monolayer permeability induced by either VEGF or thrombin (Gamble et al. 2000). Thus, in addition to the pro-angiogenic role during embryonic development, Ang-1 has barrier protective effects in vivo which are mediated through the activation of Tie-2 (Suri et al. 1996).

3.4
Nitric Oxide

NO is an important regulator of endothelial permeability (Dimmeler et al. 1999; Schubert et al. 2002). The mechanisms by which low levels of NO strengthen, while high levels of NO injure, the vascular barrier are not clear (Connelly et al. 2001). Some evidence suggests that NO can inhibit NF-κB activation and reduce leucocyte adherence to the endothelium by decreasing the expression of adhesion molecules, thereby blunting the inflammatory response (Cirino et al. 2003; Tsao et al. 1996).

In vascular endothelial cells, eNOS is localised to caveolae where caveolin-1 functions as a negative regulator of eNOS activation (Garcia-Cardena et al. 1997; Ju et al. 1997; Michel et al. 1997). The importance of caveolin-1 as a regulator of eNOS activity was shown in caveolin-1 knockout mice ($CAV1^{-/-}$; Zhao et al. 2002). $CAV1^{-/-}$ mice exhibited plasma NO levels that were fivefold higher than wild-type control mice (Zhao et al. 2002) and an increase in endothelial permeability that was reduced by the eNOS inhibitor, N^G-nitro-L-arginine (L-NAME ; Schubert et al. 2002). In contrast, endothelial-specific over-expression of caveolin-1 blocked eNOS activation and increased endothelial permeability (Bauer et al. 2005), further indicating caveolin-1 negatively regulates eNOS activity and that a basal level of NO is required to maintain a restrictive endothelial barrier. eNOS binds to caveolin-1 at residues aa 82–101, the so-called caveolin scaffolding domain (Okamoto et al. 1998; Gratton et al. 2004). In response to agonist stimulation or mechanical forces such as shear stress, increased intracellular Ca^{2+} or phosphorylation of eNOS Ser^{1179} by Akt uncouples eNOS from caveolin-1 and increases eNOS activity (Moncada et al. 1991; Dimmeler et al. 1999; Fulton et al. 1999, 2001; Gratton et al. 2004). Additional studies are needed to clarify the mechanisms that lead to a basal state of eNOS activation and the resulting barrier protective effects of NO under these conditions.

3.5
RhoGTPase Cdc42

Rho-family GTPases (Rho, Rac, Cdc42) are involved in the formation of membrane protrusions (Hall 1998). These signalling intermediates are known to induce the formation of lamellipodia (Rac) and filopodia (Cdc42) and thus may contribute to re-establishing AJ integrity (Hall 1998; Kouklis et al. 2003). As shown in Fig. 5, thrombin challenge induced the disassembly of AJs within 15 min, resulting in disruption of the endothelial barrier. This effect was reversed within 1–2 h after thrombin exposure. Kouklis et al. (2004) addressed the possible role of monomeric GTPases in the re-assembly phase. Cell lysates from naïve and thrombin-challenged cells were subjected to a pull-down assay using the GST-PAK binding domain fusion protein (GST-PBD), which binds specifically to activated Cdc42 and Rac1 (Benard et al. 1999). Only activated Cdc42 (and not Rac) bound PBD in extracts at 1 h and 2 h after thrombin exposure; i.e. at the times corresponding to re-establishment of AJ. In contrast, Cdc42 activation remained at the basal level in subconfluent, confluent untreated, or 15 min thrombin-treated endothelial cells. Thus, the monomeric GTPase Cdc42 is activated during the AJ re-assembly phase at 1–2 h after thrombin exposure when the AJs re-anneal to restore endothelial permeability. Kouklis et al. (2004) also showed that dominant-negative Cdc42 markedly interfered with AJ re-assembly at this time, further indicating Cdc42 plays a role in restoring endothelial permeability.

4
Concluding Remarks

The vascular endothelium functions as a semi-permeable barrier between the vascular compartment and the interstitium. Integrity of the endothelial cell monolayer is critical for preserving tissue homeostasis. Two general pathways describe the movement of fluid, macromolecules and leucocytes into the interstitium. The transcellular pathway utilises a gp60-activated, tyrosine kinase-dependent caveolae transport process (transcytosis) that primarily transports macromolecules across the barrier. Caveolin-1 knockout mice are providing significant insight into the role of caveolae in endothelial barrier function in that these mice, which lack the vesicle transport pathway, have to adapt or compensate to some extent for the loss of caveolae by decreasing the integrity of the interendothelial junctions to allow protein transport via the paracellular pathway. Fluid flux and PMN trafficking is thought to occur primarily by the paracellular pathway in which gaps form between endothelial cells at sites of active inflammation. Increased endothelial permeability in inflammatory states such as in acute lung injury and sepsis is dependent on the shape and configuration of pulmonary vascular endothelial cells, as determined by alterations in F-actin organisation and interendothelial junctional integrity. The increase in paracellular permeability is ultimately governed by activation of intracellular second messenger pathways, Ca^{2+}, PKC and Rho kinase that stimulate and/or prolong myosin phosphorylation, actin-myosin contraction and disruption of adherens junctions.

Starling forces govern fluid filtration from microvessels into the surrounding perimicrovascular interstitial space. The lymphatics collect the fluid and protein in the interstitium and return the fluid and dissolved solute to the vascular system. Pathophysiologic events and mediators that substantially perturb the Starling forces culminate in pulmonary oedema. Further study of these barrier-disruptive and barrier-protective mechanisms will provide insights and strategies for effective drug delivery across the barrier, oedema clearance, and the recovery of endothelial barrier properties.

References

Aberle H, Schwartz H, Kemler R (1996) Cadherin-catenin complex: protein interactions and their implications for cadherin function. J Cell Biochem 61:514–523

Abraham E (2003) Neutrophils and acute lung injury. Crit Care Med 31:S195–199

Ahn S, Maudsley S, Luttrell LM, Lefkowitz RJ, Daaka Y (1999) Src-mediated tyrosine phosphorylation of dynamin is required for β2-adrenergic receptor internalization and mitogen-activated protein kinase signaling. J Biol Chem 274:1185–1188

Ahn S, Kim J, Lucaveche CL, Reedy MC, Luttrell LM, Lefkowitz RJ, Daaka Y (2002) Src-dependent tyrosine phosphorylation regulates dynamin self-assembly and ligand-induced endocytosis of the epidermal growth factor receptor. J Biol Chem 277:26642–26651

Albelda SM, Smith CW, Ward PA (1994) Adhesion molecules and inflammatory injury. FASEB J 8:504–512

Anderson RGW (1991) Molecular motors that shape endocytic membrane. In: Steer CJ, Hanford J (eds) Intracellular trafficking of proteins. Cambridge University Press, Cambridge, pp 13–46

Anderson RGW (1998) The caveolae membrane system. Annu Rev Biochem 67:199–225

Antohe F, Dobrila L, Heltianu C, Simionescu N, Simionescu M (1993) Albumin-binding proteins function in the receptor-mediated binding and transcytosis of albumin across cultured endothelial cells. Eur J Cell Biol 60:268–275

Aoki T, Nomura R, Fujimoto T (1999) Tyrosine phosphorylation of caveolin-1 in the endothelium. Exp Cell Res 253:629–636

Azoulay E, Attalah H, Yang K, Jouault H, Schlemmer B, Brun-Buisson C, Brochard L, Harf A, Delclaux C (2002) Exacerbation by granulocyte colony-stimulating factor of prior acute lung injury: implication of neutrophils. Crit Care Med 30:2115–2122

Bauer PM, Yu J, Chen Y, Hickey R, Bernatchez PN, Looft-Wilson R, Huang Y, Giordano F, Stan RV, Sessa WC (2005) Endothelial-specific expression of caveolin-1 impairs microvascular permeability and angiogenesis. Proc Natl Acad Sci USA 102:204–209

Beckman JS (1996) Oxidative damage and tyrosine nitration from peroxynitrite. Chem Res Toxicol 9:836–844

Benard V, Bohl BP, Bokoch GM (1999) Characterization of rac and cdc42 activation in chemoattractant-stimulated human neutrophils using a novel assay for active GTPases. J Biol Chem 274:13198–13204

Bird GS, Putney JW Jr (1993) Inhibition of thapsigargin-induced calcium entry by microinjected guanine nucleotide analogues. Evidence for the involvement of a small G-protein in capacitative calcium entry. J Biol Chem 268:21486–21488

Birukova AA, Smurova K, Birukov KG, Kaibuchi K, Garcia JGN, Verin AD (2004) Role of Rho GTPases in thrombin-induced lung vascular endothelial cell barrier function. Microvasc Res 67:64–77

Brown MS, Goldstein JL (1983) Lipoprotein metabolism in the macrophage: implications for cholesterol deposition in atherosclerosis. Annu Rev Biochem 52:223–261

Bucci M, Gratton JP, Rudic RD, Acevedo L, Roviezzo F, Cirino G, Sessa WC (2000) In vivo delivery of the caveolin-1 scaffolding domain inhibits nitric oxide synthesis and reduces inflammation. Nat Med 6:1362–1367

Cao H, Courchesne WE, Mastick CC (2002) A phosphotyrosine-dependent protein interaction screen reveals a role for phosphorylation of caveolin-1 on tyrosine 14: recruitment of C-terminal Src kinase. J Biol Chem 277:8771–8774

Cao H, Orth JD, Chen J, Weller SG, Heuser JE, McNiven MA (2003) Cortactin is a component of clathrin-coated pits and participates in receptor-mediated endocytosis. Mol Cell Biol 23:2162–2170

Carson MR, Shasby S, Shasby DM (1989) Histamine and inositol phosphate accumulation in endothelium: cAMP and a G protein. Am J Physiol Lung Cell Mol Physiol 257:L259-L264

Chen Y, Norkin LC (1999) Extracellular simian virus 40 transmits a signal that promotes virus enclosure within caveolae. Exp Cell Res 246:83–90

Cirino G, Fiorucci S, Sessa WC (2003) Endothelial nitric oxide synthase: the Cinderella of inflammation? Trends Pharmacol Sci 24:91–95

Comerford KM, Lawrence DW, Synnestvedt K, Levi BP, Colgan SP (2002) Role of vasodilator-stimulated phosphoprotein in PKA-induced changes in endothelial junctional permeability. FASEB J 16:583–585

Connelly L, Palacios-Callender M, Ameixa C, Moncada S, Hobbs AJ (2001) Biphasic regulation of NF-kappa B activity underlies the pro- and anti-inflammatory actions of nitric oxide. J Immunol 166:3873–3881

Conner SD, Schmid SL (2003) Regulated portals of entry into the cell. Nature 422:37–44

Corada M, Mariotti M, Thurston G, Smith K, Kunkel R, Brockhaus M, Lampugnani MG, Martin-Padura I, Stoppacciaro A, Ruco L, McDonald DM, Ward PA, Dejana E (1999) Vascular endothelial-cadherin is an important determinant of microvascular integrity in vivo. Proc Natl Acad Sci USA 96:9815–9820

Dejana E (1996) Endothelial adherens junctions: implications in the control of vascular permeability and angiogenesis. J Clin Invest 98:1949–1953

Dejana E, Bazzoni G, Lampugnani MG (1999) Vascular endothelial (VE)-cadherin: only an intercellular glue? Exp Cell Res 252:13–19

Dimmeler S, Fleming I, Fisslthaler B, Hermann C, Busse R, Zeiher AM (1999) Activation of nitric oxide synthase in endothelial cells by Akt-dependent phosphorylation. Nature 399:601–605

Drab M, Verkade P, Elger M, Kasper M, Lohn M, Lauterbach B, Menne J, Lindschau C, Mende F, Luft FC, Schedl A, Haller H, Kurzchalia TV (2001) Loss of caveolae, vascular dysfunction, and pulmonary defects in caveolin-1 gene-disrupted mice. Science 293:2449–2452

Drazner MH, Peppel KC, Dyer S, Grant AO, Koch WJ, Lefkowitz RJ (1997) Potentiation of beta-adrenergic signaling by adenoviral-mediated gene transfer in adult rabbit ventricular myocytes. J Clin Invest 99:288–296

Dudek SM, Garcia JGN (2001) Cytoskeletal regulation of pulmonary vascular permeability. J Appl Physiol 91:1487–1500

Dvorak AM, Feng D (2001) The vesiculo-vacuolar organelle (VVO). A new endothelial cell permeability organelle. J Histochem Cytochem 49:419–432

Dvorak HF, Nagy JA, Feng D, Brown LF, Dvorak AM (1995) Vascular permeability factor/vascular endothelial growth factor, microvascular permeability, and angiogenesis. Am J Pathol 146:1029–1039

Ellis CA, Malik AB, Hamm H, Sandoval R, Voyno-Yasenetskaya T, Gilchrist A, Tiruppathi C (1999) Thrombin induces PAR-1 gene expression in endothelial cells via activation of Gi-linked Ras/mitogen-activated protein kinase pathway. J Biol Chem 274:13718–13727

Fan J, Frey RS, Rahman A, Malik AB (2002) Role of neutrophil NADPH oxidase in the mechanism of tumor necrosis factor-alpha-induced NF-kappa B activation and intercellular adhesion molecule-1 expression in endothelial cells. J Biol Chem 277:3404–3411

Fan J, Frey RS, Malik AB (2003) TLR4 signaling induces TLR2 expression in endothelial cells via neutrophil NADPH oxidase. J Clin Invest 112:1234–1243

Fasolato C, Hoth M, Penner R (1993) A GTP-dependent step in the activation mechanism of capacitative calcium influx. J Biol Chem 268:20737–20740

Fernandez I, Ying Y, Albanesi J, Anderson RG (2002) Mechanism of caveolin filament assembly. Proc Natl Acad Sci USA 99:11193–11198

Flick MR, Matthay MA (2000) Pulmonary edema and acute lung. In: Murray JF, Nadel JA (eds) Textbook of respiratory medicine, 3rd edn. WB Saunders Comp, Philadelphia, pp 1575–1629

Fujimoto T (1993) Calcium pump of the plasma membrane is localized in caveolae. J Cell Biol 120:1147–1157

Fujimoto T, Nakade S, Miyawaki A, Mikoshiba K, Ogawa K (1992) Localization of inositol 1,4,5-trisphosphate receptor-like protein in plasmalemmal caveolae. J Cell Biol 119:1507–1513

Fulton D, Gratton JP, McCabe TJ, Fontana J, Fujio Y, Walsh K, Franke TF, Papapetropoulos A, Sessa WC (1999) Regulation of endothelium-derived nitric oxide production by the protein kinase Akt. Nature 399:597–601

Fulton D, Gratton JP, Sessa WC (2001) Post-translational control of endothelial nitric oxide synthase: why isn't calcium/calmodulin enough? J Pharmacol Exp Ther 299:818–824

Gamble JR, Drew J, Trezise L, Underwood A, Parsons M, Kasminkas L, Rudge J, Yancopoulos G, Vadas MA (2000) Angiopoietin-1 is an antipermeability and anti-inflammatory agent in vitro and targets cell junctions. Circ Res 87:603–607

Gao X, Kouklis P, Xu N, Minshall RD, Sandoval R, Vogel SM, Malik AB (2000) Reversibility of increased microvessel permeability in response to VE-cadherin disassembly. Am J Physiol Lung Cell Mol Physiol 279:L1218–L1225

Garcia JG, Verin AD, Schaphorst KL (1996) Regulation of thrombin-mediated endothelial cell contraction and permeability. Semin Thromb Hemost 22:309–315

Garcia JGN, Davis HW, Patterson CE (1995) Regulation of endothelial gap formation and barrier dysfunction: role of myosin light chain phosphorylation. J Cell Physiol 163:510–522

Garcia JGN, Liu F, Verin AD, Birukova A, Dechert MA, Gerthoffer WT, Bamburg JR, English D (2001) Sphingosine 1-phosphate promotes endothelial cell barrier integrity by Edg-dependent cytoskeletal rearrangement. J Clin Invest 108:689–701

Garcia-Cardena G, Martasek P, Masters BS, Skidd PM, Couet J, Li S, Lisanti MP, Sessa WC (1997) Dissecting the interaction between nitric oxide synthase (NOS) and caveolin. Functional significance of the NOS caveolin binding domain in vivo. J Biol Chem 272:25437–25440

Ghinea N, Fixman A, Alexandru D, Popov D, Hasu M, Ghitescu I, Eskenasy M, Simionescu M, Simionescu N (1988) Identification of albumin binding proteins in capillary endothelial cells. J Cell Biol 107:231–239

Ghinea N, Eskenasy M, Simionescu M, Simionescu N (1989) Endothelial albumin binding proteins are membrane-associated components exposed on the cell surface. J Biol Chem 264:4755–4758

Ghitescu L, Fixman A, Simionescu M, Simionescu N (1986) Specific binding sites for albumin restricted to plasmalemmal vesicles of continuous capillary endothelium: receptor mediated transcytosis. J Cell Biol 102:1304–1311

Ghosh M, Peterson YK, Lanier SM, Smrcka AV (2003) Receptor- and nucleotide exchange-independent mechanisms for promoting G protein subunit dissociation. J Biol Chem 278:34747–34750

Gilbert A, Paccaud JP, Foti M, Porcheron G, Balz J, Carpentier JL (1999) Direct demonstration of the endocytic function of caveolae by a cell-free assay. J Cell Sci 112:1101–1110

Glenney JR Jr (1989) Tyrosine phosphorylation of a 22-kDa protein is correlated with transformation by Rous sarcoma virus. J Biol Chem 264:20163–20166

Goldberg RI, Smith RM, Jarett L (1987) Insulin and alpha 2-macroglobulin-methylamine undergo endocytosis by different mechanisms in rat adipocytes. I. Comparison of cell surface events. J Cell Physiol 133:203–212

Goodman RB, Pugin J, Lee JS, Matthay MA (2003) Cytokine-mediated inflammation in acute lung injury. Cytokine Growth Factor Rev 14:523–535

Goubaeva F, Ghosh M, Malik S, Yang J, Hinkle PM, Griendling KK, Neubig RR, Smrcka AV (2003) Stimulation of cellular signaling and G protein subunit dissociation by G protein betagamma subunit-binding peptides. J Biol Chem 278:19634–19641

Gratton JP, Bernatchez P, Sessa WC (2004) Caveolae and caveolins in the cardiovascular system. Circ Res 94:1408–1417

Greitz D (2002) On the active vascular absorption of plasma proteins from tissue: rethinking the role of the lymphatic system. Med Hypotheses 59:696–702

Hall A (1998) Rho GTPases and the actin cytoskeleton. Science 279:509–514

Harbeck B, Huttelmaier S, Schluter K, Jockusch BM, Illenberger S (2000) Phosphorylation of the vasodilator-stimulated phosphoprotein regulates its interaction with actin. J Biol Chem 275:30817–30825

Henley JR, Cao H, McNiven MA (1999) Participation of dynamin in the biogenesis of cytoplasmic vesicles. FASEB J 13:S243-S247

Holinstat M, Mehta D, Kozasa T, Minshall RD, Malik AB (2003) Protein kinase Calpha-induced p115RhoGEF phosphorylation signals endothelial cytoskeletal rearrangement. J Biol Chem 278:28793–28798

Horgan MJ, Ge M, Gu J, Rothlein R, Malik AB (1991) Role of ICAM-1 in neutrophil-mediated lung vascular injury following reperfusion. Am J Physiol Heart Circ Physiol 261:H1578-H1584

Issekutz AC, Rowter D, Springer TA (1999) Role of ICAM-1 and ICAM-2 and alternate CD11/CD18 ligands in neutrophil transendothelial migration. J Leukoc Biol 65:117–126

Isshiki M, Ying Y, Fujita T, Anderson RGW (2002) A molecular sensor detects signal transduction from caveolae in living cells. J Biol Chem 277:43389–43398

John TA, Vogel SM, Tiruppathi C, Malik AB, Minshall RD (2003) Quantitative analysis of albumin uptake and transport in the rat microvessel endothelial monolayer. Am J Physiol Lung Cell Mol Physiol 284:L187-L196

Jorgenson AO, Shen AC, Arnold W, Leung AT, Campbell KP (1989) Subcellular distribution of the 1,4-dihydropyridine receptor in rabbit skeletal muscle in situ: an immunofluorescence and immunocolloidal gold-labeling study. J Cell Biol 109:135–147

Ju H, Zou R, Venema VJ, Venema RC (1997) Direct interaction of endothelial nitric-oxide synthase and caveolin-1 inhibits synthase activity. J Biol Chem 272:18522–18525

Keller P, Simons K (1998) Cholesterol is required for surface transport of influenza virus hemagglutinin. J Cell Biol 140:1357–1367

Kim J, Ahn S, Guo R, Daaka Y (2003) Regulation of epidermal growth factor receptor internalization by G protein-coupled receptors. Biochemistry 42:2887–2894

Kim YN, Bertics PJ (2002) The endocytosis-linked protein dynamin associates with caveolin-1 and is tyrosine phosphorylated in response to the activation of a noninternalizing epidermal growth factor receptor mutant. Endocrinology 143:1726–1731

King GL, Johnson SM (1984) Receptor-mediated transport of insulin across endothelial cells. Science 227:1583–1586

Kirkham M, Fujita A, Chadda R, Nixon SJ, Kurzchalia TV, Sharma DK, Pagano RE, Hancock JF, Mayor S, Parton RG (2005) Ultrastructural identification of uncoated caveolin-independent early endocytic vehicles. J Cell Biol 168:465–476

Kouklis P, Konstantoulaki M, Malik AB (2003) VE-cadherin-induced Cdc42 signaling regulates formation of membrane protrusions in endothelial cells. J Biol Chem 278:16230–16236

Kouklis P, Konstantoulaki M, Vogel S, Broman M, Malik AB (2004) Cdc42 regulates the restoration of endothelial barrier function. Circ Res 94:159–166

Krueger EW, Orth JD, Cao H, McNiven MA (2003) A dynamin-cortactin-Arp2/3 complex mediates actin reorganization in growth factor-stimulated cells. Mol Biol Cell 14:1085–1096

Kubes P (1995) Nitric oxide affects microvascular permeability in the intact and inflamed vasculature. Microcirculation 2:235–244

Kuebler WM, Ying X, Singh B, Issekutz AC, Bhattacharya J (1999) Pressure is proinflammatory in lung venular capillaries. J Clin Invest 104:495–502

Kuebler WM, Ying X, Bhattacharya J (2002) Pressure-induced endothelial Ca(2+) oscillations in lung capillaries. Am J Physiol Lung Cell Mol Physiol 282:L917-L923

Kurzchalia TV, Dupree P, Parton RG, Kellner R, Virta H, Lehnert M, Simons K (1992) VIP21, a 21-kDa membrane protein is an integral component of trans-Golgi-network-derived transport vesicles. J Cell Biol 118:1003–1014

Lampugnani MG, Corada M, Caveda L, Breviario F, Ayalon O, Geiger B, Dejana E (1995) The molecular organization of endothelial cell to cell junctions: differential association of plakoglobin, beta-catenin, and alpha-catenin with vascular cadherin (VE-cadherin). J Cell Biol 129:203–217

Li S, Couet J, Lisanti MP (1996a) Src tyrosine kinases, Gα subunits, and H-Ras share a common membrane-anchored protein, caveolin. Caveolin binding negatively regulates the auto-activation of Src tyrosine kinases. J Biol Chem 271:29182–29190

Li S, Seitz R, Lisanti MP (1996b) Phosphorylation of caveolin by Src tyrosine kinases. The α-isoform of caveolin is selectively phosphorylated by vSrc in vivo. J Biol Chem 271:3863–3868

Lin HC, Duncan J, Kozasa T, Gilman AG (1998) Sequestration of the G protein beta gamma subunit complex inhibits receptor-mediated endocytosis. Proc Natl Acad Sci USA 95:5057–5060

Liu F, Verin AD, Borbiev T, Garcia JG (2001) Role of cAMP-dependent protein kinase A activity in endothelial cell cytoskeleton rearrangement. Am J Physiol Lung Cell Mol Physiol 280:L1309–L1317

Liu P, Anderson RG (1999) Spatial organization of EGF receptor transmodulation by PDGF. Biochem Biophys Res Commun 261:695–700

Liu P, Rudick M, Anderson RG (2002) Multiple functions of caveolin-1. J Biol Chem 277:41295–41298

Lo SK, Everitt J, Gu J, Malik AB (1992) Tumor necrosis factor mediates experimental pulmonary edema by ICAM-1 and CD18-dependent mechanisms. J Clin Invest 89:981–988

Lo SK, Janakidevi K, Lai L, Malik AB (1993) Hydrogen peroxide-induced increase in endothelial adhesiveness is dependent on ICAM-1 activation. Am J Physiol 264:L406–L412

Lockwich TP, Liu X, Singh BB, Jadlowiec J, Weiland S, Ambudkar IS (2000) Assembly of Trp1 in a signaling complex associated with caveolin-scaffolding lipid raft domains. J Biol Chem 275:11934–11942

Lum H, Malik AB (1994) Regulation of vascular endothelial barrier function. Am J Physiol Lung Cell Mol Physiol 267:L223–L241

Lum H, Del Vecchio PJ, Schneider AS, Goligorsky MS, Malik AB (1989) Calcium dependence of the thrombin-induced increase in endothelial albumin permeability. J Appl Physiol 66:1471–1476

Lum H, Aschner JL, Phillips PG, Fletcher PW, Malik AB (1992) Time course of thrombin-induced increase in endothelial permeability: relationship to Ca_i^{2+} and inositol polyphosphates. Am J Physiol Lung Cell Mol Physiol 263:L219–L225

Lum H, Andersen TT, Siflinger-Birnboim A, Tiruppathi C, Goligorsky MS, Fenton JW 2nd, Malik AB (1993) Thrombin receptor peptide inhibits thrombin-induced increase in endothelial permeability by receptor desensitization. J Cell Biol 120:1491–1499

Lum H, Jaffe HA, Schulz IT, Masood A, RayChaudhury A, Green RD (1999) Expression of PKA inhibitor (PKI) gene abolishes cAMP-mediated protection to endothelial barrier dysfunction. Am J Physiol 277:C580–C588

Lynch JJ, Ferro TJ, Blumenstock FA, Brockenauer AM, Malik AB (1990) Increased endothelial albumin permeability mediated by protein kinase C activation. J Clin Invest 85:1991–1998

Malik AB (1993) Endothelial cell interactions and integrins. New Horiz 1:37–51

Malik AB, Vogel SM, Minshall RD, Tiruppathi C (2000) Pulmonary circulation and regulation of fluid balance. In: Murray JF, Nadel JA (eds) Textbook of respiratory medicine, 3rd edn. WB Saunders, Philadelphia, pp 119–154

Mastick CC, Brady MJ, Saltiel AR (1995) Insulin stimulates the tyrosine phosphorylation of caveolin. J Cell Biol 129:1523–1531

McNiven MA, Kim L, Krueger EW, Orth JD, Cao H, Wong TW (2000) Regulated interactions between dynamin and the actin-binding protein cortactin modulate cell shape. J Cell Biol 151:187–198

Mehta D, Rahman A, Malik AB (2001) Protein Kinase C-α signals Rho-guanine nucleotide dissociation inhibitor phosphorylation and Rho activation and regulates the endothelial cell barrier function. J Biol Chem 276:22614–22620

Mehta D, Ahmed GU, Paria B, Holinstat M, Voyno-Yasenetskaya T, Tiruppathi C, Minshall RD, Malik AB (2003) RhoA interaction with Inositol 1,4,5-triphosphate receptor and transient receptor potential channel-1 regulates Ca^{2+} entry. Role in signaling increased endothelial permeability. J Biol Chem 278:33492–33500

Mehta D, Bhattacharya J, Matthay MA, Malik AB (2004) Integrated control of lung fluid balance. Am J Physiol Lung Cell Mol Physiol 287:L1081-L1090

Michel CC, Curry FE (1999) Microvascular permeability. Physiol Rev 79:703–761

Michel JB, Feron O, Sacks D, Michel T (1997) Reciprocal regulation of endothelial nitric-oxide synthase by Ca^{2+}-calmodulin and caveolin. J Biol Chem 272:15583–15586

Milici AJ, Watrous NE, Stukenbrok H, Palade GE (1987) Transcytosis of albumin in capillary endothelium. J Cell Biol 105:2603–2612

Mineo C, Anderson RG (2001) Potocytosis. Robert Feulgen Lecture. Histochem Cell Biol 116:109–118

Minnear FL, DeMichele MA, Moon DG, Rieder CL, Fenton JW 2nd (1989) Isoproterenol reduces thrombin-induced pulmonary endothelial permeability in vitro. Am J Physiol 257:H1613-H1623

Minshall RD, Niles WD, Tiruppathi C, Vogel SM, Gilchrist A, Hamm HE, Malik AB (2000) Endothelial cell-surface gp60 activates vesicle formation and trafficking via G(i)-coupled Src kinase signaling pathway. J Cell Biol 150:1057–1069

Minshall RD, Tiruppathi C, Vogel SM, Malik AB (2002) Vesicle formation and trafficking and its role in regulation of endothelial barrier function. Histochem Cell Biol 117:105–112

Minshall RD, Sessa WC, Stan RV, Anderson RGW, Malik AB (2003) Caveolin regulation of endothelial function. Am J Physiol Lung Cell Mol Physiol 285:L1179-L1183

Moncada S, Palmer RM, Higgs EA (1991) Nitric oxide: physiology, pathophysiology, and pharmacology. Pharmacol Rev 43:109–142

Moy AB, Van Engelenhoven J, Bodmer J, Kamath J, Keese C, Giaever C, Shasby S, Shasby DM (1996) Histamine and thrombin modulate endothelial focal adhesion through centripetal and centrifugal forces. J Clin Invest 97:1020–1027

Mundy DI, Machleidt T, Ying YS, Anderson RG, Bloom GS (2002) Dual control of caveolar membrane traffic by microtubules and the actin cytoskeleton. J Cell Sci 115:4327–4339

Murata M, Peranen J, Schreiner R, Wieland F, Kurzchalia TV, Simons K (1995) VIP21/caveolin is a cholesterol-binding protein. Proc Natl Acad Sci USA 92:10339–10343

Murthy KS, Makhlouf GM (2000) Heterologous desensitization mediated by G protein-specific binding to caveolin. J Biol Chem 275:30211–30219

Nabi IR, Le PU (2003) Caveolae/raft-dependent endocytosis. J Cell Biol 161:673–677

Niles WD, Malik AB (1999) Endocytosis and exocytosis events regulate vesicle traffic in endothelial cells. J Membr Biol 167:85–101

Oh P, McIntosh DP, Schnitzer JE (1998) Dynamin at the neck of caveolae mediates their budding to form transport vesicles by GTP-driven fission from the plasma membrane of endothelium. J Cell Biol 141:101–114

Okamoto T, Schlegel A, Scherer PE, Lisanti MP (1998) Caveolins, a family of scaffolding proteins for organizing "preassembled signaling complexes" at the plasma membrane. J Biol Chem 273:5419–5422

Parton RG, Joggerst B, Simons K (1994) Regulated internalization of caveolae. J Cell Biol 127:1199–1215

Patterson CE, Davis HW, Schaphorst KL, Garcia JG (1994) Mechanisms of cholera toxin prevention of thrombin- and PMA-induced endothelial cell barrier dysfunction. Microvasc Res 48:212–235

Patterson CE, Lum H, Schaphorst KL, Verin AD, Garcia JG (2000) Regulation of endothelial barrier function by the cAMP-dependent protein kinase. Endothelium 7:287–308

Pelkmans L, Puntener D, Helenius A (2002) Local actin polymerization and dynamin recruitment in SV40-induced internalization of caveolae. Science 296:535–539

Predescu D, Palade GE (1993) Plasmalemmal vesicles represent the large pore system of continuous microvascular endothelium. Am J Physiol 265:H725-H733

Predescu D, Horvat R, Predescu S, Palade GE (1994) Transcytosis in the continuous endothelium of the myocardial microvasculature is inhibited by N-ethylmaleimide. Proc Natl Acad Sci USA 91:3014–3018

Predescu D, Predescu S, Malik AB (2002) Transport of nitrated albumin across continuous vascular endothelium. Proc Natl Acad Sci USA 99:13932–13937

Predescu D, Vogel SM, Malik AB (2004) Functional and morphological studies of protein transcytosis in continuous endothelia. Am J Physiol Lung Cell Mol Physiol 287:L895–901

Predescu SA, Predescu DN, Palade GE (1997) Plasmalemmal vesicles function as transcytotic carriers for small proteins in the continuous microvascular endothelium. Am J Physiol 272:H937-H949

Predescu SA, Predescu DN, Palade GE (2001) Endothelial transcytotic machinery involves supramolecular protein-lipid complexes. Mol Biol Cell 12:1019–1033

Rabiet MJ, Plantier L, Rival Y, Genoux Y, Lampugnani MG, Dejana E (1996) Thrombin induced increase in endothelial permeability is associated with changes in cell-to-cell junction organization. Arterioscler Thromb Vasc Biol 16:488–496

Rahman A, Bando M, Kefer J, Anwar KN, Malik AB (1999) Protein kinase C-activated oxidant generation in endothelial cells signals intercellular adhesion molecule-1 gene transcription. Mol Pharmacol 55:575–583

Razani B, Engelman JA, Wang XB, Schubert W, Zhang XL, Marks CB, Macaluso F, Russell RG, Li M, Pestell RG, Di Vizio D, Hou H Jr, Knietz B, Lagaud G, Christ GJ, Edelmann W, Lisanti MP (2001) Caveolin-1 null mice are viable, but show evidence of hyper-proliferative and vascular abnormalities. J Biol Chem 276:38121–38138

Rizzo V, Morton C, DePaola N, Schnitzer JE, Davies PF (2003) Recruitment of endothelial caveolae into mechanotransduction pathways by flow conditioning in vitro. Am J Physiol Heart Circ Physiol 285:H1720-H1729

Rosado JA, Sage SO (2000) The actin cytoskeleton in store-mediated calcium entry. J Physiol 526:221–229

Rothberg KG, Ying YS, Kamen BA, Anderson RG (1990) Cholesterol controls the clustering of the glycophospholipid-anchored membrane receptor for 5-methyltetrahydrofolate. J Cell Biol 111:2931–2938

Rothberg KG, Heuser JE, Donzell WC, Ying YS, Glenney JR, Anderson RG (1992) Caveolin, a protein component of caveolae membrane coats. Cell 68:673–682

Sandoval R, Malik AB, Minshall RD, Kouklis P, Ellis CA, Tiruppathi C (2001) Ca^{2+} signaling and PKCα activate increased endothelial permeability by disassembly of VE-cadherin junctions. J Physiol (Lond) 533:433–445

Schnitzer JE (1992) Gp60 is an albumin-binding glycoprotein expressed by continuous endothelium involved in albumin transcytosis. Am J Physiol 262:H246-H254

Schnitzer JE, Bravo J (1993) High affinity binding, endocytosis, and degradation of conformationally modified albumins. Potential role of gp30 and gp18 as novel scavenger receptors. J Biol Chem 268:7562–7570

Schnitzer JE, Oh P (1994) Albondin-mediated capillary permeability to albumin. Differential role of receptors in endothelial transcytosis and endocytosis of native and modified albumins. J Biol Chem 269:6072–6082

Schnitzer JE, Carley WW, Palade GE (1988) Albumin interacts specifically with a 60-kDa microvascular endothelial glycoprotein. Proc Natl Acad Sci USA 85:6773–6777

Schnitzer JE, Sung A, Horvat R, Bravo J (1992) Preferential interaction of albumin-binding proteins, gp30 and gp18, with conformationally modified albumins. Presence in many cells and tissues with a possible role in catabolism. J Biol Chem 267:24544–24553

Schnitzer JE, Oh P, Pinney E, Allard J (1994) Filipin-sensitive caveolae-mediated transport in endothelium: reduced transcytosis, scavenger endocytosis, and capillary permeability of select macromolecules. J Cell Biol 127:1217–1232

Schnitzer JE, Oh P, McIntosh DP (1996) Role of GTP hydrolysis in fission of caveolae directly from plasma membranes. Science 274:239–242

Schubert W, Frank PG, Woodman SE, Hyogo H, Cohen DE, Chow CW, Lisanti MP (2002) Microvascular hyperpermeability in caveolin-1 (−/−) knock-out mice. Treatment with a specific nitric-oxide synthase inhibitor, L-name, restores normal microvascular permeability in Cav-1 null mice. J Biol Chem 277:40091–40098

Shajahan AN, Timblin BK, Sandoval RS, Malik AB, Minshall RD (2004a) Src phosphorylation of dynamin-2 regulates caveolae-mediated endocytosis and transcytosis in endothelial cells. J Biol Chem 279:20392–20400

Shajahan AN, Tiruppathi C, Smrcka AV, Malik AB, Minshall RD (2004b) Gβγ activation of Src induces caveolae-mediated endocytosis in endothelial cells. J Biol Chem 279:48055–48062

Shajahan AN, Sverdlov M, Hirth AM, Timblin BK, Tiruppathi C, Malik AB, Minshall RD (2004c) Src-dependent caveolin-1 phosphorylation destabilizes caveolin-1 oligomers and activates vesicle fission in endothelial cells (abstract). Mol Biol Cell 15:330a (1262)

Sharma DK, Brown JC, Choudhury A, Peterson TE, Holicky E, Marks DL, Simari R, Parton RG, Pagano RE (2004) Selective stimulation of caveolar endocytosis by glycosphingolipids and cholesterol. Mol Biol Cell 15:3114–3122

Siflinger-Birnboim A, Schnitzer J, Lum H, Blumenstock FA, Shen CP, Del Vecchio PJ, Malik AB (1991) Lectin binding to gp60 decreases specific albumin binding and transport in pulmonary artery endothelial monolayers. J Cell Physiol 149:575–584

Simionescu M, Simionescu N (1991) Endothelial transport of macromolecules: transcytosis and endocytosis. Cell Biol Rev 25:1–80

Singh RD, Puri V, Valiyaveettil JT, Marks DL, Bittman R, Pagano RE (2003) Selective caveolin-1-dependent endocytosis of glycosphingolipids. Mol Biol Cell 14:3254–3265

Sivasankar S, Gumbiner B, Leckband D (2001) Molecular mechanism of cadherin binding. Biophys J 80:1758–1768

Skidgel RA, Gao XP, Brovkovych V, Rahman A, Jho D, Predescu S, Standiford TJ, Malik AB (2002) Nitric oxide stimulates macrophage inflammatory protein-2 expression in sepsis. J Immunol 169:2093–2101

Smart EJ, Ying Y, Donzell WC, Anderson RG (1996) A role for caveolin in transport of cholesterol from endoplasmic reticulum to plasma membrane. J Biol Chem 271:29427–29435

Song KS, Sargiacomo M, Galbiati F, Parenti M, Lisanti MP (1997) Targeting of a G alpha subunit (Gi1 alpha) and c-Src tyrosine kinase to caveolae membranes: clarifying the role of N-myristoylation. Cell Mol Biol 43:293–303

Spiegel S, Merrill AH Jr (1996) Sphingolipid metabolism and cell growth regulation. FASEB J 10:1388–1397

Stelzner TJ, Weil JV, O'Brien RF (1989) Role of cyclic adenosine monophosphate in the induction of endothelial barrier properties. J Cell Physiol 139:157–166

Sugi H, Suzuki S, Daimon T (1982) Intracellular calcium translocation during contraction in vertebrate and invertebrate smooth muscles as studied by the pyroantimonate method. Can J Physiol Pharmacol 60:576–587

Suri C, Jones PF, Patan S, Bartunka S, Maisonpierre PC, Davis S, Sato TN, Yancopolouos GD (1996) Requisite role of angiopoietin-1, a ligand for the Tie-2 receptor, during embryonic angiogenesis. Cell 87:1171–1180

Taylor AE (1981) Capillary fluid filtration. Starling's forces and lymph flow. Circ Res 49:557–575

Taylor AE, Parker JC (1985) Pulmonary interstitial spaces and lymphatics. In: Fishman AP, Fisher AB (eds) Handbook of physiology, vol 1. American Physiological Society, Bethesda, pp 167–230

Thurston G, Suri C, Smith K, McClain J, Sato TN, Yancopoulos GD, McDonald DM (1999) Leakage-resistant blood vessels in mice transgenically overexpressing angiopoietin-1. Science 286:2511–2514

Thurston G, Rudge JS, Ioffe E, Zhou H, Ross L, Croll SD, Glazer N, Holash J, McDonald DM, Yancopoulos GD (2000) Angiopoietin-1 protects the adult vasculature against plasma leakage. Nat Med 6:460–463

Tiruppathi C, Malik AB, Del Vecchio PJ, Keese CR, Giaever I (1992) Electrical method for detection of endothelial cell shape change in real time: assessment of endothelial barrier function. Proc Natl Acad Sci USA 89:7919–7923

Tiruppathi C, Finnegan A, Malik AB (1996) Isolation and characterization of a cell surface albumin binding protein from vascular endothelial cells. Proc Natl Acad Sci USA 93:250–254

Tiruppathi C, Song W, Bergenfeldt M, Sass P, Malik AB (1997) Gp60 activation mediates albumin transcytosis in endothelial cells by a tyrosine kinase-dependent pathway. J Biol Chem 272:25968–25975

Tiruppathi C, Naqvi T, Sandoval R, Mehta D, Malik AB (2001) Synergistic effects of tumor necrosis factor-alpha and thrombin in increasing endothelial permeability. Am J Physiol Lung Cell Mol Physiol 281:L958-L968

Tiruppathi C, Freichel M, Vogel SM, Paria BC, Mehta D, Flockerzi V, Malik AB (2002) Impairment of store-operated Ca^{2+} entry in TRPC4($-/-$) mice interferes with increase in lung microvascular permeability. Circ Res 91:70–76

Tiruppathi C, Minshall RD, Paria BC, Vogel SM, Malik AB (2003) Role of Ca^{2+} signaling in the regulation of endothelial permeability. Vascul Pharmacol 39:173–185

Tiruppathi C, Naqvi T, Wu Y, Vogel SM, Minshall RD, Malik AB (2004) Albumin-mediates the transcytosis of myeloperoxidase by means of caveolae in endothelial cells. Proc Natl Acad Sci USA 101:7699–7704

Tsao PS, Buitrago R, Chan JR, Cooke JP (1996) Fluid flow inhibits endothelial adhesiveness. Nitric oxide and transcriptional regulation of VCAM-1. Circulation 94:1682–1689

Tuma PL, Hubbard AL (2003) Transcytosis: crossing cellular barriers. Physiol Rev 83:871–932

van Deurs B, Roepstorff K, Hommelgaard AM, Sandvig K (2003) Caveolae: anchored, multifunctional platforms in the lipid ocean. Trends Cell Biol 13:92–100

van Nieuw Amerongen GP, Draijer R, Vermeer MA, van Hinsbergh VWM (1998) Transient and prolonged increase in endothelial permeability induced by histamine and thrombin: role of protein kinases, calcium, and RhoA. Circ Res 83:1115–1123

van Nieuw Amerongen GP, Delft SV, Vermeer MA, Collard JG, van Hinsbergh VWM (2000) Activation of RhoA by thrombin in endothelial hyperpermeability. Role of rho kinase and protein tyrosine kinases. Circ Res 87:335–340

Vasioukhin V, Bauer C, Yin M, Fuchs E (2000) Directed actin polymerization is the driving force for epithelial cell-cell adhesion. Cell 100:209–219

Vogel SM, Gao X, Mehta D, Ye RD, John TA, Andrade-Gordon P, Tiruppathi C, Malik AB (2000) Abrogation of thrombin-induced increase in pulmonary microvascular permeability in PAR-1 knockout mice. Physiol Genomics 4:137–145

Vogel SM, Minshall RD, Pilipovic M, Tiruppathi C, Malik AB (2001a) Activation of 60 kDa albumin-binding protein (gp60) stimulates albumin transcytosis in the intact pulmonary microvessel. Am J Physiol 281:1512–1522

Vogel SM, Easington CR, Minshall RD, Niles WD, Tiruppathi C, Hollenberg SM, Parrillo JE, Malik AB (2001b) Evidence of transcellular permeability pathway in microvessels. Microvasc Res 61:87–101

Vouret-Craviari V, Boquet P, Pouyssegur J, Obberghen-Schilling EV (1998) Regulation of the actin cytoskeleton by thrombin in human endothelial cells: role of Rho proteins in endothelial barrier function. Mol Biol Cell 9:2639–2653

Wang D, Wei J, Hsu JC, Lieu MW, Chao TJ, Chen HI (1998) Effects of nitric oxide synthase inhibitors on systemic hypotension, cytokines and inducible nitric oxide synthase expression and lung injury following endotoxin administration in rats. J Biomed Sci 6:28–35

Winter MC, Kamath AM, Ries DR, Shasby SS, Chen YT, Shasby M (1999) Histamine alters cadherin-mediated sites of endothelial adhesion. Am J Physiol Lung Cell Mol Physiol 277:L988–995

Wysolmerski RB, Lagunoff D (1990) Involvement of myosin light-chain kinase in endothelial cell retraction. Proc Natl Acad Sci USA 87:16–20

Zhao YY, Liu Y, Stan RV, Fan L, Gu Y, Dalton N, Chu PH, Peterson K, Ross J, Chien KR (2002) Defects in caveolin-1 cause dilated cardiomyopathy and pulmonary hypertension in knockout mice. Proc Natl Acad Sci USA 99:11375–11380

Calcium Signalling in the Endothelium

Q.-K. Tran[1] · H. Watanabe[2] (✉)

[1]Division of Molecular Biology and Biochemistry, School of Biological Sciences, University of Missouri-Kansas City, 5007 Rockhill Road, Kansas City MO, 64110, USA

[2]Department of Clinical Pharmacology and Therapeutics, Hamamatsu University School of Medicine, 1-20-1 Handayama, 431-3192 Hamamatsu, Japan
hwat@hama-med.ac.jp

1	Introduction	146
2	Generation of Second Messengers that Release Ca^{2+} from Intracellular Ca^{2+} Stores	147
2.1	Ligand Binding and Generation of Inositol 1,4,5-Trisphosphate	147
2.2	Cyclic ADP-Ribose and Nicotinic Acid Adenine Dinucleotide Phosphate	149
3	Intracellular Ca^{2+} Stores in Endothelial Cells	150
3.1	Endoplasmic Reticulum	150
3.2	Mitochondria	150
3.3	Mechano-sensitive Ca^{2+} Stores	151
4	Ca^{2+} Channels in Endothelial Cells	152
4.1	Intracellular Ca^{2+} Channels	152
4.1.1	Ca^{2+} Release Channels on the ER	152
4.1.2	ER Ca^{2+} Uptake Channels	155
4.2	Transplasmalemmal Ca^{2+} Channels	156
4.2.1	Voltage-Dependent Ca^{2+} Channels	156
4.2.2	Non-selective Cation Channels	156
4.2.3	TRPCs as Store-Operated Ca^{2+} Entry Channels	157
5	Types of Ca^{2+} Entry in Endothelial Cells	158
5.1	Store-Operated Ca^{2+} Entry	158
5.2	Ca^{2+} Oscillations and Non-capacitative Ca^{2+} Entry	159
6	Regulation of Store-Operated Ca^{2+} Entry	161
6.1	Models for the Activation of Store-Operated Ca^{2+} Entry	161
6.1.1	Ca^{2+} Influx Factor	161
6.1.2	Conformational Coupling Model	162
6.1.3	Vesicle Secretion-Like Model	163
6.2	Factors that Regulate Ca^{2+} Entry	163
6.2.1	Ca^{2+} Store Content	163
6.2.2	Membrane Potential	164
6.2.3	Ca^{2+}-Dependent Inactivation of Ca^{2+} Entry	164
6.2.4	Roles of Protein Kinases	164
6.2.5	Mitochondria	169

7	Endothelial Nitric Oxide Synthase and Ca^{2+} Signalling	169
7.1	Nitric Oxide and Endothelial Cell Ca^{2+} Signalling	169
7.2	eNOS Affects Endothelial Cell Ca^{2+} Signalling by Limiting Intracellular Calmodulin	170
8	Conclusions	174
	References	177

Abstract Elevations in cytosolic Ca^{2+} concentration are the usual initial response of endothelial cells to hormonal and chemical transmitters and to changes in physical parameters, and many endothelial functions are dependent upon changes in Ca^{2+} signals produced. Endothelial cell Ca^{2+} signalling shares similar features with other electrically non-excitable cell types, but has features unique to endothelial cells. This chapter discusses the major components of endothelial cell Ca^{2+} signalling.

Keywords Endothelial cells · Calcium · Calcium entry · Calmodulin · Nitric oxide

1
Introduction

The endothelium has physiologically and therapeutically gone far beyond what its anatomical name would imply. It is now recognised as a multi-functional organ responsible for various physiological processes including the regulation of systemic and regional vascular tone, blood coagulation states, cell-cell adhesion, wound healing, cellular proliferation and angiogenesis. The implications of endothelial dysfunction in many pathological states have rendered modulation of endothelial functions a promising therapeutic approach.

Elevations in cytosolic Ca^{2+} concentration ($[Ca^{2+}]_i$) are the usual initial response of endothelial cells to hormonal and chemical transmitters and to changes in physical parameters, and many endothelial functions are dependent on changes in $[Ca^{2+}]_i$. Indeed, endothelial nitric oxide synthase (eNOS) that is responsible for the production of nitric oxide from endothelial cells has an absolute requirement for Ca^{2+}-calmodulin (CaM) for activation (Bredt and Snyder 1990) and appears to require Ca^{2+} entry to sustain an elevated level of activity (Lin et al. 2000). Elevations in $[Ca^{2+}]_i$ also play key roles in the production of autacoids (Crutchley et al. 1983; Kruse et al. 1994), biosynthesis of von Willebrand factor and tissue plasminogen activator, and control of intercellular permeability, cell proliferation and angiogenesis (Vischer et al. 1998).

Endothelial cells are generally viewed as electrically non-excitable, lacking functional voltage-gated Ca^{2+} channels. A major mode of Ca^{2+} entry in these cells in response to both chemical and mechanical stimuli is the so-called capacitative Ca^{2+} entry (CCE) or store-operated Ca^{2+} entry (SOCE), an entry of extracellular Ca^{2+} following depletion of intracellular Ca^{2+} stores (Putney 1990). This mode of Ca^{2+} entry is most important in non-excitable cells but also exists in all cell types. As in other cell types, one of the foci of attention in endothelial cell Ca^{2+} signalling is the yet-elusive molecular nature of SOCE. In this regard, the roles of transient receptor potential channels (TRPCs) as candidates of Ca^{2+} release-activated current (I_{CRAC}), the prototypical current of SOCE, are being extensively studied, and important factors that apparently regulate the bulk SOCE signal have been increasingly recognised. Endothelial cells, however, possess properties that apparently have positioned their Ca^{2+} signalling in a niche of its own. Among these are the multifunctional nature of endothelial cells, their constant exposure to blood shear stress and the expression of eNOS. This enzyme, in addition to its well-known role of producing the important signalling molecule nitric oxide (NO), has recently been demonstrated to be a major affector of the intracellular Ca^{2+}-CaM network. This chapter discusses the main components of endothelial cell Ca^{2+} signalling with an emphasis on factors that regulate Ca^{2+} entry, and it attempts to put in perspective factors that integrate endothelial Ca^{2+} signals in an intricate signalling environment.

2
Generation of Second Messengers that Release Ca^{2+} from Intracellular Ca^{2+} Stores

Elevations of intracellular Ca^{2+} in endothelial cells, as in other electrically non-excitable cells, are generally biphasic, initiating with Ca^{2+} release from intracellular Ca^{2+} stores and followed by Ca^{2+} entry from the extracellular milieu. In principle, this can occur (1) upon receptor activation by physiological agonists, (2) in response to mechanical stress on endothelial cells or (3) following impairment of major Ca^{2+} uptake mechanisms of intracellular Ca^{2+} stores. Experimentally, this can be achieved by treatment with physiological agonists, exposure to fluid shear stress or other mechanical forces, treatment with Ca^{2+} chelators to chelate Ca^{2+} leaking from the stores and prevent store refilling, or treatment with inhibitors of the sarcoplasmic/endoplasmic reticulum (ER) Ca^{2+}-ATPase (SERCA) or Ca^{2+} ionophores. An initial component of the responses to many physiological stimuli is the production of second messengers that trigger the release of Ca^{2+} from intracellular Ca^{2+} stores.

2.1
Ligand Binding and Generation of Inositol 1,4,5-Trisphosphate

Ligand binding to G protein (guanine nucleotide-binding protein)-coupled receptors (GPCRs), which bind to the α-subunit of the G protein q subtype ($G_{\alpha q}$)

is perhaps the best-characterised physiological mechanism leading to the release of intracellular Ca^{2+} stores in many cell types, inositol 1,4,5-trisphosphate (IP_3) being the responsible second messenger (Berridge 1993). In endothelial cells, binding to GPCRs by agonists such as bradykinin, angiotensin II, serotonin and acetylcholine causes $G_{\alpha q}$ to switch from a GDP-bound to a GTP-bound state, allowing the release of $G_{\alpha q}$ from the $G_{\beta\gamma}$ dimer. The GTP-bound $G_{\alpha q}$ subunit subsequently activates phosphoinositide phospholipase (PL)C-β, which then hydrolyses the lipid precursor phosphatidylinositol-4,5-bisphosphate (PIP_2) to yield IP_3 and diacylglycerol. The cellular response to activation of the bradykinin receptor, however, consists of both pertussis toxin-sensitive and pertussis toxin-insensitive components, implicating both G_i and G_q proteins (Liao and Homcy 1993), the former not involving increased production of IP_3 (Lambert et al. 1986). Although the heterotrimeric G proteins' classic component that activates PLC-β is a G_α, the $G_{\beta\gamma}$ dimer has been shown to be capable of activating PLC-β equally well both in vitro and in a number of cell types (Boyer et al. 1992; Camps et al. 1992). In addition, it was recently shown that the $G_{\beta\gamma}$ dimer is capable of directly activating the IP_3 receptor, causing release of intracellular Ca^{2+} stores (Zeng et al. 2003). These studies, conducted in a reconstituted system in rat pancreatic acinar cells, demonstrated that the $G_{\beta\gamma}$ dimer activates IP_3 receptors and inhibits binding of IP_3 to the receptor by allosterically modifying the IP_3 binding site or by binding directly to the IP_3 binding sites. This is supported by the observation that the activation of the IP_3 receptor by $G_{\beta\gamma}$ was abolished by heparin, a competitive inhibitor of IP_3. These latter mechanisms have yet to be demonstrated to operate in endothelial cells, although, given the ubiquitous nature of heterotrimeric G proteins, they would be predicted to do so.

On the other hand, binding to tyrosine kinase-linked receptors by growth factors such as vascular endothelial growth factor, platelet-derived growth factor, epidermal growth factor or antigens leads to autophosphorylation and hence activation the receptor β subunits' tyrosine residues, which bind phosphoinositide-specific phospholipase C-γ1 via their SH_2 domains (He et al. 1999; Meyer et al. 2003). In addition to the SH_2 and SH_3 domains unique to the PLC-γ isozymes, PLC-γ also contains a C_2 domain and two putative PH domains. These are the features shared with β family members and serve as a general mechanism in the hydrolysis of PIP_2.

Mechanical stimuli to endothelial cells, such as high shear stress, also stimulate increases in IP_3 levels. Shear stress-induced increases in IP_3 production appear to be significantly long-lasting compared to that stimulated by agonists (Prasad et al. 1993). Thus, IP_3 levels upon shear stimulation remain elevated for as long as 30 min following the onset of shear, although they eventually subside. Increases in IP_3 production are associated with decreases in phosphatidylinositol, phosphatidylethanolamine and phosphatidic acid, and with increases in diacylglycerol and free arachidonate (Bhagyalakshmi et al. 1992). Due the presence of blood-borne agonists and growth factors, it is likely that the effects of shear stress on IP_3 production in endothelial cells are mechanistically multifactorial in vivo.

2.2
Cyclic ADP-Ribose and Nicotinic Acid Adenine Dinucleotide Phosphate

In addition to IP_3, two newer important second messengers have been found to trigger release of intracellular Ca^{2+} stores in various cell types. These are the pyridine nucleotide metabolites cyclic ADP-ribose (cADPR) and nicotinic acid adenine dinucleotide phosphate (NAADP). cADPR was discovered in 1989 as a cyclised ADP-ribose having an N-glycosyl linkage between the anomeric carbon of the terminal ribose unit and the N^6-amino group of the adenine moiety from sea urchin egg extract incubated with nicotinamide adenine dinucleotide $(NAD)^+$ (Lee et al. 1989), a metabolite of which had previously been shown to trigger release of intracellular Ca^{2+} stores (Clapper et al. 1987). The enzyme responsible for cADPR synthesis is ADP-ribosyl cyclase, widespread among mammalian tissues (Lee and Aarhus 1993). cADPR was found to release Ca^{2+} from a ryanodine-sensitive pool, indicating that cADPR is an endogenous modulator of the ryanodine receptors (Galione et al. 1991). In a variety of cell types, cADPR can modulate the Ca^{2+}-induced Ca^{2+} release mechanism, and this has linked cADPR to modulation of ryanodine receptors. As discussed later, the ryanodine receptors are Ca^{2+} release channels that are found primarily in electrically excitable cells but also in endothelial cells (Sect. 4.1.1.2). Interestingly, although the ryanodine receptor appears to be a final effector, cADPR-triggered calcium release appears to have an absolute dependence on CaM (Lee et al. 1995).

NAADP is synthesised also by ADP-ribosyl cyclase, the same enzyme responsible for the cyclisation of NAD to produce cADPR (Aarhus et al. 1995). Produced by the same enzyme yet from two different substrates, which are NAD and NADP respectively, cADPR and NAADP bear little structural resemblance to each other (Lee et al. 1989). They also possess different Ca^{2+} signalling properties. Pharmacologically, Ca^{2+} release triggered by NAADP is insensitive to 8-amino-cADPR, an antagonist of the cADPR (Walseth and Lee 1993), and heparin, an antagonist of the IP_3 receptors (Lee and Aarhus 1995). From cell fractionation studies, the NAADP-sensitive Ca^{2+} stores in sea urchin eggs appear to be physically separate from those sensitive to cADPR or IP_3 (Lee and Aarhus 1995), and they appear to possess a thapsigargin-insensitive Ca^{2+}-ATPase (Genazzani and Galione 1996). NAADP is a very potent Ca^{2+}-releasing messenger, being able to release Ca^{2+} stores at nanomolar concentrations (Lee 2000).

3
Intracellular Ca^{2+} Stores in Endothelial Cells

3.1
Endoplasmic Reticulum

The ER had been considered simply the cell's main factory of protein synthesis and modification until the early 1980s, when it was found that the second messenger IP$_3$ specifically releases Ca^{2+} from the ER (Berridge and Irvine 1984; Streb et al. 1983). It is now known that the main Ca^{2+} store in non-muscle cells is the ER, while in muscle cells it is the sarcoplasmic reticulum. The ER contains large amounts of Ca^{2+}-binding proteins such as GRP 94, BiP (GRP 78), RP 60 and calreticulin, each molecule of which is able to sequester as many as 30 Ca^{2+} ions. Ca^{2+} concentration in the ER can therefore reach the millimolar range (Macer and Koch 1988). The high concentration of Ca^{2+} in the ER is in fact important for many functions of this organelle, such as vesicle trafficking, protein folding, release of stress signals and regulation of cholesterol metabolism. In endothelial cells, it has been estimated that the ER accounts for roughly 75% of the total intracellular Ca^{2+} reserve (Wood and Gillespie 1998a). Extending like a net over the entire cytoplasm, the ER is in virtually immediate contact with any intracellular Ca^{2+} signals or Ca^{2+} releasing factors (Lesh et al. 1993). The ER contributes greatly to the initiation of important Ca^{2+} signals that are involved in most other vital functions of the cell through Ca^{2+} uptake and release mechanisms to be discussed later.

3.2
Mitochondria

Mitochondria are the other important containers of intracellular Ca^{2+} in endothelial cells, accounting for approximately 25% of the Ca^{2+} reserve. The relative quantification of the Ca^{2+} storage capacity of the ER and mitochondria in endothelial cells was made by comparing the total Ca^{2+} uptake into permeabilised endothelial cells in the presence of inhibitors of mitochondria or of the ER Ca^{2+}-ATPase inhibitor thapsigargin (Wood and Gillespie 1998a). Mechanisms mobilising mitochondrial Ca^{2+} have not been as fully investigated in endothelial cells as in other cells. Mitochondrial Ca^{2+} uptake, 10–100 times kinetically slower than mitochondrial Ca^{2+} efflux (Gunter and Pfeiffer 1990), is believed to be mediated by a uniporter that facilitates the diffusion of Ca^{2+} down the electrochemical gradient across the mitochondrial membrane. Previously, mitochondria were simply considered as high-capacity, low-affinity Ca^{2+} storage pools that serve in states of Ca^{2+} overload as a life-rescuing mechanism by taking up the amount of Ca^{2+} that would otherwise overburden the ER. Recent work, however, has shown that these organelles themselves are excitable, capable of generating and conveying electrical and Ca^{2+} signals (Ichas et al. 1997). Release of Ca^{2+} from mitochondria requires Ca^{2+} to be triggered and in turn plays a critical role in forming Ca^{2+} oscillation patterns (Falcke et al. 1999). Thus, mitochondrial Ca^{2+} is released following the release of Ca^{2+} from the ER. Mitochondrial Ca^{2+}-induced Ca^{2+} release in endothelial cells is triggered during IP_3-induced Ca^{2+} mobilisation and amplifies the Ca^{2+} signals primarily emitted from the ER. Mitochondria appear in close association with regions of the ER enriched in IP_3 receptors and are particularly responsive to IP_3-induced increases in Ca^{2+}. Each mitochondrial Ca^{2+} uptake site faces multiple IP_3 receptors, a concurrent activation of which is required for optimal activation of mitochondrial Ca^{2+} uptake, and there seems to be a synaptic way of transmission of Ca^{2+} signals between mitochondria and the ER (Csordas et al. 1999). Ca^{2+} uptake by mitochondria can suppress the local positive feedback effects of Ca^{2+} on the IP_3 receptors, giving rise to subcellular heterogeneity in IP_3 sensitivity and IP_3 receptor excitability (Hajnoczky et al. 1999).

Cross-talk between mitochondrial and ER Ca^{2+} signals appears to be important in controlling the Ca^{2+} homeostasis of the cell in basal as well as in stimulated conditions. Indeed, although Ca^{2+} refilling of both the ER and mitochondria requires extracellular Ca^{2+}, in the presence of an IP_3-generating agonist, Ca^{2+} refilling of the ER appears to depend on trans-mitochondria Ca^{2+} flux; this dependence does not seem to exist in the absence of an agonist (Malli et al. 2005).

3.3
Mechano-sensitive Ca^{2+} Stores

Endothelial cells are subject to constant mechanical forces such as blood shear stress and osmotic changes. Early studies in bovine aortic endothelial cells showed that vacuum straining caused an increased in IP_3 production, as determined by immunoassays (Brophy et al. 1993). Based on this observation, it was proposed that a mechano-sensitive PLC was responsible for the observed effect, although direct PLC activity was not measured. Later studies in smooth muscle cells also showed that IP_3 levels were elevated upon stimulation of mechanical forces; this was associated with increases in PLC activity (Matsumoto et al. 1995). There has also been a pharmacological hint that PLA_2 might be involved in Ca^{2+} release activated by osmotic swelling in human umbilical vein endothelial cells, though the data suffered from the lack of specific blockers of the enzyme (Oike et al. 1994). These studies in general indicate that mechanical stimulation could trigger release of intracellular Ca^{2+} stores indirectly via the activities of mechano-sensitive, membrane-bound enzymes that catalyse the production of IP_3. However, there is also evidence that mechanical stimulation can directly activate Ca^{2+} release from internal stores in endothelial cells. A volume-sensitive, IP_3-insensitive Ca^{2+} store was proposed in a study showing that release from internal Ca^{2+} stores was still observed in response to hypotonic stress in endothelial cells permeabilised with saponin, a condition that allowed a direct effect of osmotic swelling on the ER (Jena et al. 1997). Interestingly, under these conditions, removal of external Ca^{2+} can rapidly deplete internal Ca^{2+} stores, and high concentrations of gadolinium $(Gd)^{3+}$ can block the Ca^{2+} release, suggesting that the inhibitory effects frequently seen with Gd^{3+} on Ca^{2+} entry could be due in large part to its inhibition of Ca^{2+} release in the first place and may have little to do with inhibition of the transplasmalemmal Ca^{2+} influx itself. This volume-sensitive release was not prevented by ruthenium red or prior stimulation with IP_3, indicating that the volume-sensitive storage site is distinct from mitochondria and from stores sensitive to ryanodine or IP_3. The store appears to possess Ca^{2+}-ATPase as a Ca^{2+} pump, since loading of Ca^{2+} into this pool was prevented by thapsigargin. This is perhaps the best piece of evidence available that mechanical forces can trigger Ca^{2+} release directly from intracellular organelles in endothelial cells. However, follow-up work on the molecular element responsible for such response is not available.

4
Ca^{2+} Channels in Endothelial Cells

4.1
Intracellular Ca^{2+} Channels

4.1.1
Ca^{2+} Release Channels on the ER

4.1.1.1
IP_3 Receptors

IP_3 receptors constitute the most clearly identified Ca^{2+} channels that pump Ca^{2+} from the ER. First identified in mouse and rat cerebellum as a developmentally regulated phospho-glycoprotein, the Ca^{2+} channel P400 (Furuichi et al. 1989; Mikoshiba et al. 1979), the IP_3 receptors are now known to exist in at least three isoforms (types 1, 2 and 3) in both animal and human cells (Yamada et al. 1994; Yamamoto-Hino et al. 1994). Most cells have at least one form of IP_3 receptor, and many express all three. Structurally, the IP_3 receptor channels are tetramers composed of four subunits, each containing 2,700 residues and a single IP_3-binding site. IP_3-mediated Ca^{2+} release responses are co-operative, indicating that several and perhaps all subunits are required to bind IP_3 for the channel to open (Meyer et al. 1988). A characteristic feature of IP_3 receptors is that they are regulated by both IP_3 and Ca^{2+}. High cytosolic Ca^{2+} concentration is inhibitory to IP_3 channel activity. How IP_3 and Ca^{2+} interact to regulate IP_3 channels is a challenging question to answer experimentally, due largely to the lack of high time resolution analysis of the role of IP_3 and Ca^{2+} at the single channel level. The general consensus, however, is that IP$_3$ regulates the effects of Ca^{2+} on the channel. A major model proposed that IP$_3$ channel opening depends on whether Ca^{2+} binds to the stimulatory sites or inhibitory sites. Thus, high-affinity Ca^{2+} binding to the inhibitory sites keeps the channels inactive under normal conditions, and IP$_3$ reduces Ca^{2+} binding to the inhibitory sites by reducing their affinities for Ca^{2+} and thus allowing Ca^{2+} binding to the stimulatory sites to predominate and the channel to open (Mak et al. 1998).

Another feature of IP$_3$ receptors is that they associate with a variety of molecules in the cell, and with plasmalemmal channels. There is evidence for a role of CaM in the regulation of IP$_3$ channel function, although there is significant controversy (reviewed in Taylor and Laude 2002). In addition, the probability of the IP$_3$ receptor opening is regulated by phosphorylation by non-receptor protein kinases, mostly on the cytoplasmic domain of the receptor. The IP$_3$ receptor isoform identified in endothelial cells is approximately 260 kDa, preferentially located at the perinuclear region, and both structurally and functionally analogous to that detected in neuronal tissues (Bourguignon et al. 1994). IP$_3$-induced Ca^{2+} release terminates even in the continued presence of IP$_3$, which could reflect rapid hydrolysis of IP$_3$, feedback effects of cytoplasmic and/or luminal Ca^{2+} on specific Ca^{2+} binding sites on the IP$_3$ receptors/channel complex, or intrinsic deactivation properties of IP$_3$ receptors (Oldershaw and Taylor 1993). In addition, IP$_3$-induced Ca^{2+} release appears to depend on cytosolic concentrations of monovalent cations. In permeabilised endothelial cells, the ability of different ions to allow IP$_3$-induced Ca^{2+} release was found to be K$^+$=Na$^+$>Cs$^+$>Rb$^+$>>Co^{2+}, suggesting that there is possibly a counter-ion system that controls Ca^{2+} release (Wood and Gillespie 1998b). The rate of Ca^{2+} discharge from intracellular stores apparently contributes to the regulation of cytosolic Ca^{2+} oscillations.

It should be emphasised that the ER, while being structurally and functionally the largest reservoir of cellular Ca^{2+}, is not the only organelle housing the IP$_3$ receptors. Other organelles such as the Golgi apparatus, secretory vesicles or other specialised membranes, may also function as IP$_3$-sensitive Ca^{2+} stores. A detailed discussion on these topics is beyond the scope of this chapter, and the interested reader is referred to a recent review (Vermassen et al. 2004).

4.1.1.2
Ryanodine Receptors (RyRs)

Ryanodine receptors (RyRs) constitute another family of proteins responsible for Ca^{2+} releasing channels. There is significant sequence homology between the IP$_3$ receptors and RyRs, most remarkable at the sequences that form the channel's pore (Mignery et al. 1989; Zhao et al. 1999). RyRs are found in a variety of tissues, with the highest densities in striated muscles. Three isoforms have been identified, namely RyR1, RyR2 and RyR3, predominantly in skeletal muscle, cardiac tissue and striated muscle, respectively. RyRs have also been found in endothelial cells of porcine endocardium and thoracic aorta (Lesh et al. 1993), and are more homologous to the cardiac isoform (RyR2) than to the skeletal isoform. Prestimulation of rat aortic, human aortic, human umbilical vein and bovine pulmonary endothelial cells with ryanodine significantly reduced bradykinin-induced Ca^{2+} release, suggesting that the ryanodine receptors are functional in these cells (Wang et al. 1995; Ziegelstein et al. 1994). Recent pharmacological evidence also supports the existence of RyRs in freshly isolated rabbit aortic endothelial cells, where they appear to play a role in conjunction with the Na^+-Ca^{2+} exchanger in extrusion of cytoplasmic Ca^{2+} (Liang et al. 2004). Molecular and structural characteristics of endothelial RyR channels, however, have not been determined, perhaps because IP_3 receptors are the predominant ER Ca^{2+} release channels in endothelial cells. Sequence homology between the endothelial isoform and the cardiac isoform suggests that insights into endothelial RyRs can be predicted partly based on information available for the cardiac RyR2, although the lack of functional voltage-gated Ca^{2+} channels in endothelial cells could be a factor determining the difference in RyR activity between the two tissues. From studies in non-endothelial tissues, the activity of RyRs is known to depend on a number of factors, including cytosolic and luminal Ca^{2+} concentrations, ATP and Mg^{2+}, redox status of the cell, cADPR, phosphorylation, and protein-protein interactions (Fill and Copello 2002).

4.1.1.3
Ca^{2+} Leak

In addition to Ca^{2+} release through IP_3 receptors and/or RyR channels, which requires binding of a second messenger for their activation, ER luminal Ca^{2+} is slowly but spontaneously released into the cytosol via other mechanisms. Under non-stimulated conditions, there is continuous Ca^{2+} leak from the ER. This leak is normally compensated for by Ca^{2+} uptake mechanisms and therefore is not readily observed. It is when all other known mechanisms of Ca^{2+} release and uptake are inhibited that Ca^{2+} leak manifests itself. In endothelial cells, 10-min incubation in Ca^{2+}-free medium could deplete the bradykinin-sensitive store by 60% (Paltauf-Doburzynska et al. 1999). In BHK-21 cells, Ca^{2+} leak can still be observed in the presence of EGTA in nominally Ca^{2+}-free medium, thapsigargin, heparin and ruthenium red, and therefore appears to be independent of the Ca^{2+}-ATPase, the IP_3 receptor and the ryanodine receptor (Hofer et al. 1996). In addition, basal Ca^{2+} leak in pancreatic acinar cells is not inhibited by inhibitors of IP_3 receptors, RyRs or the receptor for NAADP (Lomax et al. 2002). The cytosolic ATP concentration has been shown to regulate this Ca^{2+} leak from the ER. Thus, in permeabilised BHK-21 cells, the rate of leak can increase approximately fourfold in response to an approximate tenfold increase in ATP concentration (Hofer et al. 1996). In addition, Ca^{2+} leak has been shown to be inhibited by Ni^{2+} (Wissing et al. 2002), suggesting that the leak process can be modulated by a Ca^{2+} channel. However, no Ca^{2+} leak channel has actually been identified to date. In actuality, the physiological leak is likely to be the result of many factors, including release via the IP_3 receptor, ryanodine receptors or both, as there is certainly a low level of agonists under "resting" conditions.

4.1.2
ER Ca^{2+} Uptake Channels

A major ER surface protein is the SERCA, an ATP-dependent Ca^{2+} pump that is responsible for the sequestration of cytosolic Ca^{2+}. Three different SERCA genes are known to be expressed in vertebrates, *ATP2A1-3*, which encode for different protein isoforms of the SERCA pumps, including SERCA1a/b, SERCA2a/b and SERCA3a/b/c/d. The SERCA1 Ca21-ATPase isoform is expressed in fast skeletal muscle. SERCA2a and SERCA2b are alternative splice variants. SERCA2a is expressed in cardiac and slow skeletal muscle, while SERCA2b is expressed in smooth muscle and is found on the ER of several non-muscle cells together with SERCA3. In non-muscle cells, SERCA2b is the "house-keeping" isoform, responsible for the sequestration of cytosolic Ca^{2+}. In endothelial cells, two isoforms have been found, SERCA2b and SERCA3. In the majority of the cases, SERCA3 is found co-expressed with SERCA2b (Anger et al. 1993). Interestingly, it has been demonstrated that acetylcholine-induced Ca^{2+} signalling and endothelium-dependent relaxation of vascular smooth muscle are severely impaired in knock-out mice deficient in the SERCA3 gene (Liu et al. 1997). This suggests that the SERCA3 isoform can play a significant role in sequestering cytosolic Ca^{2+}, as does the SERCA2b isoform. In line with this, freshly isolated human umbilical vein endothelial cells were found to express only SERCA3 (Mountian et al. 1999).

The activity of the SERCA pump depends on its conformational changes into two different states, termed E1 and E2. In the E1 state, the enzyme's two Ca^{2+}-binding sites are of high affinity and face the cytoplasm. In the E2 state the Ca^{2+}-binding sites are of low affinity and face the luminal side. Either cytosolic ATP or Ca^{2+} can bind first to the E1 conformation. The $2Ca^{2+}$-E1-ATP form is phosphorylated to form $2Ca^{2+}$-E1-P. In this high-energy state, the bound Ca^{2+} ions become occluded. Conversion to the low-energy intermediate is accompanied by a major conformational change to $2Ca^{2+}$-E2-P whereby the Ca^{2+}-binding sites are converted to a low-affinity state and reorient towards the luminal face. The cycle ends with the sequential release of Ca^{2+} and phosphate and a major conformational change from the E2 to the E1 state (Wuytack et al. 2002).

Under treatment with specific SERCA pump inhibitors such as thapsigargin, dibenzohydroquinone (BHQ) or cyclopiazonic acid, ER Ca^{2+} leak is uncompensated for and the ER is depleted. In a number of cell types, including Dictyostelium and porcine aortic endothelial cells, CaM antagonists such as

W-7 and calmidazolium at micromolar concentrations also can mobilise Ca^{2+} from thapsigargin-sensitive stores, an effect apparently independent of inhibition of a CaM-dependent enzyme (Groner and Malchow 1996; Watanabe et al. 1999). In contrast to the plasma membrane Ca^{2+} ATPase pumps, SERCA pumps do not bind CaM, and CaM does not stimulate SERCA activity (Raeymaekers et al. 1983; Wibo et al. 1981). The observed Ca^{2+}-mobilising effects of CaM antagonists therefore probably reflect direct inhibitory effects on the SERCA pump rather than on CaM itself. Consistent with this, a more recent study has demonstrated that these compounds inhibit in a similar manner the Ca^{2+} ATPase activity of both sarcoplasmic reticulum vesicles and Ca^{2+} ATPase purified from smooth muscle (Khan et al. 2000). Depleting the ER's Ca^{2+} content by inhibiting the SERCA pump is not associated with increases in IP_3 production and has served as a very useful approach to investigate intracellular Ca^{2+} signalling.

4.2
Transplasmalemmal Ca^{2+} Channels

Ca^{2+} entry into endothelial cells can occur via several different mechanisms: (1) non-selective cation channels activated by a variety of agonists, inhibitors of the SERCA pump, or shear stress, (2) more selective Ca^{2+} channels activated by a store-operated mechanism, (3) a leak mechanism down the electro-chemical gradient, or, (4) in principle, an exchange mechanism such as the Na^+-Ca^{2+} exchanger. Experimentally, Ca^{2+} entry can also occur with Ca^{2+} ionophores.

4.2.1
Voltage-Dependent Ca^{2+} Channels

Endothelial cells are generally considered to be electrically non-excitable, although voltage-dependent Ca^{2+} channels have been described in a few early reports (Bossu et al. 1989, 1992). Most of the channels described were of low conductance and are generally considered to be of little functional importance (Himmel et al. 1993). Depolarisation of the plasma membrane would enhance voltage-dependent Ca^{2+} entry. However, in endothelial cells, agonist-induced Ca^{2+} entry is dramatically reduced by depolarisation of the plasma membrane with high potassium solutions, in support of a lack of functional voltage-dependent Ca^{2+} influx (Luckhoff and Busse 1990). In addition, inhibitors of voltage-dependent Ca^{2+} channels like diltiazem and verapamil did not affect agonist-induced Ca^{2+} entry in freshly isolated endothelial cells (Luckhoff and Busse 1990; Yamamoto et al. 1995).

4.2.2
Non-selective Cation Channels

A large number of Ca^{2+} entry channels described in endothelial cells are non-selective cation channels. Agonists such as thrombin, bradykinin, serotonin,

ATP and endothelin-1 activate non-selective cation channels (Brauneis et al. 1992; Colden-Stanfield et al. 1990; Groschner et al. 1994; Popp and Gogelein 1992; Zhang et al. 1994b). Inhibitors of ER Ca^{2+} ATPase as well as IP_3 applied intracellularly, all of which deplete intracellular Ca^{2+} stores, also activate non-selective cation channels (Gericke et al. 1993; Zhang et al. 1994a). Shear stress activates a non-selective cation channel that appears to be more permeable to divalent than to monovalent cation. Several groups suggested that shear stress only activates Ca^{2+} channels in the presence of ATP, and therefore the Ca^{2+} entry was attributed to activation of the Ca^{2+}-permeable purinoceptor P_2X_4 (Ando and Kamiya 1993; Davies 1995; Yamamoto et al. 2000). Other groups, however, have shown that shear stress can activate Ca^{2+} entry in the absence of any Ca^{2+}-mobilising agonists (Helmlinger et al. 1995; Kanai et al. 1995; Kwan et al. 2003; Yao et al. 2000). The basis of these discrepancies is still unclear. Under resting condition, a leak of Ca^{2+} from the extracellular medium into the cytosol has also been observed. Radio-isotopic measurements indicate that ^{45}Ca leaks into resting endothelial cells at a rate of $16\,pmol.10^{-6}\,cells.s^{-1}$ (Johns et al. 1987).

4.2.3
TRPCs as Store-Operated Ca^{2+} Entry Channels

Ca^{2+} entry activated by emptying of intracellular stores, the so-called capacitative or store-operated Ca^{2+} entry, is a most prevalent and important mode of Ca^{2+} entry in vascular endothelial cells. Electrically, store-operated Ca^{2+} entry is typically represented by the so-called Ca^{2+} release-activated Ca^{2+} entry, or I_{CRAC}, initially described in mast cells (Hoth and Penner 1992). However, despite intensive research, the molecular nature of the activated Ca^{2+} channels remains rather elusive. In this context, the mammalian homologues of the *Drosophila* TRP (transient receptor potential) protein came as valuable models for the influx channels. Trp is a *Drosophila* photoreceptor mutant incapable of maintaining a sustained potential in response to photostimulation (Cosens and Manning 1969). The fact that these receptors use a PLC signalling pathway gave the first hint that TRP might encode a component of the Ca^{2+} entry pathway (Hardie and Minke 1992, 1993). Light-induced phosphoinositide hydrolysis in *Drosophila* activates two classes of channels, one selective for Ca^{2+} and absent in the transient receptor potential mutant TRP, the other a non-selective cation channel that requires Ca^{2+} for activation. As well as being a major charge carrier for the light-induced current, Ca^{2+} entry via the TRP-dependent channels appears to be required for refilling IP_3-sensitive Ca^{2+} stores and for feedback regulation (light adaptation) of the transduction cascade. Depletion of internal Ca^{2+} stores with the SERCA inhibitor thapsigargin activates the TRP-dependent Ca^{2+} channels, suggesting that the TRP channels could be responsible for the store-operated Ca^{2+} channels (SOCC) (Vaca et al. 1994).

After more than a decade since the recognition of the TRP protein as a potential model for the Ca^{2+} entry channel, the TRP-related protein family in humans now consists of more than 20 members, classified into 3 subfamilies: the canonical TRP (TRPC) proteins, closest to the *Drosophila* TRP; the vanilloid TRP (TRPV), closely related to the vanilloid receptor; and TRPM proteins, homologous to the tumour suppressor melastatin (Montell et al. 2002). Structurally, the TRP channels consist of six predicted transmembrane-spanning helices (TM1-6), cytoplasmic N- and C-termini and a pore region between TM5 and TM6 (Clapham et al. 2001). Not all members of the three families are store-operated, however. For several members, reports differ as to whether they are store-operated, depending on the levels of expression and on the cell types studied. For example, human TRP3 forms both IP_3 receptor-dependent and receptor-independent store-operated channels in lymphocytes (Vazquez et al. 2001). Perhaps this is due to the actuality that many studies were only performed in reconstituted systems (Parekh and Putney 2005). Endothelial cells express most of the TRPC proteins; so far, six members of the TRPC family have been reported in these cells (Freichel et al. 1999). Almost all of the TRPCs expressed in endothelial cells are activated by Ca^{2+} store depletion and/or receptor activation. In particular, studies on transgenic mice have strongly implicated TRPC4 in Ca^{2+} entry and endothelium-dependent vasodilatation (Freichel et al. 2001). The TRPV4, very interestingly, is activated by different types of stimuli, including arachidonic acid, anandamide, heat and changes in cell volume. The TRPM expressed in endothelial cells, on the other hand, is activated by intracellular Ca^{2+} and has been suggested to play a role in negative feedback inhibition of all types of Ca^{2+} entry. Several TRP proteins might come together to form a cluster of Ca^{2+} channels, a possibility that would explain the functional involvement of nearly all the isoforms identified in Ca^{2+} entry in endothelial cells (Nilius et al. 2003).

5
Types of Ca^{2+} Entry in Endothelial Cells

5.1
Store-Operated Ca^{2+} Entry

A quarter of a century ago, it was observed that intracellular Ca^{2+} stores emptied by agonists are refilled very rapidly by application of extracellular Ca^{2+} (Brading and Sneddon 1980; Casteels and Droogmans 1981). The rise in intracellular Ca^{2+} concentration ($[Ca^{2+}]_i$), thus evoked in the presence of extracellular Ca^{2+}, consists of a transient component concurred or followed by a large and sustained one. Under nominally Ca^{2+}-free conditions, the response is only a small and transient rise in $[Ca^{2+}]_i$, reflecting the release of intracellular Ca^{2+} stores most commonly due to binding of IP_3 to its receptor

(Berridge 1993; Streb et al. 1983); when extracellular Ca^{2+} is reintroduced in the absence of the agonist, there is a large rise in $[Ca^{2+}]_i$, due to entry of Ca^{2+} from the extracellular medium. In endothelial cells and other non-excitable cells, this latter signal is most commonly mediated by CCE, or SOCE, a model put forward by Putney (1986, 1990), in which the opening of plasma membrane Ca^{2+} entry channels follows emptying of intracellular Ca^{2+} stores by IP_3 or other signals that release Ca^{2+} (Petersen and Cancela 1999; Vaca and Kunze 1995). The capacitative model was consolidated by the observation that agents that act as inhibitors of the smooth ER Ca^{2+} ATPase, such as thapsigargin, tert-butylhydroquinone and cyclopiazonic acid, which empty intracellular Ca^{2+} stores without increasing IP_3 production, could activate Ca^{2+} entry (Dolor et al. 1992; Thastrup et al. 1990). However, not until several years after the inception of the capacitative Ca^{2+} entry model was SOCE first measured electrically in mast cells as a "Ca^{2+} release-activated Ca^{2+} current" (I_{CRAC}) (Hoth and Penner 1992). This is a highly Ca^{2+}-selective ($P_{Ca}:P_{Na}\sim10:1$), inwardly rectifying current through very low conductance channels that are subject to feedback inhibition by intracellular Ca^{2+}. So far, CRAC is still the prototypical SOCE current. Shear stress also activates SOCE, presumably by several combined mechanisms. Blood flow transfers blood-borne agonists to the cell surface to activate PLC and increase IP_3, while the permeability of the cell membrane to extracellular Ca^{2+} increases upon exposure to blood flow and shear stress activates heterotrimeric G proteins and small G proteins, which participate in Ca^{2+} signalling. Recently, it has been proposed that a mechano-sensitive non-selective cation channel might account for shear-stimulated Ca^{2+} entry in rat aortic endothelial cells (Yao et al. 2000). This channel has relative permeability ratios of $P_{Ca}:P_{Na}:P_K=5:1:1$ and is inhibited by 8-Br-cGMP, suggesting that a protein kinase G-dependent mechanism is involved.

5.2
Ca^{2+} Oscillations and Non-capacitative Ca^{2+} Entry

The first observation of Ca^{2+} oscillations in non-excitable cells was made more than 30 years ago (Prince et al. 1972). In endothelial cells, it became obvious some years later that agonist-induced oscillations occur at low-dose agonist stimulation, whereas higher doses stimulate a sustained elevation in $[Ca^{2+}]_i$ (Jacob et al. 1988). It was proposed that the source of the oscillatory Ca^{2+} signals was the IP_3-sensitive Ca^{2+} stores and that the oscillations were dependent on the fluctuating concentrations of IP_3 produced by low concentrations of agonists (Berridge 1990; Meyer and Stryer 1988). Other models for Ca^{2+} oscillations proposed that Ca^{2+} oscillations could occur in the absence of fluctuations in IP_3 concentration, such that Ca^{2+} initially released from IP_3-sensitive stores in response to an external stimulus triggers release of Ca^{2+} via an IP_3-insensitive store based on Ca^{2+}-induced Ca^{2+} release, which serves as the source of oscillations (Goldbeter et al. 1990). In either case, refilling of IP$_3$-sensitive Ca^{2+} stores requires entry of Ca^{2+} from the extracellular space, and thus the oscillations could reflect oscillations in both the release and uptake of Ca^{2+} stores and in the transmembrane influx of Ca^{2+}. The answer to this question was provided by the demonstration in endothelial cells of large Ca^{2+} oscillations despite a constant influx rate measured by the Mn^{2+} quenching approach (Jacob 1990). Later studies in nasal secretory cells suggest that the transmembrane Ca^{2+} entry that occurs during, and thus nurtures, Ca^{2+} oscillations is itself involved in the stimulation of oscillations and is mechanistically not identical with capacitative Ca^{2+} entry (Shuttleworth and Thompson 1996a, b). This was based primarily on the observation that Ca^{2+} entry by the capacitative mechanism cannot be stimulated in nasal gland cells exhibiting oscillatory Ca^{2+} signals. Further studies in human embryonic kidney (HEK) 293 cells provided three main lines of evidence suggesting that arachidonic acid is responsible for this non-capacitative Ca^{2+} entry mechanism. Thus, exogenous administration of low concentrations of arachidonic acid (3–8 µM) induces an entry of Ca^{2+} without any detectable depletion of intracellular Ca^{2+} stores. The enzymatic machinery for arachidonic production (e.g. a cytoplasmic phospholipase A$_2$) is activated by low concentrations of agonists. Furthermore, inhibition of arachidonic production prevents Ca^{2+} entry thus triggered and yet has no effect on capacitative Ca^{2+} entry (Shuttleworth 1996; Shuttleworth and Thompson 1998). Evidence for a non-capacitative Ca^{2+} entry mechanism and mutual antagonism between the non-capacitative and capacitative mechanisms has also come from other laboratories working on different cell types (Broad et al. 1999; Luo et al. 2001; Moneer and Taylor 2002). The electrical current responsible for the arachidonate-regulated Ca^{2+} entry (I_{ARC}) has been recorded in HEK 293 cells overexpressing the muscarinic receptor M3. This current is similar to I_{CRAC} in many aspects, but is distinct in that it lacks the fast inactivation and the marked sensitivity to extracellular pH that is characteristic of I_{CRAC}. In addition, it is observed even after maximal depletion of intracellular Ca^{2+} stores (Mignen and Shuttleworth 2000). In endothelial cells it was later shown that this type of Ca^{2+} entry exists; a Ca^{2+} entry current was measured in response to arachidonic acid with similar properties to the I_{ARC} recorded in HEK 293 (Fiorio Pla and Munaron 2001). The molecular nature of this non-capacitative Ca^{2+} entry pathway is currently completely unknown. Of particular interest is a detailed study in A7R5 smooth muscle cells demonstrating that NO produced by intrinsically expressed NO synthase mediates the reciprocal regulation between non-capacitative and capacitative mechanisms (Moneer et al. 2003). Whether endothelial cell Ca^{2+} oscillations in response to agonists such as histamine, bradykinin or shear stress occur under this mechanism, and whether NO plays a role in switching between the two entry modes for Ca^{2+} are currently the subjects of investigation.

6
Regulation of Store-Operated Ca^{2+} Entry

Since the non-capacitative mechanism is relatively new and little information is available for its regulation in endothelial cells, this section only deals with the more extensively studied SOCE mechanism. As with other non-excitable cells, two major questions remain: the molecular nature of the Ca^{2+} entry channels and the signal that links the release of intracellular Ca^{2+} stores to the activation of the channels. As regards the molecular identity of store-operated channels, the discovery of the *Drosophila* transient receptor potential (TRP) gene as the candidate coding gene for an SOCE channel has led to a blooming area of research and the increasingly expanding size of the TRP family (see Sect. 4.2.3).

6.1
Models for the Activation of Store-Operated Ca^{2+} Entry

6.1.1
Ca^{2+} Influx Factor

The simplest explanation for the activation of SOCE is perhaps a soluble messenger signal linking the empty stores and the membrane Ca^{2+} entry channels. This was proposed soon after the capacitative Ca^{2+} entry model was put forward (Putney 1990). A putative calcium influx factor (CIF) was proposed when an extract of Jurkat lymphocytes-collected following store depletion by inhibition of the SERCA pumps with thapsigargin under extracellular Ca^{2+}-free conditions-was able to activate Ca^{2+} entry in macrophages, astrocytoma cells and fibroblasts (Randriamampita and Tsien 1993). The Ca^{2+} signals were inhibited significantly by econazole, an inhibitor of cytochrome P450 (CYP450), suggesting that CYP450 metabolites might be CIF. The involvement of CYP450 in the generation of a second messenger mediating capacitative Ca^{2+} entry was in fact observed in several previous studies (Alvarez et al. 1991; Montero et al. 1991, 1992). The CYP450 metabolite 5,6-epoxyeicosatrienoic acid (5,6-EET) was later proposed to be CIF in astrocytes, being able to activate Ca^{2+} entry at picomolar concentration (Rzigalinski et al. 1999). In endothelial cells, thapsigargin- and bradykinin-stimulated Ca^{2+} entry is also inhibited by CYP450 inhibitors (Takeuchi et al. 2003). Recent data in *Xenopus* oocytes have further supported the presence of CIF in smooth muscle cells, using extracts prepared from either mammalian cells in which intracellular Ca^{2+} stores were deleted by thapsigargin, or yeast in which these stores had been genetically depleted (Csutora et al. 1999). Studies on CIF in vascular smooth muscle cells also showed fluctuating Ca^{2+} responses (Trepakova et al. 2000). There is currently no study directly testing CIF prepared from endothelial cell extract.

6.1.2
Conformational Coupling Model

This model proposes that information transfer from store depletion to the plasma membrane is mediated by the IP_3 receptor functioning as the go-between of the two membrane systems. Previous versions of this model proposed that the IP_3 receptor would integrate in its cytoplasmic head information that signals capacitative Ca^{2+} entry and then transmit this information to the Ca^{2+} release-activated Ca^{2+} channels in the plasma membrane via direct protein-protein interaction (Berridge 1995). A recently modified version of this model proposes that IP_3 may either act directly to stimulate a complex formed by IP_3 receptors located in a junctional zone and Ca^{2+} entry channels in the plasma membrane, or IP_3 could act indirectly by stimulating uncoupled IP_3 receptors in the vicinity of the junctional zone to induce a localised depletion of the ER store to switch on a store-operated mechanism. At physiological agonist concentrations, the earliest Ca^{2+} response to receptor activation may be the stimulation of entry, which is then responsible for charging up the internal store to prime the IP_3 receptors for the large-scale regenerative release of Ca^{2+} that occurs during each spike (Berridge 2004).

In support of this model, Ma et al. suggested that contact of the IP_3 receptor with the plasma membrane is required both for activation and for maintenance of store-operated Ca^{2+} entry. Thus, 2-aminoethoxydiphenyl borate (APB), an inhibitor of the IP_3 receptor, almost completely blocked both agonist- and thapsigargin-stimulated Ca^{2+} entry via TRP3 channels and SOC channels (Ma et al. 2000). This inhibitor, however, while known to inhibit the IP_3 receptor, could very well be an inhibitor of the membrane channels, as suggested by the fact that it could terminate almost immediately Ca^{2+} entry during its course. Although these studies also demonstrated that 2-APB can prevent IP_3 production rather quickly if added shortly before or together with the agonist, this is not sufficient evidence to rule out a direct inhibitory effect on the influx channels. Studies from the same group subsequently demonstrated that 2-APB does not act on Ca^{2+} influx channels via inhibition of the IP_3 receptor (Ma et al. 2001). The most compelling lines of evidence that these inhibitors may not work specifically as IP_3 receptor antagonists are that, in IP_3 receptor-deficient cells, xestospongin C, an IP_3 receptor antagonist, still inhibits thapsigargin-induced Ca^{2+} entry (Castonguay and Robitaille 2002), and that in A7R5 cells, 2-APB inhibits IP_3-induced Ca^{2+} release without affecting $^3[H]IP_3$ binding to the IP_3 receptor (Missiaen et al. 2001). Oka et al. also demonstrated that xestospongin C inhibits DNP (dinitrophenol, an antigen)-induced Ca^{2+} entry, but not thapsigargin-induced SOCE (Oka et al. 2002). In neurons, the same compound is reported to empty ER Ca^{2+} stores, but does not inhibit IP_3-induced Ca^{2+} release, suggesting that xestospongin C functions as a SERCA inhibitor rather than a specific IP_3 receptor antagonist (Solovyova et al. 2002). Furthermore, in rat basophilic leukaemia (RBL)-1 cells, I_{CRAC} activity is rapidly inhibited by extracellular 2-APB, whereas intracellular 2-APB is less effective (Kukkonen et al. 2001). In the same line, 2-APB inhibits SOCE independently of the IP$_3$ receptor in human platelets and liver cells (Diver et al. 2001; Gregory et al. 2001). Detailed studies by Putney and colleagues in three different cell types also questioned the requirement for IP$_3$ or IP$_3$ receptor in the activation of SOCE (Broad et al. 2001). In vascular endothelial cells, these inhibitors have also been shown to inhibit both agonist- and thapsigargin-induced SOCE (Bishara et al. 2002); however, these were not tested if they acted as direct inhibitors of the Ca^{2+} entry channels. In addition, although there has been an indication that TRP3 and the IP$_3$ receptor interact, TRPC3 introduced into HEK 293 cells lacking all three isoforms of the IP$_3$ receptor still targets to the plasma membrane and forms functional Ca^{2+} channels (Wedel et al. 2003). Overall, it is not clear whether IP$_3$ receptor contact is responsible for activation of SOCE in most instances. Neither is it clear whether continuous contact between the IP$_3$ receptor and the plasma membrane is necessary for maintenance of Ca^{2+} entry.

6.1.3
Vesicle Secretion-Like Model

The vesicle secretion-like model proposed that the transmission of information from depleted intracellular stores resembles the secretion of vesicles to the extracellular matrix. This model was based primarily on studies in smooth muscle cells showing that stabilisation of the cortical actin network underneath the plasma membrane inhibits the activation of SOCE (Patterson et al. 1999), similar to findings in platelets (Rosado et al. 2000). Findings in support of this model in endothelial cells came from studies showing that pharmacological inhibition of vesicle transport can inhibit SOCE in corneal vascular endothelial cells (Xie et al. 2002). This model, however, has not been proved to apply in all cell types. In RBL-1 cells, this model has been seriously questioned with respect to activation of I_{CRAC}, the prototypical SOC current. In these studies, none of the treatments previously shown in smooth muscle cells to affect SOCE showed any effect on I_{CRAC} in RBL-1 cells (Bakowski et al. 2001).

6.2
Factors that Regulate Ca^{2+} Entry

6.2.1
Ca^{2+} Store Content

It is obvious that the refilling status of the ER is an important determinant of the Ca^{2+} entry signals. This seems natural, as a major function of SOCE is to refill the ER, whose many important functions depend on its Ca^{2+} content (see Sect. 2.1). In endothelial cells, SOCE is graded with the degree of store depletion (Sedova et al. 2000).

6.2.2
Membrane Potential

When the endothelial cell membrane is depolarised either by high K^+ concentration or by the K^+ channel blocker tetraethylammonium, Ca^{2+} entry in response to ATP and bradykinin is significantly diminished, while the release of Ca^{2+} from intracellular stores remains unaffected (Luckhoff and Busse 1990). Membrane potential is determined in part by K^+ and Cl^- concentrations on the two sides of the plasma membrane. Bradykinin and thapsigargin provoke Cl^- influxes that partly regulate Ca^{2+} entry (Tran et al. 1999), and Ca^{2+} influx into endothelial cells in response to histamine and ATP is sensitive to Cl^- concentration (Hosoki and Iijima 1994). Further information on the role of membrane potential on Ca^{2+} signalling in endothelial cells can be found in a recent excellent review (Adams and Hill 2004).

6.2.3
Ca^{2+}-Dependent Inactivation of Ca^{2+} Entry

Ca^{2+}-dependent inactivation is a common feature of many Ca^{2+} channels. In this mechanism, Ca^{2+} entering the cell acts in a negative feedback manner to inhibit further influx of Ca^{2+} via the channel. For a detailed discussion on this topic regarding Ca^{2+} entry channels, the reader is referred to a recent excellent review (Parekh and Putney 2005).

6.2.4
Roles of Protein Kinases

A great number of protein kinases have been implicated in controlling the bulk SOCE signal in a variety of cells. In endothelial cells, protein kinases, studied mostly with the advent of different pharmacological kinase inhibitors, have been reported to contribute to Ca^{2+} entry stimulated by a variety of stimuli such as agonists and shear stress.

6.2.4.1
Tyrosine Kinase

Conceptually, for receptor tyrosine kinase signalling, autophosphorylation or cross phosphorylation of β-subunits of the receptors is the first event linking ligand binding to downstream cascades, among them activation of PLC-γ1 (see Sect. 2.1). It would not be surprising that Ca^{2+} signals triggered by tyrosine kinase receptors can be prevented by tyrosine kinase inhibitors. However, Ca^{2+} entry stimulated by non-tyrosine kinase receptor agonists such as bradykinin and histamine, and even the SERCA inhibitor thapsigargin, can be significantly inhibited by inhibitors of tyrosine kinase such as genistein and piceatannol (Fleming and Busse 1997; Fleming et al. 1995). In other cell types, including platelets and fibroblasts, tyrosine kinase has been linked to the control of SOCE (Lee et al. 1993; Sargeant et al. 1993). In fibroblasts, it has been suggested that the non-receptor tyrosine kinase c-*src* can serve as a diffusible signal linking store depletion and Ca^{2+} entry. These studies showed that bradykinin activates c-*src* (Lee and Villereal 1996). Importantly, in fibroblasts derived from *src*⁻/*src*⁻ transgenic mice, Ca^{2+} entry stimulated by bradykinin or thapsigargin is dramatically lower than in wild-type fibroblasts. The level of capacitative Ca^{2+} entry in *src*⁻/*src*⁻ cells is restored to nearly normal levels by transfecting *src*⁻/*src*⁻ cells with chicken c-*src* (Babnigg et al. 1997). The precise mechanism by which c-*src* is linked to channel activation is unclear.

6.2.4.2
Myosin Light Chain Kinase (MLCK)

Myosin light chain kinase (MLCK) activation and the resultant phosphorylation of myosin light-chain (MLC) are key events in the initiation of smooth muscle cell contraction. In endothelial cells, MLCK and MLC are present in modest amounts (Garcia et al. 1997). Nevertheless, MLCK appears to play important roles in endothelial cell biology, including calcium signalling, endothelial barrier function, regulation of endothelium-derived relaxing factors and cell-cell interaction (Norwood et al. 2000; Tran et al. 2000; Tran and Watanabe 2003; Watanabe et al. 2001).

The first observation of MLCK's involvement in the regulation of Ca^{2+} entry in endothelial cells was made in primary cultured porcine aortic endothelial cells, where ML-9 and wortmannin, strong inhibitors of MLCK, completely inhibited the entry portion of the Ca^{2+} response provoked by both IP_3-dependent and IP_3-independent mechanisms (Fig. 1a; Watanabe et al. 1996). A number of MLCK inhibitors with different structures and specificities for MLCK, including HA 1077, wortmannin, ML-5, ML-7 and ML-9, were later shown to inhibit Ca^{2+} entry and MLC phosphorylation stimulated by bradykinin, thapsigargin and shear stress in these cells (Fig. 1b; Watanabe et al. 1998). These effects were observed in studies using different sources of endothelial cells (Norwood et al. 2000). MLCK is also implicated in agonist-induced Cl^- influx in endothelial cells (Tran et al. 1999). Involvement of MLCK in store-operated Ca^{2+} entry in endothelial cells was further consolidated with the observation that antisense oligonucleotides directed against the ATP-binding sequence of MLCK attenuate bradykinin- and thapsigargin-stimulated Ca^{2+} entry. The physiological impact of MLCK inhibition is demonstrated by a drastic reduction of agonist-stimulated NO production from primary cultured endothelial cells (Fig. 1c), perhaps via blockade of SOCE, and of acetylcholine-stimulated hyperpolarisation of endothelium-intact smooth muscle cells in mesenteric arteries (Fig. 1d; Watanabe et al. 2001). Whether inhibition of MLCK affects release from intracellular stores is controversial, based on different data from different laboratories using different sources for endothelial cells (Norwood et al. 2000; Watanabe et al. 1996). Ion channels and gene expression, even in the same cell type, are highly variable depending on cell isolation, culture and growth conditions, and thus controversial data observed from different endothelia with different methods of isolation and culture seems unavoidable (Tran and Watanabe 2003). In human platelets, wortmannin inhibited significantly thrombin-induced Ca^{2+} entry and MLC phosphorylation without affecting intracellular store release (Hashimoto et al. 1993), and in human monocytes, MLCK inhibitors also inhibit Ca^{2+} entry but not Ca^{2+} release from intracellular stores (Tran et al. 2001).

There is thus plenty of evidence for the involvement of MLCK in SOCE in many cell types including human platelets, endothelial cells and human monocytes. Similar observations have also been made in A7R5 smooth muscle cells and HEK 293 cells (Q.K. Tran, personal observations). What, then, is the precise mechanism whereby MLCK modulates Ca^{2+} entry? Since the inhibition of Ca^{2+} entry is well correlated with inhibition of MLC phosphorylation, the simplest explanation that has been suggested is that MLCK inhibitors prevent the reorganisation of the cytoskeleton around Ca^{2+} entry channels that is associated with activation of Ca^{2+} entry (Fig. 1e, f; Watanabe et al. 1998). Direct changes in the cytoskeleton have often been associated with changes in SOCE signals, although the effects vary with the type and dose of pharmacological agent used (Holda and Blatter 1997; Patterson et al. 1999). Although a clear change in cell morphology is not observed following treatment with MLCK inhibitors, changes in acto-myosin complex formation could be predicted based on the observed changes in MLC phosphorylation. However,

Fig. 1 a–f MLCK as an important regulatory input for endothelial cell Ca^{2+} entry and endothelium-dependent vasodilatation. **a, b** MLCK inhibition prevents Ca^{2+} entry stimulated by agonist (**a**) or fluid flow (**b**). Primary porcine aortic endothelial cells loaded with fura-2/AM (**a**, 2 µM) or indo-1/AM (**b**, 10 µM) were stimulated with bradykinin (**a**, 10 nM) with (*open circles*) or without (*closed circles*) treatment with the specific MLCK inhibitor ML-9 (100 µM) or exposed to fluid flow (**b**, 5 dynes/cm^2) in the presence or absence of ML-9 (100 µM) as indicated. Note the complete inhibition by ML-9 of the plateau phases of Ca^{2+} entry stimulated by bradykinin or laminar fluid flow. (Reproduced from Watanabe et al. 1996 and Watanabe et al. 1998, with permission). **c, d** MLCK inhibition prevents both production of nitric oxide (**c**) and endothelium-dependent vasodilatation (**d**). Nitrite and nitrate were detected using an HPLC system, and acetylcholine (Ach)-induced hyperpolarisation of smooth muscle cell membrane in rat mesenteric artery was measured as described in Watanabe et al. (2001). *Ach*, acetylcholine; *AS*, antisense directed against the ATP-binding domain of MLCK; *BK*, 10 nM bradykinin; *TG*, 1 µM thapsigargin; *Wort*, 100 µM wortmannin. (Reproduced from Watanabe et al. 2001, with permission). **e, f** MLCK inhibition prevents Ca^{2+} entry in correlation with phosphorylation of myosin light chain. ML-9 inhibits MLC phosphorylation in a dose-dependent manner (**e**), which correlates with inhibition of Ca^{2+} entry (**f**). *MLC-UP*, *MLC-P* and *MLC-PP*, represent, respectively, un-phosphorylated, mono-phosphorylated and di-phosphorylated myosin light chain. (Reproduced from Watanabe et al. 1998, with permission)

Calcium Signalling in the Endothelium 167

since actin and myosin do not bind transmembrane proteins, a direct link between MLC and Ca^{2+} entry channels seemed unlikely. Recent studies in pulmonary endothelial cells demonstrated that thapsigargin triggered a selective store-operated Ca^{2+} entry current in endothelial cells (I_{SOC}), in addition to a non-selective cation channels. This current contributed approximately 50% of the total thapsigargin-stimulated Ca^{2+} entry signal and was completely inhibited by specific disruption of the spectrin-protein 4.1 interaction (Wu

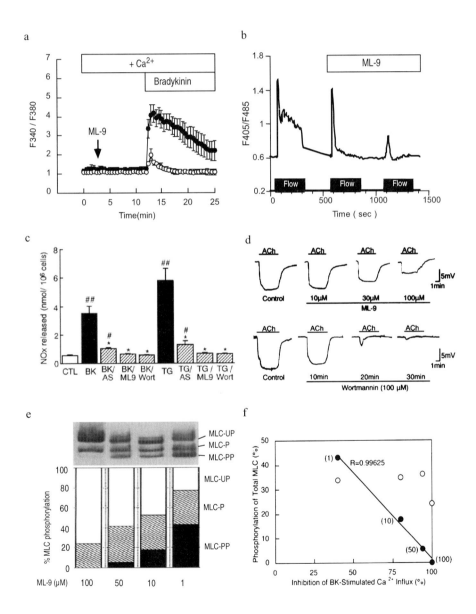

et al. 2001). Further studies suggested an interaction between protein 4.1 and TRPC4 in pulmonary artery endothelial cells (Cioffi et al. 2003). Based on this line of evidence, it is tempting to speculate that MLCK activation could be linked to Ca^{2+} entry via the interaction between spectrin and protein 4.1, the latter being capable of interacting directly with TRPC4, a candidate of SOCC in endothelial cells. However, in the many studies described above, MLCK inhibitors almost completely blocked Ca^{2+} entry, while I_{SOC} can contribute to at most roughly 50% of the bulk SOCE signal. There are several possible explanations for this. First, the observations that application of MLCK inhibitors during Ca^{2+} entry abolishes the ion influx almost immediately, and that their removal from the medium instantly restores Ca^{2+} entry, suggest that MLCK inhibitors may affect membrane potential or may also have direct effects on non-selective cation channels. Indeed, thapsigargin-induced Ba^{2+} and Sr^{2+} entry in endothelial cells and HEK 293 cells can be prevented completely by these inhibitors (Q.K. Tran, personal observations). In addition, Mn^{2+} influx can also be inhibited, albeit not completely, by MLCK inhibitors (Takahashi et al. 1997). Second, a link between cytoskeletal reorganisation and non-selective cation channels in endothelial cells cannot be ruled out. Until the molecular identity of non-selective cation channels in endothelial cells is clarified, this is by no means a simple undertaking.

6.2.4.3
Protein Kinase G

Several groups have reported the involvement of PKG in SOCE in endothelial cells. Yao and colleagues showed that stretch- and shear stress-activated Ca^{2+} channels in vascular endothelial cells are inhibited by active PKG (Yao et al. 2000; Kwan et al. 2000). This was attributed to an autocrine effect of NO, which will be discussed in Sect. 7.1.

6.2.4.4
Protein Kinase C

Effects of PKC on SOCE appear to be cell-type specific. In thyroid cells and human neutrophils, phorbol ester was found to reduce receptor- or thapsigargin-stimulated Ca^{2+} entry (Montero et al. 1994; Tornquist 1993), and PKC was suggested to accelerate the inactivation of I_{CRAC} in rat basophilic leukaemia (RBL) cells and Jurkat T cells (Parekh and Penner 1995). In porcine aortic endothelial cells, inhibitors of PKC such as bisindolylmaleimide I and staurosporine appear to have little or no effect on SOCE stimulated by bradykinin or thapsigargin (Watanabe et al. 1998); however, down-regulation of conventional PKC isoforms by long-term treatment with phorbol ester reduces agonist- and thapsigargin-evoked Ca^{2+} signals in bovine aortic endothelial cells and HEK 293 cells (Q.K. Tran, personal observations).

6.2.5
Mitochondria

Mitochondrial Ca^{2+} uptake can modulate SOCE signals either by affecting the ER refilling process (see Sect. 2.2), or by generating subplasmalemmal microdomains of low Ca^{2+} that sustain SOCE (Malli et al. 2003). In addition, there has been some indication that mitochondrial Ca^{2+} uptake can stimulate production of NO within mitochondria, an action that can affect SOCE (see Sect. 7.1) and the ER stress response (Xu et al. 2004).

7
Endothelial Nitric Oxide Synthase and Ca^{2+} Signalling

7.1
Nitric Oxide and Endothelial Cell Ca^{2+} Signalling

Nitric oxide (NO) produced by eNOS is a potent vasodilator that relaxes smooth muscle cells by increasing cytosolic cGMP (Moncada and Higgs 1993). eNOS has an absolute requirement for Ca^{2+}-CaM for its activation (Bredt and Snyder 1990), and Ca^{2+} entry appears to be required for sustained activation of the enzyme (Lin et al. 2000). A significant amount of work has been done to test the hypothesis that NO could act in a negative feedback manner to inhibit Ca^{2+} entry. In smooth muscle cells there appear to be several mechanisms by which NO could act to inhibit Ca^{2+} entry. NO could inhibit L-type Ca^{2+} channels directly, or do so indirectly by changing membrane potential via activation of Ca^{2+}-dependent K^+ channels (Blatter and Wier 1994; Bolotina et al. 1994). Both of these effects are apparently due to increases in cGMP concentrations. In addition, NO may inhibit Ca^{2+} entry by promoting sarcoplasmic reticulum Ca^{2+} uptake via SERCA activity. This was deduced from experiments in which NO gas at low concentrations (10^{-10}–10^{-6} M) reduced agonist-induced but not thapsigargin-induced Ca^{2+} entry in smooth muscle cells and platelets (Cohen et al. 1999; Trepakova et al. 1999). In endothelial cells, several groups have reported inhibitory effects of NO on Ca^{2+} entry, and increased SERCA pump activity by NO has also been suggested (Dedkova and Blatter 2002; Takeuchi et al. 2004). In addition, high concentrations of NO appear to reduce eNOS protein expression, due possibly to cleavage of the eNOS protein (Takeuchi et al. 2004). Thus, there appears to be plenty of evidence for a negative feedback activity of NO on Ca^{2+} entry. However, a potential difficulty with many of these results is that NO donors have been used at high concentrations, usually in the high micromolar range. These doses are obviously non-physiological, and the inhibitory effects of NO produced under physiological conditions on Ca^{2+} entry in endothelial cells might be more subtle. Testing the effect of inhibition of intrinsic NO production, e.g. by treatment with N^G-nitro-L-arginine methyl ester (L-NAME), on Ca^{2+} entry would appear to be a more physiological paradigm. Such reports are few and gave modest effects in endothelial cells as compared to smooth muscle cells (Wang et al. 1996).

7.2
eNOS Affects Endothelial Cell Ca^{2+} Signalling by Limiting Intracellular Calmodulin

The Ca^{2+}-binding protein calmodulin (CaM) is a ubiquitous transducer of intracellular Ca^{2+} signals in all cell types. It is involved in many cellular functions via its many target proteins, including adenylyl cyclases and phosphodi-

Fig. 2 a–d Competition for limiting intracellular CaM as a novel coupling mechanism for disparate CaM targets-the example of eNOS and the PMCA in endothelial cells is shown. **a** Manipulation of eNOS phosphorylation and CaM binding: forskolin (*FSK*, 50 µM) and 3-isobutyl-1-methylxanthine (*IBMX*, 0.5 mM). Immunoblots of anti-eNOS immunoprecipitates were performed using anti-eNOS and anti-CaM antibodies, and those of whole cell homogenates were performed using phospho-specific antibodies for Thr^{497} and Ser^{1179}. *Columns* represent densitometric values for immunoblots of control (*cross-hatched*) and treated (*filled*) samples, respectively (*n*=5). Measurements are performed in the presence of the NOS inhibitor L-NAME (100 µM). **b** Treatment to increase CaM binding to eNOS results in substantial reduction in free Ca^{2+}-CaM concentration in BAECs. Time courses of free Ca^{2+}-CaM (*upper panel*) and Ca^{2+} (*lower panel*) simultaneously determined in control (*open circles*) and FSK/IBMX-treated cells (*filled circles*). Measurements are performed in the presence of L-NAME (100 µM). **c** Enhanced CaM binding to eNOS is associated with reduced responses of other CaM targets during SOCE. Concurrent measurements of $Ca^{2+}{}_i$ (*lower panel*) and the response of $BSCaM_2$, a FRET-based biosensor that is constructed based on the CaM-binding domain of MLCK and that therefore functions as an analogue for MLCK in terms of CaM binding (*upper panel*). ER Ca^{2+} stores are first emptied by thapsigargin (TG, 1 µM) in Ca^{2+}-free buffer and store-operated Ca^{2+} entry is triggered by the addition of small amounts of Ca^{2+} and terminated by addition of 1 mM BAPTA. Following wash out of BAPTA and addition of FSK-IBMX, which increases CaM binding to eNOS by roughly threefold (Fig. 2a), subsequent similar SOCE signal evokes a significantly reduced response of $BSCaM_2$. All measurements are performed in the presence of the NOS inhibitor L-NAME (100 µM). **d** Phosphorylation-dependent increases in CaM binding to eNOS are associated with ∼40% reduction in the CaM-dependent activity of plasma membrane Ca^{2+}-ATPase (*PMCA*) in BAECs. *Upper panel*, protocol for in-cell determination of PMCA activity. PMCA activity is reflected in the time course of Ca^{2+} extrusion following removal of extracellular Ca^{2+} at peak Ca^{2+} entry in the absence of extracellular Na^+. The Ca^{2+} extrusion time courses of cells with similar $Ca^{2+}{}_i$ values at peak of Ca^{2+} entry are fitted to a monoexponential and the τ values obtained are an inverse measure of PMCA activity. *TGN*, thapsigargin (1 µM); Na^+-free, cell buffer in which Na^+ is replaced by equimolar *N*-methyl-D-glucamine. *Lower panel*, PMCA activity in BAECs transfected with or without fluorescent CaM biosensors in the presence (*filled columns*) or absence (*cross-hatched columns*) of pretreatment with FSK-IBMX. $BSCaM_2$ and $BSCaM_{0.3}$, fluorescent CaM indicators with apparent K_d for CaM ∼2 nM and 0.3 nM, respectively. *Asterisks* indicate statistical significance. (Reproduced from Tran et al. 2003, with permission)

esterases (Gu and Cooper 1999), numerous protein kinases (Nairn and Picciotto 1994), the protein phosphatase calcineurin (Aramburu et al. 2000), NO synthase (Bredt and Snyder 1990), the plasma membrane Ca^{2+}-ATPase (PMCA) (Vincenzi and Larsen 1980), and several ion channels (Levitan 1999).

Up to 40% of the total cellular CaM is associated with proteins from which it is virtually inseparable regardless of the free Ca^{2+} concentration, such as phosphorylase kinase (Picton et al. 1983), inducible NO synthase (Cho et al.

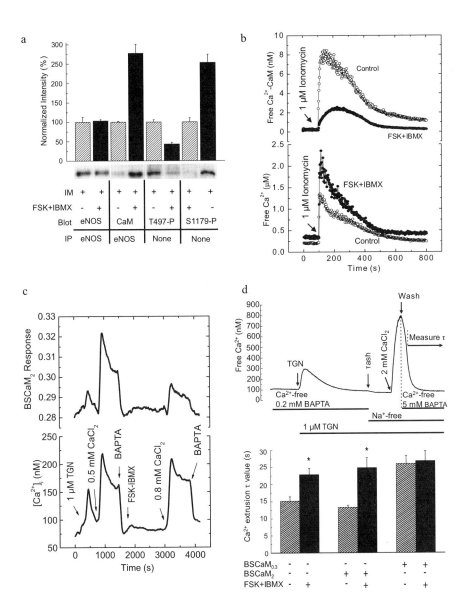

1992) and several unconventional myosins (Mooseker and Cheney 1995). The available CaM concentration therefore is significantly smaller than the total cellular CaM concentration.

In commercial bovine aortic endothelial cells (BAECs), eNOS can bind as much as 25% of the total cellular CaM (Tran et al. 2003), and in primary endothelial cells this number appears to be significantly higher (Q.K. Tran, unpublished observations). This observation suggested that eNOS can control the intracellular Ca^{2+}-CaM concentration via changes in its CaM binding status. A key determinant of the CaM binding status of eNOS is the phosphorylation status of Thr^{497} (Thr^{495} in the human sequence) in the CaM-binding domain. Under basal conditions, this residue is phosphorylated, which results in apparently no CaM binding to the synthase. In response to Ca^{2+}-elevating agonists or Ca^{2+} ionophores, dephosphorylation of Thr^{497} is associated with substantial increases in CaM binding to eNOS. As shown in Fig. 2a, treatment with forskolin (FSK, 50 μM) and 3-isobutyl-1-methylxanthine (IBMX), which dephosphorylates Thr^{497} eNOS and phosphorylates Ser^{1179}, is associated with a an approximately threefold increase in CaM binding to eNOS (Tran et al. 2003). Concurrent measurements of Ca^{2+} and Ca^{2+}-CaM concentrations revealed that this is associated with substantial reduction in the free intracellular Ca^{2+}-CaM concentration produced by ionomycin in BAECs (Fig. 2b; Tran et al. 2003). Experiments using the SOCE paradigm further demonstrate that the response of a fluorescent CaM biosensor constructed based on the CaM-binding domain of MLCK ($BSCaM_2$, apparent K_d for CaM ~2 nM) is substantially reduced following treatment that increases eNOS CaM binding, despite an SOCE signal manipulated to be similar with that obtained under control conditions (Fig. 2c; Q.K. Tran, unpublished data). Thus, the phosphorylation-dependent increases in CaM binding to eNOS can substantially reduce the free intracellular Ca^{2+}-CaM produced in endothelial cells. The physiological aspect of this effect is an associated approximate 40% reduction in the activity of the plasma membrane Ca^{2+}-ATPase (PMCA) in wild-type BAECs or BAECs expressing $BSCaM_2$ (Fig. 2d; Tran et al. 2003). Most importantly, this effect disappears in cells in which intracellular CaM has been buffered with a very high-affinity CaM binding protein $BSCaM_{0.3}$ (apparent K_d for CaM ~0.3 nM; Fig. 2d), confirming that it is due to increased CaM buffering by eNOS. These effects are totally independent of NO, as sufficient doses of L-NAME were applied throughout, and have been confirmed in a reconstituted system, HEK 293 cells expressing eNOS and CaM biosensors (Tran et al. 2005).

These studies have provided the first direct evidence that competition for limiting CaM can be a general mechanism coupling the activities of CaM targets in the cell. Further evidence in support of this mechanism later came from studies in neurons, in which phosphorylation-dependent changes in the CaM binding status of a novel CaM-binding protein can affect the functions of calcineurin and L-type Ca^{2+} channels in similar manners in which eNOS affects PMCA activity in endothelial cells (Rakhilin et al. 2004). In principle, the effects of limiting CaM on Ca^{2+} signalling can be pervasive; an excellent discussion can be found in Persechini and Stemmer (2002). In endothelial cells, the effect of eNOS to limit CaM availability apparently also affects SOCE. Indeed, while PMCA activity is substantially inhibited in BAECs treated with FSK-IBMX, which apparently accounts for the increased Ca^{2+} levels in ionomycin-treated cells (Fig. 2b, lower panel; Tran et al. 2003), after treatment with FSK-IBMX, higher extracellular Ca^{2+} is required to trigger an SOCE signal with a peak similar to a previous one stimulated under control condition. The suppressed PMCA activity appears to result in a more sustained Ca^{2+} response after the peak (Fig. 2c; Q.K. Tran, unpublished data). In addition, CaM binding to the more C-terminal CaM-binding site of TRPC1 has been suggested to regulate Ca^{2+}-dependent feedback inhibition of SOCE (Singh et al. 2002), and SOCE in non-excitable RBL-1 cells is impaired either following over-expression of a dominant-negative CaM mutant or following whole-cell dialysis with a CaM inhibitory peptide (Moreau et al. 2005).

Not only can eNOS limit the magnitudes of CaM-binding responses of proteins involved in Ca^{2+} signalling such as the PMCA and MLCK (BSCaM$_2$ is similar in its interaction with CaM to MLCK and therefore can be considered as a target analogue of MLCK), it can also limit the CaM-binding time courses of these targets due to differences in the kinetics of their interactions with CaM (Tran et al. 2005). Indeed, in vitro studies showed that eNOS binds $(Ca^{2+})_4$-CaM with a K_d value of 0.2 nM and an association rate constant of approximately 1.3×10^5 $M^{-1}s^{-1}$. These values are respectively 10- and 100-fold smaller than the corresponding values for the MLCK analogue BSCaM$_2$. As a result, when Ca^{2+} is added to a mixture of CaM, MLCK analogue and eNOS in vitro, a large fluorescence transient is observed as $(Ca^{2+})_4$-CaM is rapidly bound to the analogue and then slowly captured by the higher-affinity synthase (Fig. 3a).

In vivo, a rapid and sustained increase in free Ca^{2+} concentration in cells expressing both the cytoplasmic MLCK analogue and membrane-targeted eNOS only elicits a transient MLCK analogue response as opposed to a plateau response in cells expressing only the MLCK analogue. Increased CaM binding to eNOS with FSK-IBMX further enhances these effects (Fig. 3b). Transient responses are not observed in cells co-expressing the fluorescent analogue and a mutant T497D synthase unable to bind CaM (Tran et al. 2005). These data clearly demonstrate that eNOS can limit both the magnitudes and time courses of CaM responses of lower-affinity, less-abundant targets. The experimental protocol used in these studies was intended to rapidly elevate the intracellular free Ca^{2+} concentration to a sustained high level, allowing observation of the redistribution of CaM between eNOS and the MLCK analogue in the absence of potential complications associated with sub-maximal Ca^{2+} transients, especially spatial heterogeneity. A more detailed discussion of the situation during transient Ca^{2+} signals can be found in Tran et al. (2005).

8
Conclusions

This chapter has aimed to address the major components of Ca^{2+} signalling in the endothelial cell. Due to space constraints, many important issues such as spatial and temporal aspects of endothelial Ca^{2+} signals have been reluctantly left out. Four stages of Ca^{2+} signalling in these cells can be summarised (Fig. 4):

1. Stimuli-ER Ca^{2+} depletion: Physiological agonists or mechanical stimuli can deplete ER Ca^{2+} content by increasing IP_3 production and activation of IP_3 receptor, a process involving activation of trimeric G protein and activation of phospholipase C; ER Ca^{2+} content can also be pharmacologically emptied by inhibition of the ER Ca^{2+}-ATPase.

2. ER Ca^{2+} depletion-Activation of Ca^{2+} entry: Proposed coupling mechanisms (Fig. 4, diagonal box) include conformational coupling, Ca^{2+} influx factor or vesicle secretion-like coupling. There are many regulatory inputs (Fig. 4, large shaded arrow), including ER Ca^{2+} content, mitochondrial sequestration and interaction with ER, protein kinases such as MLCK and tyrosine kinase (PTK), possibly NO, membrane potential, and Ca^{2+}-dependent inactivation. Ca^{2+} entry channels are being identified, represented by members the transient receptor potential (TRP) protein family.

Fig. 3 a,b eNOS, as an abundant and high-affinity CaM-binding protein, can affect not only the magnitudes but also the time courses of activities of other Ca^{2+}-CaM targets in cells. **a** In vitro $(Ca^{2+})_4$-CaM redistribution kinetics in mixtures of eNOS and BSCaM$_2$. *Upper time course*: 250 nM BSCaM$_2$, 250 nM eNOS and 450 nM CaM; *Lower time course*: 250 nM BSCaM$_2$, 1 µM eNOS and 1.1 µM CaM. The eNOS association rate constant (k_f) and total concentration (E_t) values of 1.6×10^5 $M^{-1}s^{-1}$ and 230 nM *(upper curve)*, and 1.1×10^5 $M^{-1}s^{-1}$, 1.1 µM *(lower curve)* were determined (Tran et al. 2005). In these calculations, the K_d values for the $(Ca^{2+})_4$-CaM complexes with eNOS and BSCaM$_2$ were fixed at 0.2 and 1.4 nM. Measured k_f and k_r values for BSCaM$_2$ are 3.3×10^7 $M^{-1}s^{-1}$ and 0.06 s^{-1} (data not shown). **b** The presence of eNOS limits both the magnitude and time course of BSCaM$_2$ response in cells. *Upper panel*: fractional response of the fluorescent CaM biosensor BSCaM$_2$, a CaM-binding analogue of MLCK, in HEK 293 cells expressing only the biosensor *(open circles)*, or co-expressing BSCaM$_2$ and a fusion of bovine eNOS and the fluorescent protein DsRed2 after treatment with either FSK-IBMX *(filled circles)* or an equal volume of Me$_2$SO$_4$ *(open squares*, vehicle medium for FSK-IBMX). A saturated indo-1 response, corresponding in these cells to a free Ca^{2+} concentration above ∼4 µM, was produced by addition of 10 µM ionomycin and 10 mM CaCl$_2$, as indicated. *Lower panel*: indo-1 responses determined concurrently. (Reproduced from Tran et al. 2005, with permission)

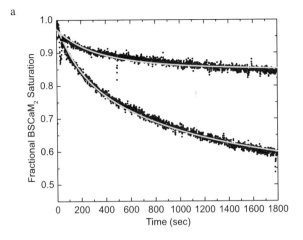

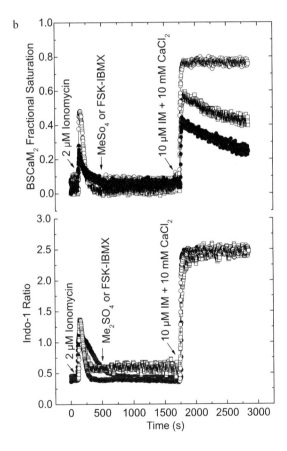

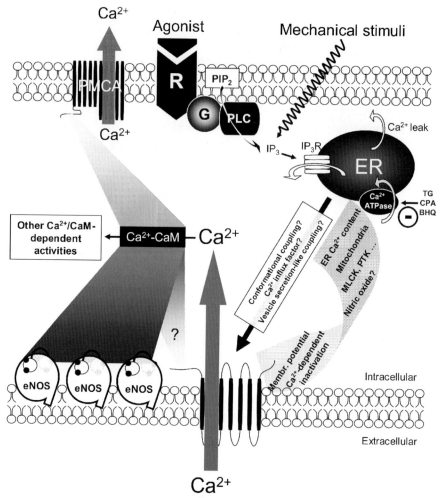

Fig. 4 Several key components of endothelial cell Ca^{2+} signalling. See text (Sect. 8). *BHQ*, dibenzohydroquinone; *CPA*, cyclopiazonic acid; *ER*, endoplasmic reticulum; *G*, trimeric G proteins; *IP$_3$R*, IP$_3$ receptor; *PIP$_2$*, phosphatidyl inositol 4,5-bisphosphate; *PLC*, phospholipase C, *IP$_3$*, inositol 1,4,5-trisphosphate; *PMCA*, plasma membrane Ca^{2+}-ATPase; *R*, surface receptor for agonists; *TGN*, thapsigargin

3. Transduction of intracellular Ca^{2+} signals: Calmodulin is the most important transducer of Ca^{2+} signals. eNOS, being abundant and having high affinity for CaM, can limit the available Ca^{2+}-CaM for other CaM-binding proteins involved in Ca^{2+} signalling. Competition for limiting CaM likely represents a pervasive coupling mechanism (see Sect. 7.2).

4. Mechanisms for removal of cytoplasmic Ca^{2+}: These comprise PMCA, the ER Ca^{2+}-ATPase, mitochondria and the Na^+-Ca^{2+} exchange.

In general, several aspects of endothelial cell Ca^{2+} signalling are distinct from that in other non-excitable cells, including among other things the constant exposure to shear stress and the presence of eNOS, which can substantially affect the activities of other Ca^{2+} signalling components via its role as a dominant affector of the intracellular CaM network. The signalling events from ligand-receptor binding to activation of intracellular Ca^{2+} channels and store depletion are fairly well clarified. Although many TRPC channels have been identified with significant functional impact in endothelial cells, the search for a clear identification of Ca^{2+} entry channels as well as their regulatory mechanisms is still underway.

References

Aarhus R, Graeff RM, Dickey DM, Walseth TF, Lee HC (1995) ADP-ribosyl cyclase and CD38 catalyze the synthesis of a calcium-mobilizing metabolite from NADP. J Biol Chem 270:30327–30333

Adams DJ, Hill MA (2004) Potassium channels and membrane potential in the modulation of intracellular calcium in vascular endothelial cells. J Cardiovasc Electrophysiol 15:598–610

Alvarez J, Montero M, Garcia-Sancho J (1991) Cytochrome P-450 may link intracellular Ca2+ stores with plasma membrane Ca2+ influx. Biochem J 274:193–197

Ando J, Kamiya A (1993) Blood flow and vascular endothelial cell function. Front Med Biol Eng 5:245–264

Anger M, Samuel JL, Marotte F, Wuytack F, Rappaport L, Lompre AM (1993) The sarco-(endo)plasmic reticulum Ca(2+)-ATPase mRNA isoform, SERCA 3, is expressed in endothelial and epithelial cells in various organs. FEBS Lett 334:45–48

Aramburu J, Rao A, Klee CB (2000) Calcineurin: from structure to function. Curr Top Cell Regul 36:237–295

Babnigg G, Bowersox SR, Villereal ML (1997) The role of pp60c-src in the regulation of calcium entry via store-operated calcium channels. J Biol Chem 272:29434–29437

Bakowski D, Glitsch MD, Parekh AB (2001) An examination of the secretion-like coupling model for the activation of the Ca2+ release-activated Ca2+ current I(CRAC) in RBL-1 cells. J Physiol 532:55–71

Berridge M (2004) Conformational coupling: a physiological calcium entry mechanism. Sci STKE 2004:pe33

Berridge MJ (1990) Calcium oscillations. J Biol Chem 265:9583–9586

Berridge MJ (1993) Inositol trisphosphate and calcium signalling. Nature 361:315–325

Berridge MJ (1995) Capacitative calcium entry. Biochem J 312:1–11

Berridge MJ, Irvine RF (1984) Inositol trisphosphate, a novel second messenger in cellular signal transduction. Nature 312:315–321

Bhagyalakshmi A, Berthiaume F, Reich KM, Frangos JA (1992) Fluid shear stress stimulates membrane phospholipid metabolism in cultured human endothelial cells. J Vasc Res 29:443–449

Bishara NB, Murphy TV, Hill MA (2002) Capacitative Ca(2+) entry in vascular endothelial cells is mediated via pathways sensitive to 2 aminoethoxydiphenyl borate and xestospongin C. Br J Pharmacol 135:119–128

Blatter LA, Wier WG (1994) Nitric oxide decreases [Ca2+]i in vascular smooth muscle by inhibition of the calcium current. Cell Calcium 15:122–131

Bolotina VM, Najibi S, Palacino JJ, Pagano PJ, Cohen RA (1994) Nitric oxide directly activates calcium-dependent potassium channels in vascular smooth muscle. Nature 368:850–853

Bossu JL, Feltz A, Rodeau JL, Tanzi F (1989) Voltage-dependent transient calcium currents in freshly dissociated capillary endothelial cells. FEBS Lett 255:377–380

Bossu JL, Elhamdani A, Feltz A (1992) Voltage-dependent calcium entry in confluent bovine capillary endothelial cells. FEBS Lett 299:239–242

Bourguignon LY, Iida N, Sobrin L, Bourguignon GJ (1994) Identification of an IP3 receptor in endothelial cells. J Cell Physiol 159:29–34

Boyer JL, Waldo GL, Harden TK (1992) Beta gamma-subunit activation of G-protein-regulated phospholipase C. J Biol Chem 267:25451–25456

Brading AF, Sneddon P (1980) Evidence for multiple sources of calcium for activation of the contractile mechanism of guinea-pig taenia coli on stimulation with carbachol. Br J Pharmacol 70:229–240

Brauneis U, Gatmaitan Z, Arias IM (1992) Serotonin stimulates a Ca2+ permeant nonspecific cation channel in hepatic endothelial cells. Biochem Biophys Res Commun 186:1560–1566

Bredt DS, Snyder SH (1990) Isolation of nitric oxide synthetase, a calmodulin-requiring enzyme. Proc Natl Acad Sci U S A 87:682–685

Broad LM, Cannon TR, Taylor CW (1999) A non-capacitative pathway activated by arachidonic acid is the major Ca2+ entry mechanism in rat A7r5 smooth muscle cells stimulated with low concentrations of vasopressin. J Physiol 517:121–134

Broad LM, Braun FJ, Lievremont JP, Bird GS, Kurosaki T, Putney JW Jr (2001) Role of the phospholipase C-inositol 1,4,5-trisphosphate pathway in calcium release-activated calcium current and capacitative calcium entry. J Biol Chem 276:15945–15952

Brophy CM, Mills I, Rosales O, Isales C, Sumpio BE (1993) Phospholipase C: a putative mechanotransducer for endothelial cell response to acute hemodynamic changes. Biochem Biophys Res Commun 190:576–581

Camps M, Carozzi A, Schnabel P, Scheer A, Parker PJ, Gierschik P (1992) Isozyme-selective stimulation of phospholipase C-beta 2 by G protein beta gamma-subunits. Nature 360:684–686

Casteels R, Droogmans G (1981) Exchange characteristics of the noradrenaline-sensitive calcium store in vascular smooth muscle cells or rabbit ear artery. J Physiol 317:263–279

Castonguay A, Robitaille R (2002) Xestospongin C is a potent inhibitor of SERCA at a vertebrate synapse. Cell Calcium 32:39–47

Cho HJ, Xie QW, Calaycay J, Mumford RA, Swiderek KM, Lee TD, Nathan C (1992) Calmodulin is a subunit of nitric oxide synthase from macrophages. J Exp Med 176:599–604

Cioffi DL, Wu S, Stevens T (2003) On the endothelial cell I(SOC). Cell Calcium 33:323–336

Clapham DE, Runnels LW, Strubing C (2001) The TRP ion channel family. Nat Rev Neurosci 2:387–396

Clapper DL, Walseth TF, Dargie PJ, Lee HC (1987) Pyridine nucleotide metabolites stimulate calcium release from sea urchin egg microsomes desensitized to inositol trisphosphate. J Biol Chem 262:9561–9568

Cohen RA, Weisbrod RM, Gericke M, Yaghoubi M, Bierl C, Bolotina VM (1999) Mechanism of nitric oxide-induced vasodilatation: refilling of intracellular stores by sarcoplasmic reticulum Ca2+ ATPase and inhibition of store-operated Ca2+ influx. Circ Res 84:210–219

Colden-Stanfield M, Schilling WP, Possani LD, Kunze DL (1990) Bradykinin-induced potassium current in cultured bovine aortic endothelial cells. J Membr Biol 116:227–238

Cosens DJ, Manning A (1969) Abnormal electroretinogram from a Drosophila mutant. Nature 224:285–287

Crutchley DJ, Ryan JW, Ryan US, Fisher GH (1983) Bradykinin-induced release of prostacyclin and thromboxanes from bovine pulmonary artery endothelial cells. Studies with lower homologs and calcium antagonists. Biochim Biophys Acta 751:99–107

Csordas G, Thomas AP, Hajnoczky G (1999) Quasi-synaptic calcium signal transmission between endoplasmic reticulum and mitochondria. EMBO J 18:96–108

Csutora P, Su Z, Kim HY, Bugrim A, Cunningham KW, Nuccitelli R, Keizer JE, Hanley MR, Blalock JE, Marchase RB (1999) Calcium influx factor is synthesized by yeast and mammalian cells depleted of organellar calcium stores. Proc Natl Acad Sci U S A 96:121–126

Davies PF (1995) Flow-mediated endothelial mechanotransduction. Physiol Rev 75:519–560

Dedkova EN, Blatter LA (2002) Nitric oxide inhibits capacitative Ca2+ entry and enhances endoplasmic reticulum Ca2+ uptake in bovine vascular endothelial cells. J Physiol 539:77–91

Diver JM, Sage SO, Rosado JA (2001) The inositol trisphosphate receptor antagonist 2-aminoethoxydiphenylborate (2-APB) blocks Ca2+ entry channels in human platelets: cautions for its use in studying Ca2+ influx. Cell Calcium 30:323–329

Dolor RJ, Hurwitz LM, Mirza Z, Strauss HC, Whorton AR (1992) Regulation of extracellular calcium entry in endothelial cells: role of intracellular calcium pool. Am J Physiol 262:C171–C181

Falcke M, Hudson JL, Camacho P, Lechleiter JD (1999) Impact of mitochondrial Ca2+ cycling on pattern formation and stability. Biophys J 77:37–44

Fill M, Copello JA (2002) Ryanodine receptor calcium release channels. Physiol Rev 82:893–922

Fiorio Pla A, Munaron L (2001) Calcium influx, arachidonic acid, and control of endothelial cell proliferation. Cell Calcium 30:235–244

Fleming I, Busse R (1997) Tyrosine phosphorylation and bradykinin-induced signaling in endothelial cells. Am J Cardiol 80:102A–109A

Fleming I, Fisslthaler B, Busse R (1995) Calcium signaling in endothelial cells involves activation of tyrosine kinases and leads to activation of mitogen-activated protein kinases. Circ Res 76:522–529

Freichel M, Schweig U, Stauffenberger S, Freise D, Schorb W, Flockerzi V (1999) Store-operated cation channels in the heart and cells of the cardiovascular system. Cell Physiol Biochem 9:270–283

Freichel M, Suh SH, Pfeifer A, Schweig U, Trost C, Weissgerber P, Biel M, Philipp S, Freise D, Droogmans G, Hofmann F, Flockerzi V, Nilius B (2001) Lack of an endothelial store-operated Ca2+ current impairs agonist-dependent vasorelaxation in TRP4−/− mice. Nat Cell Biol 3:121–127

Furuichi T, Yoshikawa S, Miyawaki A, Wada K, Maeda N, Mikoshiba K (1989) Primary structure and functional expression of the inositol 1,4,5-trisphosphate-binding protein P400. Nature 342:32–38

Galione A, Lee HC, Busa WB (1991) Ca(2+)-induced Ca2+ release in sea urchin egg homogenates: modulation by cyclic ADP-ribose. Science 253:1143–1146

Garcia JG, Lazar V, Gilbert-McClain LI, Gallagher PJ, Verin AD (1997) Myosin light chain kinase in endothelium: molecular cloning and regulation. Am J Respir Cell Mol Biol 16:489–494

Genazzani AA, Galione A (1996) Nicotinic acid-adenine dinucleotide phosphate mobilizes Ca2+ from a thapsigargin-insensitive pool. Biochem J 315:721–725

Gericke M, Droogmans G, Nilius B (1993) Thapsigargin discharges intracellular calcium stores and induces transmembrane currents in human endothelial cells. Pflugers Arch 422:552–557

Goldbeter A, Dupont G, Berridge MJ (1990) Minimal model for signal-induced Ca2+ oscillations and for their frequency encoding through protein phosphorylation. Proc Natl Acad Sci U S A 87:1461–1465

Gregory RB, Rychkov G, Barritt GJ (2001) Evidence that 2-aminoethyl diphenylborate is a novel inhibitor of store-operated Ca2+ channels in liver cells, and acts through a mechanism which does not involve inositol trisphosphate receptors. Biochem J 354:285–290

Groner M, Malchow D (1996) Calmodulin-antagonists inhibit vesicular Ca2+ uptake in Dictyostelium. Cell Calcium 19:105–111

Groschner K, Graier WF, Kukovetz WR (1994) Histamine induces K+, Ca2+, and Cl− currents in human vascular endothelial cells. Role of ionic currents in stimulation of nitric oxide biosynthesis. Circ Res 75:304–314

Gu C, Cooper DM (1999) Calmodulin-binding sites on adenylyl cyclase type VIII. J Biol Chem 274:8012–8021

Gunter TE, Pfeiffer DR (1990) Mechanisms by which mitochondria transport calcium. Am J Physiol 258:C755–C786

Hajnoczky G, Hager R, Thomas AP (1999) Mitochondria suppress local feedback activation of inositol 1,4,5-trisphosphate receptors by Ca2+. J Biol Chem 274:14157–14162

Hardie RC, Minke B (1992) The trp gene is essential for a light-activated Ca2+ channel in Drosophila photoreceptors. Neuron 8:643–651

Hardie RC, Minke B (1993) Novel Ca2+ channels underlying transduction in Drosophila photoreceptors: implications for phosphoinositide-mediated Ca2+ mobilization. Trends Neurosci 16:371–376

Hashimoto Y, Watanabe T, Kinoshita M, Tsukamoto K, Togo M, Horie Y, Matsuda Y, Kurokawa K (1993) Ca2+ entry pathways activated by the tumor promoter thapsigargin in human platelets. Biochim Biophys Acta 1220:37–41

He H, Venema VJ, Gu X, Venema RC, Marrero MB, Caldwell RB (1999) Vascular endothelial growth factor signals endothelial cell production of nitric oxide and prostacyclin through flk-1/KDR activation of c-Src. J Biol Chem 274:25130–25135

Helmlinger G, Berk BC, Nerem RM (1995) Calcium responses of endothelial cell monolayers subjected to pulsatile and steady laminar flow differ. Am J Physiol 269:C367–C375

Himmel HM, Whorton AR, Strauss HC (1993) Intracellular calcium, currents, and stimulus-response coupling in endothelial cells. Hypertension 21:112–127

Hofer AM, Curci S, Machen TE, Schulz I (1996) ATP regulates calcium leak from agonist-sensitive internal calcium stores. Faseb J 10:302–308

Holda JR, Blatter LA (1997) Capacitative calcium entry is inhibited in vascular endothelial cells by disruption of cytoskeletal microfilaments. FEBS Lett 403:191–196

Hosoki E, Iijima T (1994) Chloride-sensitive Ca2+ entry by histamine and ATP in human aortic endothelial cells. Eur J Pharmacol 266:213–218

Hoth M, Penner R (1992) Depletion of intracellular calcium stores activates a calcium current in mast cells. Nature 355:353–356

Ichas F, Jouaville LS, Mazat JP (1997) Mitochondria are excitable organelles capable of generating and conveying electrical and calcium signals. Cell 89:1145–1153

Jacob R (1990) Agonist-stimulated divalent cation entry into single cultured human umbilical vein endothelial cells. J Physiol 421:55–77

Jacob R, Merritt JE, Hallam TJ, Rink TJ (1988) Repetitive spikes in cytoplasmic calcium evoked by histamine in human endothelial cells. Nature 335:40–45

Jena M, Minore JF, O'Neill WC (1997) A volume-sensitive, IP3-insensitive Ca2+ store in vascular endothelial cells. Am J Physiol 273:C316–C322

Johns A, Lategan TW, Lodge NJ, Ryan US, Van Breemen C, Adams DJ (1987) Calcium entry through receptor-operated channels in bovine pulmonary artery endothelial cells. Tissue Cell 19:733–745

Kanai AJ, Strauss HC, Truskey GA, Crews AL, Grunfeld S, Malinski T (1995) Shear stress induces ATP-independent transient nitric oxide release from vascular endothelial cells, measured directly with a porphyrinic microsensor. Circ Res 77:284–293

Khan SZ, Longland CL, Michelangeli F (2000) The effects of phenothiazines and other calmodulin antagonists on the sarcoplasmic and endoplasmic reticulum Ca(2+) pumps. Biochem Pharmacol 60:1797–1806

Kruse HJ, Grunberg B, Siess W, Weber PC (1994) Formation of biologically active autacoids is regulated by calcium influx in endothelial cells. Arterioscler Thromb 14:1821–1828

Kukkonen JP, Lund PE, Akerman KE (2001) 2-Aminoethoxydiphenyl borate reveals heterogeneity in receptor-activated Ca(2+) discharge and store-operated Ca(2+) influx. Cell Calcium 30:117–129

Kwan HY, Huang Y, Yao X (2000) Store-operated calcium entry in vascular endothelial cells is inhibited by cGMP via a protein kinase G-dependent mechanism. J Biol Chem 275:6758–6763

Kwan HY, Leung PC, Huang Y, Yao X (2003) Depletion of intracellular Ca2+ stores sensitizes the flow-induced Ca2+ influx in rat endothelial cells. Circ Res 92:286–292

Lambert TL, Kent RS, Whorton AR (1986) Bradykinin stimulation of inositol polyphosphate production in porcine aortic endothelial cells. J Biol Chem 261:15288–15293

Lee HC (2000) NAADP: an emerging calcium signaling molecule. J Membr Biol 173:1–8

Lee HC, Aarhus R (1993) Wide distribution of an enzyme that catalyzes the hydrolysis of cyclic ADP-ribose. Biochim Biophys Acta 1164:68–74

Lee HC, Aarhus R (1995) A derivative of NADP mobilizes calcium stores insensitive to inositol trisphosphate and cyclic ADP-ribose. J Biol Chem 270:2152–2157

Lee HC, Walseth TF, Bratt GT, Hayes RN, Clapper DL (1989) Structural determination of a cyclic metabolite of NAD+ with intracellular Ca2+-mobilizing activity. J Biol Chem 264:1608–1615

Lee HC, Aarhus R, Graeff RM (1995) Sensitization of calcium-induced calcium release by cyclic ADP-ribose and calmodulin. J Biol Chem 270:9060–9066

Lee KM, Villereal ML (1996) Tyrosine phosphorylation and activation of pp60c-src and pp125FAK in bradykinin-stimulated fibroblasts. Am J Physiol 270:C1430–C1437

Lee KM, Toscas K, Villereal ML (1993) Inhibition of bradykinin- and thapsigargin-induced Ca2+ entry by tyrosine kinase inhibitors. J Biol Chem 268:9945–9948

Lesh RE, Marks AR, Somlyo AV, Fleischer S, Somlyo AP (1993) Anti-ryanodine receptor antibody binding sites in vascular and endocardial endothelium. Circ Res 72:481–488

Levitan IB (1999) It is calmodulin after all! Mediator of the calcium modulation of multiple ion channels. Neuron 22:645–648

Liang W, Buluc M, van Breemen C, Wang X (2004) Vectorial Ca2+ release via ryanodine receptors contributes to Ca2+ extrusion from freshly isolated rabbit aortic endothelial cells. Cell Calcium 36:431–443

Liao JK, Homcy CJ (1993) The G proteins of the G alpha i and G alpha q family couple the bradykinin receptor to the release of endothelium-derived relaxing factor. J Clin Invest 92:2168–2172

Lin S, Fagan KA, Li KX, Shaul PW, Cooper DM, Rodman DM (2000) Sustained endothelial nitric-oxide synthase activation requires capacitative Ca2+ entry. J Biol Chem 275:17979–17985

Liu LH, Paul RJ, Sutliff RL, Miller ML, Lorenz JN, Pun RY, Duffy JJ, Doetschman T, Kimura Y, MacLennan DH, Hoying JB, Shull GE (1997) Defective endothelium-dependent relaxation of vascular smooth muscle and endothelial cell Ca2+ signaling in mice lacking sarco(endo)plasmic reticulum Ca2+-ATPase isoform 3. J Biol Chem 272:30538–30545

Lomax RB, Camello C, Van Coppenolle F, Petersen OH, Tepikin AV (2002) Basal and physiological Ca(2+) leak from the endoplasmic reticulum of pancreatic acinar cells. Second messenger-activated channels and translocons. J Biol Chem 277:26479–26485

Luckhoff A, Busse R (1990) Calcium influx into endothelial cells and formation of endothelium-derived relaxing factor is controlled by the membrane potential. Pflugers Arch 416:305–311

Luo D, Broad LM, Bird GS, Putney JW Jr (2001) Mutual antagonism of calcium entry by capacitative and arachidonic acid-mediated calcium entry pathways. J Biol Chem 276:20186–20189

Ma HT, Patterson RL, van Rossum DB, Birnbaumer L, Mikoshiba K, Gill DL (2000) Requirement of the inositol trisphosphate receptor for activation of store-operated Ca2+ channels. Science 287:1647–1651

Ma HT, Venkatachalam K, Li HS, Montell C, Kurosaki T, Patterson RL, Gill DL (2001) Assessment of the role of the inositol 1,4,5-trisphosphate receptor in the activation of transient receptor potential channels and store-operated Ca2+ entry channels. J Biol Chem 276:18888–18896

Macer DR, Koch GL (1988) Identification of a set of calcium-binding proteins in reticuloplasm, the luminal content of the endoplasmic reticulum. J Cell Sci 91:61–70

Mak DO, McBride S, Foskett JK (1998) Inositol 1,4,5-trisphosphate activation of inositol trisphosphate receptor Ca2+ channel by ligand tuning of Ca2+ inhibition. Proc Natl Acad Sci U S A 95:15821–15825

Malli R, Frieden M, Osibow K, Zoratti C, Mayer M, Demaurex N, Graier WF (2003) Sustained Ca2+ transfer across mitochondria is Essential for mitochondrial Ca2+ buffering, soreoperated Ca2+ entry, and Ca2+ store refilling. J Biol Chem 278:44769–44779

Malli R, Frieden M, Trenker M, Graier WF (2005) The role of mitochondria for Ca2+ refilling of the endoplasmic reticulum. J Biol Chem 280:12114–12122

Matsumoto H, Baron CB, Coburn RF (1995) Smooth muscle stretch-activated phospholipase C activity. Am J Physiol 268:C458–C465

Meyer RD, Latz C, Rahimi N (2003) Recruitment and activation of phospholipase Cgamma1 by vascular endothelial growth factor receptor-2 are required for tubulogenesis and differentiation of endothelial cells. J Biol Chem 278:16347–16355

Meyer T, Stryer L (1988) Molecular model for receptor-stimulated calcium spiking. Proc Natl Acad Sci U S A 85:5051–5055

Meyer T, Holowka D, Stryer L (1988) Highly cooperative opening of calcium channels by inositol 1,4,5-trisphosphate. Science 240:653–656

Mignen O, Shuttleworth TJ (2000) I(ARC), a novel arachidonate-regulated, noncapacitative Ca(2+) entry channel. J Biol Chem 275:9114–9119

Mignery GA, Sudhof TC, Takei K, De Camilli P (1989) Putative receptor for inositol 1,4,5-trisphosphate similar to ryanodine receptor. Nature 342:192–195

Mikoshiba K, Huchet M, Changeux JP (1979) Biochemical and immunological studies on the P400 protein, a protein characteristic of the Purkinje cell from mouse and rat cerebellum. Dev Neurosci 2:254–275

Missiaen L, Callewaert G, De Smedt H, Parys JB (2001) 2-Aminoethoxydiphenyl borate affects the inositol 1,4,5-trisphosphate receptor, the intracellular Ca2+ pump and the non-specific Ca2+ leak from the non-mitochondrial Ca2+ stores in permeabilized A7r5 cells. Cell Calcium 29:111–116

Moncada S, Higgs A (1993) The L-arginine-nitric oxide pathway. N Engl J Med 329:2002–2012

Moneer Z, Taylor CW (2002) Reciprocal regulation of capacitative and non-capacitative Ca2+ entry in A7r5 vascular smooth muscle cells: only the latter operates during receptor activation. Biochem J 362:13–21

Moneer Z, Dyer JL, Taylor CW (2003) Nitric oxide co-ordinates the activities of the capacitative and non-capacitative Ca2+-entry pathways regulated by vasopressin. Biochem J 370:439–448

Montell C, Birnbaumer L, Flockerzi V, Bindels RJ, Bruford EA, Caterina MJ, Clapham DE, Harteneck C, Heller S, Julius D, Kojima I, Mori Y, Penner R, Prawitt D, Scharenberg AM, Schultz G, Shimizu N, Zhu MX (2002) A unified nomenclature for the superfamily of TRP cation channels. Mol Cell 9:229–231

Montero M, Alvarez J, Garcia-Sancho J (1991) Agonist-induced Ca2+ influx in human neutrophils is secondary to the emptying of intracellular calcium stores. Biochem J 277:73–79

Montero M, Alvarez J, Garcia-Sancho J (1992) Control of plasma-membrane Ca2+ entry by the intracellular Ca2+ stores. Kinetic evidence for a short-lived mediator. Biochem J 288:519–525

Montero M, Garcia-Sancho J, Alvarez J (1994) Phosphorylation down-regulates the store-operated Ca2+ entry pathway of human neutrophils. J Biol Chem 269:3963–3967

Mooseker MS, Cheney RE (1995) Unconventional myosins. Annu Rev Cell Dev Biol 11:633–675

Moreau B, Straube S, Fisher RJ, Putney JW Jr, Parekh AB (2005) Ca2+-calmodulin-dependent facilitation and Ca2+ inactivation of Ca2+ release-activated Ca2+ channels. J Biol Chem 280:8776–8783

Mountian I, Manolopoulos VG, De Smedt H, Parys JB, Missiaen L, Wuytack F (1999) Expression patterns of sarco/endoplasmic reticulum Ca(2+)-ATPase and inositol 1,4,5-trisphosphate receptor isoforms in vascular endothelial cells. Cell Calcium 25:371–380

Nairn AC, Picciotto MR (1994) Calcium/calmodulin-dependent protein kinases. Semin Cancer Biol 5:295–303

Nilius B, Droogmans G, Wondergem R (2003) Transient receptor potential channels in endothelium: solving the calcium entry puzzle? Endothelium 10:5–15

Norwood N, Moore TM, Dean DA, Bhattacharjee R, Li M, Stevens T (2000) Store-operated calcium entry and increased endothelial cell permeability. Am J Physiol Lung Cell Mol Physiol 279:L815–L824

Oike M, Droogmans G, Nilius B (1994) Mechanosensitive Ca2+ transients in endothelial cells from human umbilical vein. Proc Natl Acad Sci U S A 91:2940–2944

Oka T, Sato K, Hori M, Ozaki H, Karaki H (2002) Xestospongin C, a novel blocker of IP3 receptor, attenuates the increase in cytosolic calcium level and degranulation that is induced by antigen in RBL-2H3 mast cells. Br J Pharmacol 135:1959–1966

Oldershaw KA, Taylor CW (1993) Luminal Ca2+ increases the affinity of inositol 1,4,5-trisphosphate for its receptor. Biochem J 292:631–633

Paltauf-Doburzynska J, Frieden M, Graier WF (1999) Mechanisms of Ca2+ store depletion in single endothelial cells in a Ca(2+)-free environment. Cell Calcium 25:345–353

Parekh AB, Penner R (1995) Depletion-activated calcium current is inhibited by protein kinase in RBL-2H3 cells. Proc Natl Acad Sci U S A 92:7907–7911

Parekh AB, Putney JW Jr (2005) Store-operated calcium channels. Physiol Rev 85:757–810

Patterson RL, van Rossum DB, Gill DL (1999) Store-operated Ca2+ entry: evidence for a secretion-like coupling model. Cell 98:487–499

Persechini A, Stemmer PM (2002) Calmodulin is a limiting factor in the cell. Trends Cardiovasc Med 12:32–37

Petersen OH, Cancela JM (1999) New Ca2+-releasing messengers: are they important in the nervous system? Trends Neurosci 22:488–495

Picton C, Shenolikar S, Grand R, Cohen P (1983) Calmodulin as an integral subunit of phosphorylase kinase from rabbit skeletal muscle. Methods Enzymol 102:219–227

Popp R, Gogelein H (1992) A calcium and ATP sensitive nonselective cation channel in the antiluminal membrane of rat cerebral capillary endothelial cells. Biochim Biophys Acta 1108:59–66

Prasad AR, Logan SA, Nerem RM, Schwartz CJ, Sprague EA (1993) Flow-related responses of intracellular inositol phosphate levels in cultured aortic endothelial cells. Circ Res 72:827–836

Prince WT, Berridge MJ, Rasmussen H (1972) Role of calcium and adenosine-3′:5′-cyclic monophosphate in controlling fly salivary gland secretion. Proc Natl Acad Sci U S A 69:553–557

Putney JW Jr (1986) A model for receptor-regulated calcium entry. Cell Calcium 7:1–12

Putney JW Jr (1990) Capacitative calcium entry revisited. Cell Calcium 11:611–624

Raeymaekers L, Wuytack F, Eggermont J, De Schutter G, Casteels R (1983) Isolation of a plasma-membrane fraction from gastric smooth muscle. Comparison of the calcium uptake with that in endoplasmic reticulum. Biochem J 210:315–322

Rakhilin SV, Olson PA, Nishi A, Starkova NN, Fienberg AA, Nairn AC, Surmeier DJ, Greengard P (2004) A network of control mediated by regulator of calcium/calmodulin-dependent signaling. Science 306:698–701

Randriamampita C, Tsien RY (1993) Emptying of intracellular Ca2+ stores releases a novel small messenger that stimulates Ca2+ influx. Nature 364:809–814

Rosado JA, Jenner S, Sage SO (2000) A role for the actin cytoskeleton in the initiation and maintenance of store-mediated calcium entry in human platelets. Evidence for conformational coupling. J Biol Chem 275:7527–7533

Rzigalinski BA, Willoughby KA, Hoffman SW, Falck JR, Ellis EF (1999) Calcium influx factor, further evidence it is 5, 6-epoxyeicosatrienoic acid. J Biol Chem 274:175–182

Sargeant P, Farndale RW, Sage SO (1993) ADP- and thapsigargin-evoked Ca2+ entry and protein-tyrosine phosphorylation are inhibited by the tyrosine kinase inhibitors genistein and methyl-2,5-dihydroxycinnamate in fura-2-loaded human platelets. J Biol Chem 268:18151–18156

Sedova M, Klishin A, Huser J, Blatter LA (2000) Capacitative Ca2+ entry is graded with degree of intracellular Ca2+ store depletion in bovine vascular endothelial cells. J Physiol 523:549–559

Shuttleworth TJ (1996) Arachidonic acid activates the noncapacitative entry of Ca2+ during [Ca2+]i oscillations. J Biol Chem 271:21720–21725

Shuttleworth TJ, Thompson JL (1996a) Ca2+ entry modulates oscillation frequency by triggering Ca2+ release. Biochem J 313:815–819

Shuttleworth TJ, Thompson JL (1996b) Evidence for a non-capacitative Ca2+ entry during [Ca2+] oscillations. Biochem J 316:819–824

Shuttleworth TJ, Thompson JL (1998) Muscarinic receptor activation of arachidonate-mediated Ca2+ entry in HEK293 cells is independent of phospholipase C. J Biol Chem 273:32636–32643

Singh BB, Liu X, Tang J, Zhu MX, Ambudkar IS (2002) Calmodulin regulates Ca(2+)-dependent feedback inhibition of store-operated Ca(2+) influx by interaction with a site in the C terminus of TrpC1. Mol Cell 9:739–750

Solovyova N, Fernyhough P, Glazner G, Verkhratsky A (2002) Xestospongin C empties the ER calcium store but does not inhibit InsP3-induced Ca2+ release in cultured dorsal root ganglia neurones. Cell Calcium 32:49–52

Streb H, Irvine RF, Berridge MJ, Schulz I (1983) Release of Ca2+ from a nonmitochondrial intracellular store in pancreatic acinar cells by inositol-1,4,5-trisphosphate. Nature 306:67–69

Takahashi R, Watanabe H, Zhang XX, Kakizawa H, Hayashi H, Ohno R (1997) Roles of inhibitors of myosin light chain kinase and tyrosine kinase on cation influx in agonist-stimulated endothelial cells. Biochem Biophys Res Commun 235:657–662

Takeuchi K, Watanabe H, Tran QK, Ozeki M, Uehara A, Katoh H, Satoh H, Terada H, Ohashi K, Hayashi H (2003) Effects of cytochrome P450 inhibitors on agonist-induced Ca2+ responses and production of NO and PGI2 in vascular endothelial cells. Mol Cell Biochem 248:129–134

Takeuchi K, Watanabe H, Tran QK, Ozeki M, Sumi D, Hayashi T, Iguchi A, Ignarro LJ, Ohashi K, Hayashi H (2004) Nitric oxide: inhibitory effects on endothelial cell calcium signaling, prostaglandin I2 production and nitric oxide synthase expression. Cardiovasc Res 62:194–201

Taylor CW, Laude AJ (2002) IP3 receptors and their regulation by calmodulin and cytosolic Ca2+. Cell Calcium 32:321–334

Thastrup O, Cullen PJ, Drobak BK, Hanley MR, Dawson AP (1990) Thapsigargin, a tumor promoter, discharges intracellular Ca2+ stores by specific inhibition of the endoplasmic reticulum Ca2(+)-ATPase. Proc Natl Acad Sci U S A 87:2466–2470

Tornquist K (1993) Modulatory effect of protein kinase C on thapsigargin-induced calcium entry in thyroid FRTL-5 cells. Biochem J 290:443–447

Tran QK, Watanabe H (2003) Myosin light chain kinase in endothelial cell calcium signaling and endothelial functions. In: Pierce GN, Nagano M, Zahradka P, Dhalla NS (eds) Progress in experimental cardiology, vol 8. Kluwer Academic/Plenum Publishers, New York pp 163–174

Tran QK, Watanabe H, Zhang XX, Takahashi R, Ohno R (1999) Involvement of myosin light-chain kinase in chloride-sensitive Ca2+ influx in porcine aortic endothelial cells. Cardiovasc Res 44:623–631

Tran QK, Ohashi K, Watanabe H (2000) Calcium signalling in endothelial cells. Cardiovasc Res 48:13–22

Tran QK, Watanabe H, Le HY, Pan L, Seto M, Takeuchi K, Ohashi K (2001) Myosin light chain kinase regulates capacitative ca(2+) entry in human monocytes/macrophages. Arterioscler Thromb Vasc Biol 21:509–515

Tran QK, Black DJ, Persechini A (2003) Intracellular coupling via limiting calmodulin. J Biol Chem 278:24247–24250

Tran QK, Black DJ, Persechini A (2005) Dominant affectors in the calmodulin network shape the time courses of target responses in the cell. Cell Calcium 37:541–553

Trepakova ES, Cohen RA, Bolotina VM (1999) Nitric oxide inhibits capacitative cation influx in human platelets by promoting sarcoplasmic/endoplasmic reticulum Ca2+-ATPase-dependent refilling of Ca2+ stores. Circ Res 84:201–209

Trepakova ES, Csutora P, Hunton DL, Marchase RB, Cohen RA, Bolotina VM (2000) Calcium influx factor directly activates store-operated cation channels in vascular smooth muscle cells. J Biol Chem 275:26158–26163

Vaca L, Kunze DL (1995) IP3-activated Ca2+ channels in the plasma membrane of cultured vascular endothelial cells. Am J Physiol 269:C733–C738

Vaca L, Sinkins WG, Hu Y, Kunze DL, Schilling WP (1994) Activation of recombinant trp by thapsigargin in Sf9 insect cells. Am J Physiol 267:C1501–C1505

Vazquez G, Lievremont JP, St J Bird G, Putney JW Jr (2001) Human Trp3 forms both inositol trisphosphate receptor-dependent and receptor-independent store-operated cation channels in DT40 avian B lymphocytes. Proc Natl Acad Sci U S A 98:11777–11782

Vermassen E, Parys JB, Mauger JP (2004) Subcellular distribution of the inositol 1,4,5-trisphosphate receptors: functional relevance and molecular determinants. Biol Cell 96:3–17

Vincenzi FF, Larsen FL (1980) The plasma membrane calcium pump: regulation by a soluble Ca2+ binding protein. Fed Proc 39:2427–2431

Vischer UM, Lang U, Wollheim CB (1998) Autocrine regulation of endothelial exocytosis: von Willebrand factor release is induced by prostacyclin in cultured endothelial cells. FEBS Lett 424:211–215

Walseth TF, Lee HC (1993) Synthesis and characterization of antagonists of cyclic-ADP-ribose-induced Ca2+ release. Biochim Biophys Acta 1178:235–242

Wang X, Lau F, Li L, Yoshikawa A, van Breemen C (1995) Acetylcholine-sensitive intracellular Ca2+ store in fresh endothelial cells and evidence for ryanodine receptors. Circ Res 77:37–42

Wang Y, Shin WS, Kawaguchi H, Inukai M, Kato M, Sakamoto A, Uehara Y, Miyamoto M, Shimamoto N, Korenaga R, Ando J, Toyo-oka T (1996) Contribution of sustained Ca2+ elevation for nitric oxide production in endothelial cells and subsequent modulation of Ca2+ transient in vascular smooth muscle cells in coculture. J Biol Chem 271:5647–5655

Watanabe H, Takahashi R, Zhang XX, Kakizawa H, Hayashi H, Ohno R (1996) Inhibition of agonist-induced Ca2+ entry in endothelial cells by myosin light-chain kinase inhibitor. Biochem Biophys Res Commun 225:777–784

Watanabe H, Takahashi R, Zhang XX, Goto Y, Hayashi H, Ando J, Isshiki M, Seto M, Hidaka H, Niki I, Ohno R (1998) An essential role of myosin light-chain kinase in the regulation of agonist- and fluid flow-stimulated Ca2+ influx in endothelial cells. FASEB J 12:341–348

Watanabe H, Takahashi R, Tran QK, Takeuchi K, Kosuge K, Satoh H, Uehara A, Terada H, Hayashi H, Ohno R, Ohashi K (1999) Increased cytosolic Ca(2+) concentration in endothelial cells by calmodulin antagonists. Biochem Biophys Res Commun 265:697–702

Watanabe H, Tran QK, Takeuchi K, Fukao M, Liu MY, Kanno M, Hayashi T, Iguchi A, Seto M, Ohashi K (2001) Myosin light-chain kinase regulates endothelial calcium entry and endothelium-dependent vasodilation. FASEB J 15:282–284

Wedel BJ, Vazquez G, McKay RR, St J Bird G, Putney JW Jr (2003) A calmodulin/inositol 1,4,5-trisphosphate (IP3) receptor-binding region targets TRPC3 to the plasma membrane in a calmodulin/IP3 receptor-independent process. J Biol Chem 278:25758–25765

Wibo M, Morel N, Godfraind T (1981) Differentiation of Ca2+ pumps linked to plasma membrane and endoplasmic reticulum in the microsomal fraction from intestinal smooth muscle. Biochim Biophys Acta 649:651–660

Wissing F, Nerou EP, Taylor CW (2002) A novel Ca2+-induced Ca2+ release mechanism mediated by neither inositol trisphosphate nor ryanodine receptors. Biochem J 361:605–611

Wood PG, Gillespie JI (1998a) Evidence for mitochondrial Ca(2+)-induced Ca2+ release in permeabilised endothelial cells. Biochem Biophys Res Commun 246:543–548

Wood PG, Gillespie JI (1998b) In permeabilised endothelial cells IP3-induced Ca2+ release is dependent on the cytoplasmic concentration of monovalent cations. Cardiovasc Res 37:263–270

Wu S, Sangerman J, Li M, Brough GH, Goodman SR, Stevens T (2001) Essential control of an endothelial cell ISOC by the spectrin membrane skeleton. J Cell Biol 154:1225–1233

Wuytack F, Raeymaekers L, Missiaen L (2002) Molecular physiology of the SERCA and SPCA pumps. Cell Calcium 32:279–305

Xie Q, Zhang Y, Zhai C, Bonanno JA (2002) Calcium influx factor from cytochrome P-450 metabolism and secretion-like coupling mechanisms for capacitative calcium entry in corneal endothelial cells. J Biol Chem 277:16559–16566

Xu W, Liu L, Charles IG, Moncada S (2004) Nitric oxide induces coupling of mitochondrial signalling with the endoplasmic reticulum stress response. Nat Cell Biol 6:1129–1134

Yamada N, Makino Y, Clark RA, Pearson DW, Mattei MG, Guenet JL, Ohama E, Fujino I, Miyawaki A, Furuichi T, et al (1994) Human inositol 1,4,5-trisphosphate type-1 receptor, InsP3R1: structure, function, regulation of expression and chromosomal localization. Biochem J 302:781–790

Yamamoto K, Korenaga R, Kamiya A, Ando J (2000) Fluid shear stress activates Ca(2+) influx into human endothelial cells via P2X4 purinoceptors. Circ Res 87:385–391

Yamamoto N, Watanabe H, Kakizawa H, Hirano M, Kobayashi A, Ohno R (1995) A study on thapsigargin-induced calcium ion and cation influx pathways in vascular endothelial cells. Biochim Biophys Acta 1266:157–162

Yamamoto-Hino M, Sugiyama T, Hikichi K, Mattei MG, Hasegawa K, Sekine S, Sakurada K, Miyawaki A, Furuichi T, Hasegawa M, et al (1994) Cloning and characterization of human type 2 and type 3 inositol 1,4,5-trisphosphate receptors. Receptors Channels 2:9–22

Yao X, Kwan HY, Chan FL, Chan NW, Huang Y (2000) A protein kinase G-sensitive channel mediates flow-induced Ca(2+) entry into vascular endothelial cells. Faseb J 14:932–938

Zeng W, Mak DO, Li Q, Shin DM, Foskett JK, Muallem S (2003) A new mode of Ca2+ signaling by G protein-coupled receptors: gating of IP3 receptor Ca2+ release channels by Gbetagamma. Curr Biol 13:872–876

Zhang H, Inazu M, Weir B, Buchanan M, Daniel E (1994a) Cyclopiazonic acid stimulates Ca2+ influx through non-specific cation channels in endothelial cells. Eur J Pharmacol 251:119–125

Zhang H, Inazu M, Weir B, Daniel E (1994b) Endothelin-1 inhibits inward rectifier potassium channels and activates nonspecific cation channels in cultured endothelial cells. Pharmacology 49:11–22

Zhao M, Li P, Li X, Zhang L, Winkfein RJ, Chen SR (1999) Molecular identification of the ryanodine receptor pore-forming segment. J Biol Chem 274:25971–25974

Ziegelstein RC, Spurgeon HA, Pili R, Passaniti A, Cheng L, Corda S, Lakatta EG, Capogrossi MC (1994) A functional ryanodine-sensitive intracellular Ca2+ store is present in vascular endothelial cells. Circ Res 74:151–156

Eicosanoids and the Vascular Endothelium

K. Egan · G. A. FitzGerald (✉)

Institute for Translational Medicine and Therapeutics, School of Medicine,
University of Pennsylvania, 153 Johnson Pavilion, Philadelphia PA, 19104, USA
garret@spirit.gcrc.upenn.edu

1	Introduction	190
2	The COX Pathway	191
2.1	Molecular Biology of COXs	192
2.2	COX Expression in the Cardiovascular System	192
2.3	PG Isomerase/Synthases	193
2.4	Prostaglandin Receptors	193
2.5	Prostacyclin	194
2.5.1	PGI_2 Effects	195
2.5.2	The IP	196
2.5.3	Studies in IP Transgenic Mice	196
2.6	Thromboxane A_2	197
2.6.1	TXA_2 Effects	198
2.6.2	TXA_2 Receptor	198
2.6.3	Studies in Transgenic Mice	199
2.6.4	Pharmacologic Agents that Act at the TP	199
2.7	Other Prominent Eicosanoids	200
2.7.1	PGD_2	200
2.7.2	PGE_2	200
2.7.3	$PGF_{2\alpha}$	201
3	COX Inhibitors	201
	References	204

Abstract Cyclooxygenase (COX) enzymes catalyse the biotransformation of arachidonic acid to prostaglandins which subserve important functions in cardiovascular homeostasis. Prostacyclin (PGI_2) and prostaglandin (PG)E_2, dominant products of COX activity in macro- and microvascular endothelial cells, respectively, in vitro, modulate the interaction of blood cells with the vasculature and contribute to the regulation of blood pressure. COXs are the target for inhibition by nonsteroidal anti-inflammatory drugs (NSAIDs—which include those selective for COX-2) and for aspirin. Modulation of the interaction between COX products of the vasculature and platelets underlies both the cardioprotection afforded by aspirin and the cardiovascular hazard which characterises specific inhibitors of COX-2.

Keywords Endothelium · Vascular · Prostacyclin · Thromboxane · Cyclooxygenase

1
Introduction

Arachidonic acid (AA) is a 20-carbon fatty acid containing four double bonds (Δ 5, 8, 11, 14: C 20:4) that circulates in plasma in both free and esterified forms. AA is derived from dietary linoleic and linolenic acids. It is esterified in the *sn-2* position of phospholipids from which it is liberated by phospholipases (PL). While these include PLA_1, PLA_2, PLC and PLD, a cytosolic PLA_2 has a particular affinity for AA as a substrate for cleavage (Leslie 1997). PLA_2s hydrolyse the *sn-2* ester bond of membrane phospholipids (particularly phosphatidyl choline and phosphatidyl ethanolamine) with the release of arachidonate (Lin et al. 1992). Multiple additional phospholipase A_2s—group IIa secretory ($sPLA_2$), group V $sPLA_2$, group VI calcium-independent ($iPLA_2$) and group X $sPLA_2$—have been characterised. The cyclooxygenase (COX) enzyme utilizes AA as its preferred substrate to catalyse the formation of prostaglandins (PGs) and thromboxane (TX) (Fig. 1).

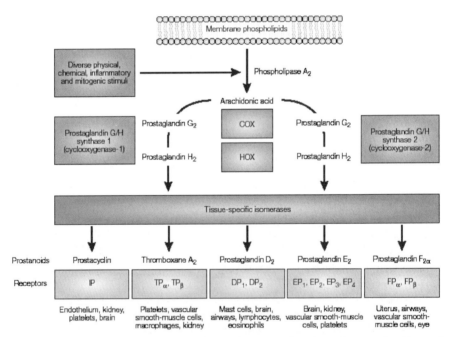

Fig. 1 The cyclooxygenase (*COX*) cascade: production and actions of prostaglandins and thromboxane. Arachidonic acid (AA) is converted by cytosolic prostaglandin G/H synthases, which have both COX and hydroperoxidase (*HOX*) activity, to the unstable intermediate prostaglandin H_2 (PGH_2). PGH_2 is converted by tissue-specific isomerases to multiple prostaglandins that activate specific cell-membrane receptors of the superfamily of G protein-coupled receptors. Some of the tissues in which individual prostanoids exert prominent effects are indicated. (Reproduced with permission from FitzGerald 2003)

2
The COX Pathway

Although commonly referred to as COXs, the PG G/H synthases have two distinct enzyme activities, namely an endoperoxide synthase activity (COX) that oxygenates and cyclises the unesterified precursor fatty acid, and a hydroperoxidase activity (HOX) that sequentially forms the cyclic endoperoxides G (PGG) and H (PGH). PGG_2 and PGH_2 are chemically unstable, but are rapidly transformed by the downstream prostaglandin isomerases to produce the various PGs (Fig. 1). The PGs are bisenoic products that contain two double bonds, denoted by a subscript 2 (e.g. PGE_2). They belong to the larger family of products of AA termed eicosanoids as they contain 20 carbon atoms (Greek: eikosi/εικωσι=20). Products formed from substrates with different numbers of double bonds differ in their subscript. Thus, PGs formed from eicosapentaenoic acid (C20:5), which has one more double bond than arachidonic acid, form PGE_3. The isoeicosanoids, a family of free radical catalysed isomers, are formed by non-enzymatic, direct peroxidation of AA in situ in cell membranes (Pratico et al. 2004).

The COX enzyme exists in two forms—COX-1 and COX-2 (Smith et al. 1996)—and occurs as a dimer (Picot et al. 1994; Garavito et al. 1995; Loll et al. 1995), monotopically inserted into the endoplasmic reticulum membrane. Recent evidence suggests that COX-1 and COX-2 may also heterodimerize (Yu et al. 2006). AA can access the active site in the body of the enzyme via a hydrophobic tunnel from the endoplasmic reticulum-bound surface of the enzyme. Access to the active site is more accommodating in COX-2, although many of the key residues for catalysis are conserved between the two isoforms. A side pocket in the tunnel affords the structural opportunity for the synthesis of inhibitors which inhibit selectively COX-2 (Smith et al. 1996; FitzGerald and Loll 2001). The crystal structures of COX-1 and COX-2 are similar, with a 61% amino acid identity, and both enzymes have similar kinetics for AA (Smith et al. 1996). COX-1 is localised predominantly to the endoplasmic reticulum, whereas COX-2 is present in both the endoplasmic reticulum and the nuclear membrane (Spencer et al. 1998). COX-1 and COX-2 may use different pools of arachidonate that are mobilised in response to different cellular stimuli for PG synthesis (Spencer et al. 1998) and may display differential affinity for downstream synthases, at least in heterologous systems. These preferences may be governed by spatiotemporal associations and quantitative provision of substrate (Ueno et al. 2001). Traditionally, COX-1 is viewed as a constitutive enzyme with housekeeping functions, such as gastric epithelial cytoprotection and haemostasis, whereas COX-2 has been deemed inducible, particularly at sites of inflammation. It is now apparent that this is an oversimplification of biological reality. For example, expression of both enzymes (COX-1 and COX-2) is evident in the brain and kidney (FitzGerald and Patrono 2001) and both are upregulated in the synovia of inflamed joints (Iniguez et al.

1998). Specifically, with reference to the present chapter, COX-2 is induced in endothelial cells by laminar shear, suggesting haemodynamic induction of the enzyme in endothelium in vivo (Topper et al. 1996).

2.1
Molecular Biology of COXs

The COX-1 gene is located on chromosome 9, whereas COX-2 is located on chromosome 1 (Smith et al. 1996). Characteristic features of the COX-1 gene are consistent with its being suitable for rapid transcription and messenger RNA (mRNA) processing, thus producing a continuously transcribed, stable message (Smith et al. 1996). This provides a constant level of enzyme in most cell types to synthesise PGs responsible for homeostatic functions. In contrast, the features of the COX-2 gene are those of an "immediate-early" gene that is not always present, but is highly regulated—for example by cytokines or mitogens. The COX-2 gene is smaller than that of COX-1, with exons 1 and 2 of COX-1 (containing the translation site and original peptide) condensed into a single exon in COX-2. Additionally, the introns of COX-2 are smaller than those of COX-1, and COX-2 has a TATA box promoter, which is lacking in COX-1. Lastly, the mRNA of COX-2 contains long 3′ untranslated regions, which exhibit several different polyadenylation signals and multiple "AUUUA" instability sequences that act to mediate rapid degradation of the transcript. Insight into the mechanisms which regulate expression of the COXs continues to emerge. Thus, both post-transcriptional and post-translational mechanisms appear to converge with transcriptional regulation in determining the altered expression of COX-2 (Dixon 2004; Cok et al. 2003).

2.2
COX Expression in the Cardiovascular System

In cardiovascular tissues, COX-1 is constitutively expressed in cultured endothelial (ECs) and vascular smooth muscle cells (VSMCs) under static conditions in vitro, while the expression of COX-2 is increased by growth factors, cytokines, phorbol esters and lipopolysaccharide in many cell types, including those of the vasculature (Herschman 1996). Laminar shear upregulates COX-2 expression in ECs in vitro (Topper et al. 1996), while disturbed shear, designed to mimic the disordered laminar shear at sites of atherogenesis, fails to have this effect. The offset kinetics of shear-induced COX-2 expression in endothelial cells is unknown. Thus, it is perhaps unsurprising that COX-2 has been variably detected in endothelial cells ex vivo after tissues have been harvested and stained for immunodetection or in situ analysis. Experiments in humans suggest that COX-2 is the dominant source of PGI_2 produced even under physiological conditions (McAdam et al. 1999; Catella-Lawson et al. 1999). Expression of vascular COX-2 is elevated in response to injury in vivo

(Connolly et al. 2002) and expression of both COX-1 and COX-2 is increased in foam cells and in VSMC in atherosclerotic plaque (Schonbeck et al. 1999).

2.3
PG Isomerase/Synthases

Cell-specific isomerases and synthases occur in different tissues and determine the terminal PG produced—usually one or two products by a particular cell (see Fig. 1). For example, prostacyclin synthase (PGIS) renders PGI_2 the dominant PG of macrovascular ECs (Gryglewski et al. 1986; Spisni et al. 1995). In contrast, platelets produce TXA_2 because TX synthase predominates in those cells. As mentioned, the COX isoforms preferentially co-localise with a particular synthase, at least in heterologous systems. Thus, COX-1 couples preferentially with TXA_2 and PGF synthases (Ueno et al. 2001) and the cytosolic PGE synthase (PGES) isozymes (Tanioka et al. 2000). COX-2 preferentially couples with PGIS (Ueno et al. 2001) and the microsomal PGES isozymes (Murakami et al. 2000), which are also induced by cytokines and tumour promoters. Co-localisation of the respective COX and PGES enzymes is evident in vivo during development in zebrafish (Pini et al. 2005). However, this may not reflect completely the situation in vivo. For example, both COX enzymes contribute to microsomal PGE synthase (mPGES)-1-derived PGE_2 in vivo (Cheng et al 2006).

2.4
Prostaglandin Receptors

PGs have short half-lives (seconds to minutes at physiological pH) and, as such, act as autacoids at nearby membrane G protein-coupled receptors (GPCR), rather than as circulating hormones (FitzGerald et al. 1981; Breyer et al. 2001). While formed intracellularly, PGs may diffuse through cell membranes or be actively transported (Bao et al. 2002) to activate membrane GPCRs. PGs may also undergo active transport intracellularly for catabolism (Vezza et al. 2001). There are nine PG receptor subtypes which are conserved in mammals from mouse to human: PGD_2 receptors (D prostanoid (DP)1 and CRTH2 or DP2), PGE_2 receptor (EP; EP1, EP2, EP3 and EP4), the $PGF_{2\alpha}$ receptor (FP), the PGI_2 receptor (IP) and the TXA_2 receptor (TP) (Narumiya et al. 1999). Each receptor is encoded by a separate gene, although splice variants of each gene may occur, as exemplified amongst the EP3, FP and TP receptors in which C terminal variants have been reported. All PGs act at GPCRs and transduce their diverse responses through second messenger systems such as adenylate cyclase or phospholipase C. All derive from an ancestral EP, with the exception of the DP2, which is a member of the N-formyl-L-methionyl-L-leucyl-L-phenylalanine (fMLP) receptor superfamily. PGs may also activate nuclear receptors (Lim and Dey 2002). However, it remains speculative that sufficient PG concentrations are attained in vivo to effect these responses (Bell-Parikh et al. 2003).

2.5
Prostacyclin

The final step in PGI$_2$ synthesis is the isomerisation of PGH$_2$ by PGIS. The gene encoding PGIS, a cytochrome P450 haemoprotein, is approximately 70 kb long. Expression of PGIS mRNA is upregulated by several cytokines and hormones (e.g. oestrogen) and downregulated by acrolein (N. Volkel, personal communication). Peroxynitrite selectively inhibits PGIS by post-translational modification of the enzyme (Zou et al. 1997). PGI$_2$ synthesis occurs most notably in ECs and VSMCs (Moncada et al. 1976), but it also occurs in heart, kidney, gastric mucosa, macrophages, lung, brain and small intestine. Inhibition of urinary PGI$_2$ metabolite (PGI-M) excretion by structurally distinct selective inhibitors of COX-2—rofecoxib and celecoxib—is indistinguishable from that by structurally distinct mixed inhibitors—ibuprofen and indomethacin (Catella-Lawson et al. 1999; McAdam et al. 1999). However, while this indicates that COX-2 is likely to be the dominant source of endothelial PGI$_2$ in vivo, COX-1 may also contribute to EC biosynthesis. Selective inhibition of COX-2 in mice accelerates the response to a thrombogenic stimulus in vivo (Cheng et al. 2006).

PGI$_2$ has a double-ring structure; a cyclopentenone ring and a second ring formed by an oxygen bridge between carbons 6 and 9 (Fig. 2). It is hydrol-

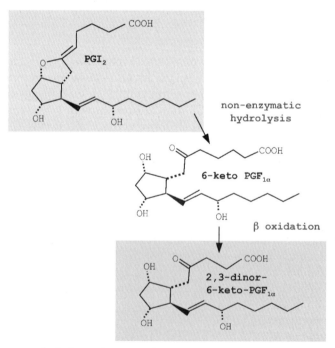

Fig. 2 Spontaneous hydrolysis of prostacyclin (PGI$_2$) to form the 6-keto hydrolysis product and enzymatic formation of the 2,3-dinor metabolite

ysed non-enzymatically ($t_{1/2}$=3 min) to the inactive 6-keto-PGF$_{1\alpha}$. The major route of elimination of PGI$_2$ is in the urine. Measurement of a major urinary metabolite (Brash et al. 1983), such as 2,3 dinor 6-keto-PGF$_{1\alpha}$, provides a time-integrated, non-invasive index of the total biosynthesis of PGI$_2$ in vivo (FitzGerald et al. 1983; Pratico et al. 2000a).

2.5.1
PGI$_2$ Effects

Isolated from vascular tissue initially, PGI$_2$ is a potent vasodilator and inhibitor of platelet aggregation in vitro (Moncada et al. 1976; Moncada et al. 1977). It was shown also to promote inflammation and regulate salt and water handling by the kidney, suggesting a pervasive role in cardiovascular dysfunction (Moncada and Vane 1978). The recognition of these properties prompted consideration of the use of PGI$_2$ or more stable analogues as therapeutic agents, initially in patients with peripheral obstructive arterial disease (Szczeklik et al. 1979). More recently, intravenous, oral and inhaled PGI$_2$ analogues have established a place in the therapy of pulmonary hypertension (Gibbs et al. 2004). PGI$_2$ also appears to modulate the pulmonary response to viral infection. Overexpression of PGIS constrains the response to respiratory syncytial virus (RSV) infection in mice, and weight loss, cytokine response and delayed viral clearance in response to RSV are all exacerbated in mice lacking the IP (Hashimoto et al. 2004).

Recently, interest has developed in the anti-oxidant effects of PGI$_2$. Thus, an analogue has been shown to modulate the oxidant stress caused by doxorubicin in cardiomyocytes in vitro (Adderley and Fitzgerald 1999) and in vivo (Dowd et al. 2001). Similarly, infusion of an analogue into patients with pulmonary hypertension—a disease in which both platelet activation and oxidant stress have been implicated—depressed isoprostane generation, but not TX metabolite excretion (Robbins et al. 2005). Deletion of the IP exacerbates the oxidant injury of ischaemia/reperfusion (Xiao et al. 2001). More recently, Egan and colleagues (2004) have implicated loss of an antioxidant effect mediated by the IP in the accelerated atherogenesis in female mice lacking both the low-density lipoprotein (LDL) receptor and the IP (Egan et al. 2004). Indeed, COX-2-derived PGI$_2$ may contribute substantially to the antioxidant effects of oestrogen in this model. Infusion of PGI$_2$ reduces blood pressure, but paradoxically elevates renin (FitzGerald et al. 1979). Excretion of PGI-M is markedly elevated in human pregnancy, a state of constitutive hypotension and elevated renin. Interestingly, the elevation of PGI-M is less pronounced from the first trimester in those destined to develop pregnancy-induced hypertension, a low renin condition manifest typically in the late second or third trimester (Fitzgerald et al. 1987).

Anti-mitogenic effects of PGI$_2$ have been demonstrated in smooth muscle cells in vitro (Zucker et al. 1998) and studies in mice deficient in the IP have demonstrated the antiproliferative effect of PGI_2 in the response to vascular injury (Cheng et al. 2002). Similarly, viral delivery of the PGIS enzyme ameliorates the response to vascular injury in rats, preventing intimal hyperplasia following balloon carotid injury (Todaka et al. 1999). Finally, PGI_2 enhances reverse cholesterol transport from vascular cells in vitro by modulating cholesterol ester hydrolase (Hajjar et al. 1982) and impairs cellular adhesion to the vessel wall (Kobayashi et al. 2004). The antiproliferative effect of high-density lipoprotein on VSMC in vitro is mediated via its apoE moiety that induces COX-2-dependent PGI_2, acting via the IP to inhibit cyclin A (Kothapalli et al. 2004).

2.5.2
The IP

A single prostanoid receptor, the IP, has been identified and cloned (Namba et al. 1994; Boie et al. 1994). The human IP gene encodes a GPCR protein consisting of 386 amino acid residues. It is located on chromosome 19 and spans a total of 7 kb. The IP typically couples to Gs and thus elevates cyclic adenosine monophosphate (cAMP). In addition, the IP may also couple to G_q and thus activate phospholipase C. Elevated cAMP stimulates ATP-sensitive K^+ channels to cause hyperpolarisation of the cell membrane and inhibit development of contraction of vascular smooth muscle. Elevated cAMP levels also decrease cytosolic Ca^{2+}, inhibiting contractile machinery (Smyth and FitzGerald 2002). Wilson and colleagues have demonstrated that the IP can undergo homodimerisation and heterodimerisation with the TP, with consequent alterations in ligand affinity and signalling patterns (Wilson et al. 2004).

IP mRNA is abundantly expressed in mouse megakaryocytes and arterial smooth muscle, consistent with the actions of PGI_2 in the cardiovascular system. In addition, IP mRNA can be found in the thymus, kidney, heart, liver, spinal column and particularly in neurons of the dorsal root ganglion, indicating a role for the IP in the mediation of pain.

2.5.3
Studies in IP Transgenic Mice

IP knockout mice are viable, normotensive and reproduce normally (Murata et al. 1997). These mice revealed the importance of the IP in mediating both pain and inflammation (Murata et al. 1997), properties shared with the EP1 and the EP3 (Minami et al. 2001). Similarly, the IP, this time acting in concert with the EP2 and EP4, contributes to joint inflammation in collagen-induced arthritis in the mouse (Honda et al. 2006). As expected, IP knockout mice are more sensitive to thrombogenic stimuli (Murata et al. 1997) and exhibit an increased proliferative response to vascular injury and an augmentation in the attendant platelet activation. This last phenotype is rescued by coincident deletion of the TP (Cheng et al. 2002). Deletion of the IP elevates blood pressure on some backgrounds and increases the response to salt loading (Francois et al. 2005). It is, in this phenotype, reminiscent of deletion of the EP2 and EP4 receptors (Tilley et al. 1999; Audoly et al. 1999). However, consistent with its effect on renin, deletion of the IP reduces blood pressure in hyper-reninaemic mice in which the renal artery has been clipped (Fujino et al. 2004). More recently, the atheroprotective effect of the IP has been demonstrated in both the apoE and LDL knockout mouse models of atherosclerosis, where IP deletion accelerated atherogenesis (Egan et al. 2004; Kobayashi et al. 2004). A synthesis of the information in these papers suggests that acceleration of interactions between neutrophils, platelets and the vessel wall and, in particular, the attendant increase in oxidant stress, mediates the impact of IP deletion on initiation and early development of atherogenesis. These observations are concordant with the failure of turbulent shear stress in vitro to upregulate COX-2 in ECs (Topper et al. 1996), mimicking the potential functional deficiency in COX-2-derived PGI_2 formation secondary to the disturbed laminar shear that pertains at vascular sites prone to the initial development of atherosclerosis in humans. Disruption of this pathway may contribute to the time-dependent increase in cardiovascular hazard and risk transformation which appears to complicate extended dosing with selective inhibitors of COX-2 (*vide infra*).

2.6
Thromboxane A_2

The final step in the synthesis of TXA_2 is the isomerisation of PGH_2 to TXA_2 by TX synthase (Needleman et al. 1976). The TX synthase gene is found on chromosome 7 and spans greater than 150 kb with 13 exons (Tanabe et al. 1993). TXA_2 is the principal metabolite of COX-1-derived metabolism of AA in platelets (Hamberg et al. 1975). Either COX may generate TX as a principal product. Thus, COX-2 expressed in macrophages may contribute to TXA_2 biosynthesis and has been speculated to contribute to the syndrome of aspirin resistance (Patrignani 2003). Recently, Evangelista et al. (2006) have demonstrated that de novo synthesis of COX-1 by platelets, at least in vitro, may undermine the sustained and complete inhibition of platelet COX-1 derived TxA_2 by aspirin. In humans, TX synthase mRNA is found in platelets, lung, placenta, kidney, spleen, thymus, prostate gland and peripheral blood leucocytes. Similar to other enzymes in the AA biosynthetic cascade, TX synthase is subject to mechanism-based inactivation (Fitzpatrick et al. 2004).

TXs have a six-member oxirane ring, differing from the cyclopentenone ring in conventional PGs (Fig. 3). TXA_2 breaks down non-enzymatically into the stable inactive hydrolysis product, thromboxane B_2 (TXB_2) with a half-life of roughly 30 s at physiological pH (Bhagwat et al. 1985). Urinary excretion of products of the two major pathways of TX disposition (Roberts et al. 1977), 2,3 dinor TXB_2 and 11-dehydro TXB_2, reflect biosynthesis of TX in vivo (Lawson et al. 1985; Catella et al. 1986).

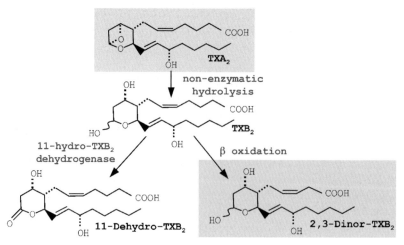

Fig. 3 Spontaneous hydrolysis of thromboxane A_2 (TXA$_2$) to form TXB$_2$ and subsequent metabolism to 11-dehydro- and 2,3-dinor-TXB$_2$

2.6.1
TXA$_2$ Effects

TXA$_2$ activates platelets, but also acts to amplify the response to more potent agonists, such as thrombin (FitzGerald 1991). TXA$_2$ is a potent vasoconstrictor (Dorn et al. 1987) and causes proliferation (Pakala et al. 1997) and hypertrophy (Ali et al. 1993) of VSMC in vitro.

2.6.2
TXA$_2$ Receptor

The human TXA$_2$ receptor (TP) was the first receptor of the prostaglandin/eicosanoid pathway to be cloned (Hirata et al. 1991). The human TP is derived from a single gene located on chromosome 19 that spans 15 kb and has 3 exons divided by 2 introns (Nusing et al. 1993). Splice variants occur of the human, but not the mouse, TP that differ particularly in their cytoplasmic tails and thus their subsequent G protein signalling specificities. TPα is probably the sole isoform expressed as a protein in platelets (Habib et al. 1999). TPβ, originally cloned from ECs, appears to limit angiogenesis in part by disrupting the actions of vascular endothelial growth factor (VEGF) (Ashton and Ware 2004). Additional to its natural ligands, TXA$_2$ and PGH$_2$, the TP can be ligated and activated by infusion of the isoprostanes, iPF$_{2\alpha}$-III and iPE$_2$-III in vivo (Audoly et al. 2000). It is unknown whether sufficient endogenous concentrations of these iPs, or indeed other lipid peroxidation products (Li et al 2006), accumulate in settings of disease to activate the TP or other eicosanoid receptors (Kunapuli et al. 1997). The TP isoforms can heterodimerise in vitro, increasing their affinity for activation by iPF$_{2\alpha}$-III. Heterodimerisation with the IP converts classical G$_q$-dependent signalling to a preference for G$_s$. The TP can also signal via G11, G12, G13 and the G protein/tissue transglutaminase, Gh (Warumiya et al. 1999; Zhang et al. 2003).

TP mRNA is expressed in tissues rich in vasculature, such as the lung, kidney and heart, as well as in thymus and spleen and in spinal chord. Immature thymocytes express the TP at a density as high as that in platelets (Namba et al. 1992). TP activation modulates acquired immunity in vivo by negatively regulating dendritic cell–T cell interactions (Kabashima et al. 2003). A naturally occurring mutation in the first intracellular loop of the TP is associated with a mild bleeding disorder and platelet resistance to TP agonists (Hirata et al. 1996), while a polymorphism in the TP has been linked to bronchodilator resistance in asthma (Unoki et al. 2000).

2.6.3
Studies in Transgenic Mice

Only the TPα isoform of the TP is expressed in mice. TP knockout mice exhibit a mild bleeding tendency and are resistant to platelet aggregation by TP agonists (Thomas et al. 1998). Similarly, deletion of TX synthase results in a mild bleeding disorder and resistance to AA-induced sudden death (Yu et al. 2004). TP antagonism or deletion decreases the vascular proliferative response to catheter-induced injury in the mouse, while directed vascular overexpression of TPβ augments the response to injury (Cheng et al. 2002). Overexpression of TPβ also results in a syndrome reminiscent of intrauterine growth retardation (IUGR), probably secondary to placental ischaemia (Rocca et al. 2000), and biosynthesis of TXA$_2$ is markedly elevated in patients with severe pregnancy-induced hypertension and with IUGR (Fitzgerald et al. 1990). TP antagonism or deletion retards atherogenesis in murine models of atherosclerosis (Cayatte et al. 2000; Egan et al. 2005). Finally, in addition to its effects on dendritic cell–T lymphocyte interactions, deletion of the TP modulates the immune-mediated inflammatory response to tissue transplantation (Thomas et al. 2003).

2.6.4
Pharmacologic Agents that Act at the TP

TP antagonists were developed by several companies and shown to be well tolerated with a modest effect on cutaneous bleeding time and inhibition of TP-dependent platelet aggregation ex vivo. Unfortunately, their introduction into clinical trials coincided with the emergence of evidence for the efficacy of low-dose aspirin in the secondary prevention of heart attack and stroke. Thus, most programmes were abandoned for economic reasons—"a more expensive aspirin"—particularly when two clinical trials failed to demonstrate superiority of TP antagonists over aspirin in prevention of delayed cardiovascular events or radiological evidence of restenosis in patients who underwent angioplasty. However, the emergence of information suggesting that lipid peroxidation products may activate TPs and that suppression of PGI_2 may have adverse cardiovascular consequences in vivo has renewed interest in this therapeutic approach (Pratico et al. 2000a; Dogne et al. 2004; Morrow 2006). Thus, new programmes of development are being considered where oxidant stress and COX activation coincide, such as in the treatment of atherosclerosis and transplant rejection, and in settings where the depression of PGI_2 by even low doses of aspirin (Clarke et al. 1991) may be undesirable. This would include the use of antagonists as adjunctive cardioprotective therapy with selective inhibitors of COX-2 or where pharmacodynamic interactions might preclude aspirin use for cardioprotection, such as during chronic therapy with ibuprofen-like drugs (Catella-Lawson et al. 2001). Molecules which share TP antagonism with other properties, such as inhibition of TX synthase or antagonism of DP2, are also under consideration (Ishizuka et al. 2003, 2004; Hanson et al. 2005).

2.7
Other Prominent Eicosanoids

2.7.1
PGD_2

A major prostaglandin product of mast cells (Roberts et al. 1980), PGD_2 is released upon their activation due to allergic and other stimuli (Sladek et al. 1991). PGD_2 is also a COX product in platelets, albeit a minor one. However, albumin possesses a PGD isomerase activity and can enhance PGD formation in platelets treated with TX synthase inhibitors (Patscheke 1985). PGD_2 formation is increased (along with other PGs) in platelets of TX synthase-deleted mice (Yu et al. 2004). PGD_2 acts on the DP1 via Gs to elevate platelet cAMP (Wright et al. 1998). Expression of the DP2 is evident in eosinophils, basophils and T lymphocytes, where its ligation results in an elevation in intracellular calcium (Nagata and Hirai 2003). However, it is less potent than PGI_2 and PGE_2 as an inhibitor of platelet activation (Moncada et al. 1977). Nonetheless, the relevance of PGD_2 to cardiovascular biology in vivo is largely unknown. The emergence of selective agonists and antagonists for DP1 and DP2 and the availability of mice lacking these receptors promise to clarify this situation.

2.7.2
PGE_2

PGE_2 mediates pain and inflammation as well as the febrile response (Hata and Breyer 2004). Along with PGI_2, it induces diuresis and natriuresis (Fleming et al. 1998). PGE_2 is the predominant COX product of microvascular endothelial cells in vitro (Gerritsen 1987) and like PGI_2, is a dominantly product of COX-2 under physiological conditions in vivo (Murphey et al. 2004).

PGE$_2$ activates four receptor subtypes; EP1 and EP3 are coupled via G$_q$ to elevation of intracellular calcium while EP2 and EP4 signal predominantly via G$_s$. Activation of these receptors by varied concentrations of PGE$_2$ may exert contrasting biological effects—vasodilatation via EP2 and vasoconstriction via EP1. Although much information has been derived from mice lacking these receptors (Narumiya and FitzGerald 2001), the phenotypic response to gene deletion can be quite strikingly conditioned by genetic background (Austin and FitzGerald 1999). PGE$_2$ may also activate other prostanoid receptors. Thus, low concentrations of PGE$_2$ activate platelets via EP3 and perhaps EP1, while higher concentrations inhibit platelet aggregation by ligating the IP (Fabre et al. 2001). Activation of EP3 receptors causes contraction of intestinal smooth muscle, inhibition of gastric acid secretion, increased gastric mucus secretion, inhibition of lipolysis, inhibition of autonomic neurotransmitter release and stimulation of contraction of the pregnant uterus (Narumiya and FitzGerald 2001). Both EP1 and EP3 mediate the febrile response to administered lipopolysaccharide and turpentine in mice, but do not contribute to the circadian variation in body temperature (Oka et al. 2003), despite recent evidence that PGE$_2$ may contribute to regulation of the molecular clock (Tsuchiya et al. 2005). Both EP1 and EP3 receptors mediate vasoconstriction (Jadhav et al. 2004) and recently the EP3 gene has been associated with a severe phenotype (i.e. progression to surgery) of peripheral vascular disease, prompting development of an EP3 antagonist for this disease. Deletion of the EP4 delays closure of the ductus arteriosus (Nguyen et al. 1997). Activation of the EP4 has been implicated in both atherosclerotic plaque progression (Cipollone et al. 2004) and destabilisation (Takayama et al. 2002).

2.7.3
PGF$_{2\alpha}$

Activation of vascular FP by PGF$_{2\alpha}$ elevates blood pressure and induces VSMC proliferation (Fujino et al. 2002). However, although FP-deficient mice have nicely delineated the role of this receptor in parturition (Narumiya and FitzGerald 2001), little information is available as to its importance in vascular biology. Variants of the FP have been described (Vielhauer et al. 2004).

3
COX Inhibitors

non-steroidal anti-inflammatory drugs (NSAIDs) and aspirin target the COX enzymes for inhibition. The pharmacology of these drugs has been reviewed elsewhere (Marnett et al. 1999; Patrono et al. 2004; Burke et al. 2005). Briefly, NSAIDs are traditional competitive active site inhibitors that have a reversible inhibitory effect on the enzymes, while aspirin irreversibly targets Ser[529] close to the active site of the enzyme, obstructing access of the substrate, AA (Funk et al. 1991). Aspirin reduces the secondary incidence of myocardial infarction and stroke in men and women via inhibition of platelet COX-1-derived TXA_2, predominantly acting in the presystemic circulation (Pedersen and FitzGerald 1984). Low doses of aspirin used for cardioprotection (<100 mg/day) preferentially inhibit COX-1, while higher, anti-inflammatory doses inhibit both COX enzymes. A direct randomised comparison of the cardioprotective effects of aspirin has not been performed, although indirect comparisons suggest an inverse dose-related cardioprotective effect (Antithrombotic Trialists' Collaboration 2002) and a direct, dose-related increase in adverse gastrointestinal effects (Patrono et al. 2004). Aspirin is also effective in the primary prevention of myocardial infarction in men and of stroke in women (Ridker et al. 2005; Physicians' Health Study 1989). This apparent gender-specific distinction may merely reflect the relative incidence of myocardial infarction and stroke amongst the sexes, with the impact of aspirin being detectable—when the absolute incidence of any events is so low—only in the more prevalent condition. Aspirin does not appear to differentially inhibit platelet function in men vs women (Becker DM et al. 2006). In both cases, the small benefit is offset by a roughly twofold increase in the incidence of serious gastrointestinal bleeds (Patrono et al. 2004).

Given that TXA_2 is only one of several platelet agonists, it is unsurprising that some patients taking aspirin suffer myocardial infarction or stroke. Such treatment failures have commonly been grouped in a syndrome of "aspirin resistance" (Patrignani 2003; Mason et al. 2004). However, at present there is little evidence that integrates some stable biochemical, genetic or functional measurement of "resistance" to clinical outcome (Hennekens et al. 2004).

NSAIDs are used widely for relief of pain and inflammation, but gastrointestinal complications have limited their efficacy. These have been ascribed to inhibition of COX-1-derived PGE_2 and PGI_2, which afford cytoprotection in gastroduodenal endothelium, and predisposition to bleeding consequent to inhibition of platelet COX-1-derived TXA_2. The PGs which predominantly mediate pain and inflammation (PGE_2 and PGI_2) are assumed to derive predominantly from COX-2, providing the rationale for development of specific inhibitors of COX-2, such as the coxibs (FitzGerald and Patrono 2001).

While these drugs have never been tested to determine if they afford greater (or less) efficacy than traditional (t)NSAIDs, two coxibs, rofecoxib and lumiracoxib, have been shown to result in a reduced incidence of serious adverse gastrointestinal effects at doses which are equi-efficacious with tNSAIDs (Bombardier et al. 2000; Schnitzer et al. 2004). However, five placebo-controlled trials with three members of this class—valdecoxib, celecoxib and rofecoxib—have demonstrated that they elevate the incidence of heart attack and stroke (Wong et al. 2005). This appears to result from depression of COX-2-derived PGI_2 without a concomitant effect on platelet TXA_2, as mature platelets do not contain COX-2 (McAdam et al. 1999; FitzGerald 2004). Selectivity for COX-2 is relative rather than absolute, and some tNSAIDs, such as diclofenac and meloxicam, exhibit selectivity (at least in vitro) similar to that of celecoxib, the least selective of the purpose-designed COX-2 inhibitors. Placebo-controlled trials have not been performed to assess the effects of such tNSAIDs on the cardiovascular system.

A further complication with tNSAID therapy may result from drug–drug interactions. Prior occupancy of the active site of platelet COX-1 by ibuprofen prevents access of aspirin to afford the sustained inhibition of platelet TXA_2 thought to be intrinsic to its property of cardioprotection (Catella-Lawson et al. 2001). Some clinical evidence consistent with this interaction—which probably involves similar drugs, such as flurbiprofen and indomethacin—has emerged (Schnitzer et al. 2004). Epidemiological analysis of the impact of ibuprofen alone suggests that it neither increases nor decreases the risk of myocardial infarction (Garcia-Rodriguez et al. 2004). Naproxen, by contrast, appears to afford some protection, albeit with a less pronounced signal than for aspirin. This may reflect the variable pharmacokinetics of naproxen, which in some, but not all, individuals sustains inhibition of TXA_2 formation to result in platelet inhibition throughout the dosing interval (Capone et al. 2004). Recently, evidence has emerged to suggest a similar pharmacodynamic interaction between naproxen and low-dose aspirin as previously observed for ibuprofen (Capone et al. 2005). Given the potential protective effect in some, but not all, patients treated chronically with naproxen, the implications of such an interaction would be less pronounced than for ibuprofen. Both ibuprofen and naproxen preferentially inhibit COX-1 in vitro (FitzGerald and Patrono 2001), consistent with their failure to exhibit a cardiovascular hazard in epidemiological studies when they are used without the complicating feature of co-therapy with aspirin.

Currently, rofecoxib and valdecoxib have been withdrawn from the market and Celebrex is subject to substantial restriction in the United States. The recognition of hazard for these drugs seems rational, given its biological plausibility (FitzGerald 2003) and evidence of cardiovascular risk from structurally distinct COX-2 inhibitors from placebo-controlled, randomised clinical trials. Indeed, there is a remarkable congruence between the mechanistic data derived from studies in humans and mice and the nature of the information that has emerged from studies in large populations—both through pharmacoepidemiology and randomised clinical trials (Grosser et al. 2006).

Acknowledgements Dr. FitzGerald is the Bobst Professor of Pharmacology and is supported by grants (HL 70128, HL 62250, HL 54500 and RR 00040) from the National Institutes of Health. Dr. Egan was supported by a grant from the American Heart Association.

References

Adderley SR, Fitzgerald DJ (1999) Oxidative damage of cardiomyocytes is limited by extracellular regulated kinases 1/2-mediated induction of cyclooxygenase-2. J Biol Chem 274:5038–5046

Ali S, Davis MG, Becker MW, et al (1993) Thromboxane A2 stimulates vascular smooth muscle hypertrophy by up-regulating the synthesis and release of endogenous basic fibroblast growth factor. J Biol Chem 268:17397–17403

Antithrombotic Trialists, Collaboration (2002) Collaborative meta-analysis of randomised trials of antiplatelet therapy for prevention of death, myocardial infarction, and stroke in high risk patients. BMJ 324:71–86

Ashton AW, Ware JA (2004) Thromboxane A2 receptor signaling inhibits vascular endothelial growth factor-induced endothelial cell differentiation and migration. Circ Res 95:372–379

Audoly LP, Tilley SL, Goulet J, et al (1999) Identification of specific EP receptors responsible for the hemodynamic effects of PGE2. Am J Physiol 277:H924–H930

Audoly LP, Rocca B, Fabre JE, et al (2000) Cardiovascular responses to the isoprostanes iPF2alpha-III and iPE2-III are mediated via the thromboxane A2 receptor in vivo. Circulation 101:2833–2840

Austin SA, FitzGerald GA (1999) Not a mouse stirring: deletion of the EP2 and love's labor's lost. J Clin Invest 103:1481–1482

Bao Y, Pucci ML, Chan BS, et al (2002) Prostaglandin transporter PGT is expressed in cell types that synthesize and release prostanoids. Am J Physiol Renal Physiol 282:F1103–F1110

Becker DM, Segal J, Vaidya D, Yanek LR, Herrera Galeano JE, Bray PF, Moy TF, Becker LC, Faraday N (2006) Sex differences in platelet reactivity and response to low-dose aspirin therapy. JAMA 295:1420–1427

Bell-Parikh C, Ide T, Lawson JA, et al (2003) Biosynthesis of 15-deoxy-delta12,14-PGJ2 and the ligation of PPARgamma. J Clin Invest 112:945–955

Bhagwat SS, Hamann PR, Still WC, et al (1985) Synthesis and structure of the platelet aggregation factor thromboxane A2. Nature 315:511–513

Boie Y, Rushmore TH, Darmon-Goodwin A, et al (1994) Cloning and expression of a cDNA for the human prostanoid IP receptor. J Biol Chem 1994 269:12173–12178

Bombardier C, Laine L, Reicin A, et al (2000) Comparison of upper gastrointestinal toxicity of rofecoxib and naproxen in patients with rheumatoid arthritis. VIGOR Study Group. N Engl J Med 343:1520–1528

Brash AR, Jackson EK, Saggese CA, et al (1983) Metabolic disposition of prostacyclin in humans. J Pharmacol Exp Ther 226:78–87

Breyer RM, Bagdassarian CK, Myers SA, et al (2001) Prostanoid receptors: subtypes and signaling. Annu Rev Pharmacol Toxicol 41:661–690

Burke A, Smyth EM, FitzGerald GA (2005) Analgesic-anti-pyretic and anti-inflammatory agents and drugs employed in the treatment of gout. In: Brunton LL, Lazo JS, Parker KL, et al (eds) Goodman and Gilman's the pharmacological basis of therapeutics, 11th edn. McGraw-Hill, New York, 673–715

Capone M, Tacconelli S, Sciulli MG, et al (2004) Clinical pharmacology of platelet, monocyte, and vascular cyclooxygenase inhibition by naproxen and low-dose aspirin in healthy subjects. Circulation 109:1468–1471

Capone M, Sciulli MG, Tacconelli S, et al (2005) Pharmacodynamic interaction of naproxen with low-dose aspirin in healthy subjects. J Am Coll Cardiol 45:1295–1301

Catella F, Healy D, Lawson JA, et al (1986) 11-Dehydrothromboxane B2: a quantitative index of thromboxane A2 formation in the human circulation. Proc Natl Acad Sci U S A 83:5861–5865

Catella-Lawson F, McAdam B, Morrison BW, et al (1999) Effects of specific inhibition of cyclooxygenase-2 on sodium balance, hemodynamics, and vasoactive eicosanoids. J Pharmacol Exp Ther 289:735–741

Catella-Lawson F, Reilly MP, Kapoor SC, et al (2001) Cyclooxygenase inhibitors and the antiplatelet effects of aspirin. N Engl J Med 345:1809–1817

Cayatte AJ, Du Y, Oliver-Krasinski J, et al (2000) The thromboxane receptor antagonist S18886 but not aspirin inhibits atherogenesis in apo E-deficient mice: evidence that eicosanoids other than thromboxane contribute to atherosclerosis. Arterioscler Thromb Vasc Biol 20:1724–1728

Cheng Y, Austin SC, Rocca B, et al (2002) Role of prostacyclin in the cardiovascular response to thromboxane A2. Science 296:539–541

Cheng Y, Wang M, Yu Y, Lawson J, Funk C, FitzGerald GA (2006) Cyclooxygenases, microsomal prostaglandind E synthase-1 and cardiovascular fcunction. J Clin Invest 116:1391-1399

Cipollone F, Rocca B, Patrono C (2004) Cyclooxygenase-2 expression and inhibition in atherothrombosis. Arterioscler Thromb Vasc Biol 24:246–255

Clarke RJ, Mayo G, Price P, et al (1991) Suppression of thromboxane A2 but not of systemic prostacyclin by controlled-release aspirin. N Engl J Med 325:1137–1141

Cok SJ, Acton SJ, Morrison AR (2003) The proximal region of the 3′-untranslated region of cyclooxygenase-2 is recognized by a multimeric protein complex containing HuR, TIA-1, TIAR, and the heterogeneous nuclear ribonucleoprotein U. J Biol Chem 278:36157–36162

Connolly E, Bouchier-Hayes DJ, Kaye E, et al (2002) Cyclooxygenase isozyme expression and intimal hyperplasia in a rat model of balloon angioplasty. J Pharmacol Exp Ther 300:393–398

Dixon DA (2004) Dysregulated post-transcriptional control of COX-2 gene expression in cancer. Curr Pharm Des 10:635–646

Dogne JM, Hanson J, de Leval X, et al (2004) New developments on thromboxane modulators. Mini Rev Med Chem 4:649–657

Dorn GW, Sens D, Chaikhouni A, et al (1987) Cultured human vascular smooth muscle cells with functional thromboxane A_2 receptors: measurement of U46619-induced ^{45}calcium efflux. Circ Res 60:952–956

Dowd NP, Scully M, Adderley SR, et al (2001) Inhibition of cyclooxygenase-2 aggravates doxorubicin-mediated cardiac injury in vivo. J Clin Invest 108:585–590

Egan KM, Lawson JA, Fries S, et al (2004) COX-2-derived prostacyclin confers atheroprotection on female mice. Science 306:1954–1957

Egan KM, Wang M, Lucitt MB, et al (2005) Cyclooxygenases, thromboxane, and atherosclerosis: plaque destabilization by cyclooxygenase-2 inhibition combined with thromboxane receptor antagonism. Circulation 111:334–342

Evangelista V, Manarini S, DiSanto A, Capone ML, Riciotti E, et al (2006) De novo synthesis of cyclooxygenase-1 counteracts the suppression of platelet thromboxane biosynthesis by aspirin. Circ Res 98:593–595

Fabre JE, Nguyen M, Athirakul K, et al (2001) Activation of the murine EP3 receptor for PGE2 inhibits cAMP production and promotes platelet aggregation. J Clin Invest 107:603–610

Fitzgerald DJ, Entman SS, Mulloy K, et al (1987) Decreased prostacyclin biosynthesis preceding the clinical manifestation of pregnancy-induced hypertension. Circulation 75:956–963

Fitzgerald DJ, Rocki W, Murray R, et al (1990) Thromboxane A2 synthesis in pregnancy-induced hypertension. Lancet 335:751–754

FitzGerald GA (1991) Mechanisms of platelet activation: thromboxane A2 as an amplifying signal for other agonists. Am J Cardiol 68:11B–15B

FitzGerald GA (2003) COX-2 and beyond: approaches to prostaglandin inhibition in human disease. Nat Rev Drug Discov 11:879–890

FitzGerald GA (2004) Coxibs and cardiovascular disease. N Engl J Med 351:1709–1711

FitzGerald GA, Loll P (2001) COX in a crystal ball: current status and future promise of prostaglandin research. J Clin Invest 107:1335–1337

FitzGerald GA, Patrono C (2001) The coxibs, selective inhibitors of cyclooxygenase-2. N Engl J Med 345:433–442

FitzGerald GA, Friedman LA, Miyamori I, et al (1979) A double blind placebo controlled crossover study of prostacyclin in man. Life Sci 25:665–672

FitzGerald GA, Brash AR, Falardeau P, et al (1981) Estimated rate of prostacyclin secretion into the circulation of normal man. J Clin Invest 68:1272–1276

FitzGerald GA Pedersen AK, Patrono C (1983) Analysis of prostacyclin and thromboxane biosynthesis in cardiovascular disease. Circulation 67:1174–1177

Fitzpatrick FA (2004) Cyclooxygenase enzymes: regulation and function. Curr Pharm Des 10:577–588

Fleming EF, Athirakul K, Oliverio MI, et al (1998) Urinary concentrating function in mice lacking EP3 receptors for prostaglandin E2. Am J Physiol 275:F955–F961

Francois H, Atirakul K, Howell D, Dash R, Mao L, Kim HS, Rockman HA, FitzGerald GA, Koller BH, Coffman TC (2005) Prostacyclin protects against elevated blood pressure and cardiac fibrosis. Cell Metab 2:201–207

Fujino T, Yuhki K, Yamada T, et al (2002) Effects of the prostanoids on the proliferation or hypertrophy of cultured murine aortic smooth muscle cells. Br J Pharmacol 136:530–539

Fujino T, Nakagawa N, Yuhki K, et al (2004) Decreased susceptibility to renovascular hypertension in mice lacking the prostaglandin I2 receptor IP. J Clin Invest 114:805–812

Funk CD, Funk LB, Kennedy ME, et al (1991) Human platelet/erythroleukemia cell prostaglandin G/H synthase: cDNA cloning, expression, and gene chromosomal assignment. Faseb J 5:2304–2312

Garavito RM, Picot D, Loll PJ (1995) The 3.1 A X-ray crystal structure of the integral membrane enzyme prostaglandin H2 synthase-1. Adv Prostaglandin Thromboxane Leukot Res 23:99–103

Garcia-Rodriguez LA, Varas-Lorenzo C, Maguire A, et al (2004) Nonsteroidal antiinflammatory drugs and the risk of myocardial infarction in the general population. Circulation 109:3000–3006

Gerritsen ME (1987) Eicosanoid production by the coronary microvascular endothelium. Fed Proc 46:47–53

Gibbs JS, Broberg CS, Gatzoulis MA (2004) Idiopathic pulmonary arterial hypertension: current state of play and new treatment modalities. Int J Cardiol 97 Suppl 1:7–10

Grosser T, Fries S, FitzGerald GA (2006) Biological basis for the cardiovascular consequences of COX-2 inhibition: therapeutic challenges and opportunities. J Clin Invest 116:4–15

Gryglewski RJ, Moncada S, Palmer RM (1986) Bioassay of prostacyclin and endothelium-derived relaxing factor (EDRF) from porcine aortic endothelial cells. Br J Pharmacol 87:685–694

Habib A, FitzGerald GA, Maclouf J (1999) Phosphorylation of the thromboxane receptor alpha, the predominant isoform expressed in human platelets. J Biol Chem 274:2645–2651

Hajjar D, Weksler BB, Falcone DJ, et al (1982) Prostacyclin modulates cholesteryl ester hydrolytic activity by its effect on cyclic adenosine monophosphate in rabbit aortic smooth muscle cells. J Clin Invest 70:479–488

Hamberg M, Svensson J, Samuelsson B (1975) Thromboxanes: a new group of biologically active compounds derived from prostaglandin endoperoxides. Proc Natl Acad Sci U S A 72:2994–2998

Hanson J, Rolin S, Reynaud D, et al (2005) In vitro and in vivo pharmacological characterization of BM-613 [N-n-Pentyl-N'-[2-(4'-methylphenylamino)-5-nitrobenzenesulfonyl]urea], a novel dual thromboxane synthase inhibitor and thromboxane receptor antagonist. J Pharmacol Exp Ther 313:293–301

Hashimoto K, Graham BS, Geraci MW, et al (2004) Signaling through the prostaglandin I_2 receptor IP protects against respiratory syncytial virus-induced illness. J Virol 78:10303–10309

Hata AN, Breyer RM (2004) Pharmacology and signaling of prostaglandin receptors: multiple roles in inflammation and immune modulation. Pharmacol Ther 103:147–166

Hennekens C, Schror K, Weisman S, et al (2004) Terms and conditions: semantic complexity and aspirin resistance. Circulation 110:1706–1708

Herschman HR (1996) Prostaglandin synthase 2. Biochim Biophys Acta 1299:125–140

Hirata M, Hayashi Y, Ushikubi F, et al (1991) Cloning and expression of cDNA for a human thromboxane A_2 receptor. Nature 349:617–620

Hirata T, Ushikubi F, Kakizuka A, et al (1996) Two thromboxane A_2 receptor isoforms in human platelets. Opposite coupling to adenylyl cyclase with different sensitivity to Arg60 to Leu mutation. J Clin Invest 97:949–956

Honda T, Segi-Nishida E, Miyachi Y, Narumiya S (2006) Prostacyclin-IP signaling and prostaglandin E2-EP2/EP4 signaling both mediate joint inflammation in mouse collagen-induced arthritis. J Exp Med 203:325–335

Iniguez MA, Pablos JL, Carreira PE, et al (1998) Detection of COX-1 and COX-2 isoforms in synovial fluid cells from inflammatory joint diseases. Br J Rheumatol 37:773–778

Ishizuka T, Matsui T, Kurita A, et al (2003) Ramatroban, a TP receptor antagonist, improves vascular responses to acetylcholine in hypercholesterolemic rabbits in vivo. Eur J Pharmacol 468:27–35

Ishizuka T, Matsui T, Okamoto Y et al (2004) Ramatroban (BAY u 3405): a novel dual antagonist of TXA2 receptor and CRTh2, a newly identified prostaglandin D2 receptor. Cardiovasc Drug Rev 22:71–90

Jadhav V, Jabre A, Lin SZ, et al (2004) EP1- and EP3-receptors mediate prostaglandin E2-induced constriction of porcine large cerebral arteries. J Cereb Blood Flow Metab 24:1305–1316

Kabashima K, Murata T, Tanaka H, et al (2003) Thromboxane A2 modulates interaction of dendritic cells and T cells and regulates acquired immunity. Nat Immunol 4:694–701

Kobayashi T, Tahara Y, Matsumoto M, et al (2004) Roles of thromboxane A(2) and prostacyclin in the development of atherosclerosis in apoE-deficient mice. J Clin Invest 114:784–794

Kothapalli D, Fuki I, Ali K, et al (2004) Antimitogenic effects of HDL and APOE mediated by Cox-2-dependent IP activation. J Clin Invest 113:609–618

Kunapuli P, Lawson JA, Rokach J, et al (1997) Functional characterization of the ocular PGF2alpha receptor. Activation by the isoprostane, 12-iso-PGF2alpha. J Biol Chem 272:27147–27154

Lawson JA, Brash AR, Doran J, et al (1985) Measurement of urinary 2,3-dinor-thromboxane B2 and thromboxane B2 using bonded-phase phenylboronic acid columns and capillary gas chromatography—negative-ion chemical ionization mass spectrometry. Anal Biochem 150:463–470

Leslie CC (1997) Properties and regulation of cytosolic phospholipase A2. J Biol Chem 272:16709–16712

Li R, Mouillesseaux KP, Montoya D, Cruz D, Gharavi N, Dun M, Koroniak L, Berliner JA (2006) Identification of prostaglandin E2 receptor subtype 2 as a receptor activated by OxPAPC. Circ Res 98:642–650

Lim H, Dey SK (2002) A novel pathway of prostacyclin signaling-hanging out with nuclear receptors. Endocrinology 143:3207–3210

Lin LL, Lin AY, DeWitt DL (1992) Interleukin-1 alpha induces the accumulation of cytosolic phospholipase A2 and the release of prostaglandin E2 in human fibroblasts. J Biol Chem 267:23451–23454

Loll PJ, Picot D, Garavito RM (1995) The structural basis of aspirin activity inferred from the crystal structure of inactivated prostaglandin H2 synthase. Nat Struct Biol 2:637–643

Marnett LJ, Rowlinson SW, Goodwin DC, et al (1999) Arachidonic acid oxygenation by COX-1 and COX-2. Mechanisms of catalysis and inhibition. J Biol Chem 274:22903–22906

Mason PJ, Freedman JE, Jacobs AK (2004) Aspirin resistance: current concepts. Rev Cardiovasc Med 5:156–163

McAdam B, Catella-Lawson F, Mardini IA, et al (1999) Systemic biosynthesis of prostacyclin by cyclooxygenase (COX)-2: the human pharmacology of a selective inhibitor of COX-2. Proc Natl Acad Sci U S A 96:272–277

Minami T, Nakano H, Kobayashi T, et al (2001) Characterization of EP receptor subtypes responsible for prostaglandin E2-induced pain responses by use of EP1 and EP3 receptor knockout mice. Br J Pharmacol 133:438–444

Moncada S, Vane JR (1978) Pharmacology and endogenous roles of prostaglandin endoperoxides, thromboxane A2, and prostacyclin. Pharmacol Rev 30:293–331

Moncada S, Gryglewski R, Bunting S, et al (1976) An enzyme isolated from arteries transforms prostaglandin endoperoxides to an unstable substance that inhibits platelet aggregation. Nature 263:663–665

Moncada S, Vane JR, Whittle BJ (1977) Relative potency of prostacyclin, prostaglandin E1 and D2 as inhibitors of platelet aggregation in several species. J Physiol 273:2P–4P

Morrow JD (2006) The isoprostanes—unique products of arachidonate peroxidation: their role as mediators of oxidant stress. Curr Pharm Des 12:895–902

Murakami M, Naraba H, Tanioka T, et al (2000) Regulation of prostaglandin E2 biosynthesis by inducible membrane-associated prostaglandin E2 synthase that acts in concert with cyclooxygenase-2. J Biol Chem 275:32783–32792

Murata T, Ushikubi F, Matsuoka T, et al (1997) Altered pain perception and inflammatory response in mice lacking prostacyclin receptor. Nature 388:678–682

Murphey LJ, Williams MK, Sanchez SC, et al (2004) Quantification of the major urinary metabolite of PGE2 by a liquid chromatographic/mass spectrometric assay: determination of cyclooxygenase-specific PGE2 synthesis in healthy humans and those with lung cancer. Anal Biochem 334:266–275

Nagata K, Hirai H (2003) The second PGD(2) receptor CRTH2: structure, properties, and functions in leukocytes. Prostaglandins Leukot Essent Fatty Acids 69:169–177

Namba T, Sugimoto Y, Hirata M, et al (1992) Mouse thromboxane A2 receptor: cDNA cloning, expression and northern blot analysis. Biochem Biophys Res Commun 184:1197–1203

Namba T, Oida H, Sugimoto Y, et al (1994) cDNA cloning of a mouse prostacyclin receptor. Multiple signaling pathways and expression in thymic medulla. J Biol Chem 269:9986–9992

Narumiya S, FitzGerald GA (2001) Genetic and pharmacological analysis of prostanoid receptor function. J Clin Invest 108:25–30

Narumiya S, Sugimoto Y, Ushikubi F (1999) Prostanoid receptors: structures, properties, and functions. Physiol Rev 79:1193–1226

Needleman P, Moncada S, Bunting S, et al (1976) Identification of an enzyme in platelet microsomes which generates thromboxane A2 from prostaglandin endoperoxides. Nature 261:558–560

Nguyen M, Camenisch T, Snouwaert JN, et al (1997) The prostaglandin receptor EP4 triggers remodelling of the cardiovascular system at birth. Nature 390:78–81

Nusing RM, Hirata M, Kakizuka A, et al (1993) Characterization and chromosomal mapping of the human thromboxane A_2 receptor gene. J Biol Chem 268:25253–25259

Oka T, Oka K, Kobayashi T, et al (2003) Characteristics of thermoregulatory and febrile responses in mice deficient in prostaglandin EP1 and EP3 receptors. J Physiol 551:945–954

Pakala R, Willerson JT, Benedict CR, et al (1997) Effect of serotonin, thromboxane A_2, and specific receptor antagonists on vascular smooth muscle cell proliferation. Circulation 96:2280–2286

Patrignani P (2003) Aspirin insensitive eicosanoid biosynthesis in cardiovascular disease. Thromb Res 110:281–286

Patrono C, Coller B, FitzGerald GA, et al (2004) Platelet-active drugs: the relationships among dose, effectiveness, and side effects: the Seventh ACCP Conference on Antithrombotic and Thrombolytic Therapy. Chest 126:234S–264S

Patscheke H (1985) Thromboxane synthase inhibition potentiates washed platelet activation by endogenous and exogenous arachidonic acid. Biochem Pharmacol 1985 34:1151–1156

Pedersen AK, FitzGerald GA (1984) Dose-related kinetics of aspirin. Presystemic acetylation of platelet cyclooxygenase. N Engl J Med 311:1206–1211

Physician's Health Study (1989) Aspirin and primary prevention of coronary heart disease. N Engl J Med 321:1825–1828

Picot D, Loll PJ, Garavito RM (1994) The X-ray crystal structure of the membrane protein prostaglandin H2 synthase-1. Nature 1994 367:243–249

Pini B, Grosser T, Lawson JA, et al (2005) Prostaglandin E synthases in zebrafish. Arterioscler Thromb Vasc Biol 25:315–320

Pratico D, Cyrus T, Li H, et al (2000) Endogenous biosynthesis of thromboxane and prostacyclin in 2 distinct murine models of atherosclerosis. Blood 96:3823–3826

Pratico D, Cheng Y, FitzGerald GA (2000) TP or not TP: primary mediators in a close runoff? Arterioscler Thromb Vasc Biol 20:1695–1698

Pratico D, Rokach J, Lawson J, et al (2004) F2-isoprostanes as indices of lipid peroxidation in inflammatory diseases. Chem Phys Lipids 128:165–171

Ridker PM, Cook NR, Lee IM, et al (2005) A randomized trial of low-dose aspirin in the primary prevention of cardiovascular disease in women. N Engl J Med 352:1293–1304

Robbins IM, Morrow JD, Christman BW (2005) Oxidant stress but not thromboxane decreases with epoprostenol therapy. Free Radic Biol Med 38:568–574

Roberts LJ II, Sweetman BJ, Payne NA, et al (1977) Metabolism of thromboxane B_2 in man. Identification of the major urinary metabolite. J Biol Chem 252:7415–7417

Roberts LJ II, Sweetman BJ, Lewis RA, et al (1980) Increased production of prostaglandin D2 in patients with systemic mastocytosis. N Engl J Med 303:1400–1404

Rocca B, Loeb AL, Strauss JF III, et al (2000) Directed vascular expression of the thromboxane A2 receptor results in intrauterine growth retardation. Nat Med 6:219–221

Schnitzer TJ, Burmester GR, Mysler E, et al (2004) Comparison of lumiracoxib with naproxen and ibuprofen in the Therapeutic Arthritis Research and Gastrointestinal Event Trial (TARGET), reduction in ulcer complications: randomised controlled trial. Lancet 364:665–674

Schonbeck U, Sukhova GK, Graber P, et al (1999) Augmented expression of cyclooxygenase-2 in human atherosclerotic lesions. Am J Pathol 155:1281–1291

Sladek K, Sheller JR, FitzGerald GA, et al (1991) Formation of PGD2 after allergen inhalation in atopic asthmatics. Adv Prostaglandin Thromboxane Leukot Res 21A:433–436

Smith WL, Garavito RM, DeWitt DL (1996) Prostaglandin endoperoxide H synthases (cyclooxygenases)-1 and -2. J Biol Chem 271:33157–33160

Smyth EM, FitzGerald GA (2002) Human prostacyclin receptor. Vitam Horm 65:149–165

Spencer AG, Woods JW, Arakawa T, et al (1998) Subcellular localization of prostaglandin endoperoxide H synthases-1 and -2 by immunoelectron microscopy. J Biol Chem 273:9886–9893

Spisni E, Bartolini G, Orlandi M, et al (1995) Prostacyclin (PGI2) synthase is a constitutively expressed enzyme in human endothelial cells. Exp Cell Res 219:507–513

Szczeklik A, Nizankowski R, Skawinski S, et al (1979) Successful therapy of advanced arteriosclerosis obliterans with prostacyclin. Lancet 1:1111–1114

Takayama K, Garcia-Cardena G, Sukhova GK, et al (2002) Prostaglandin E2 suppresses chemokine production in human macrophages through the EP4 receptor. J Biol Chem 277:44147–44154

Tanabe T, Yokoyama C, Miyata A, et al (1993) Molecular cloning and expression of human thromboxane synthase. J Lipid Mediat 6:139–144

Tanioka T, Nakatani Y, Semmyo N, et al (2000) Molecular identification of cytosolic prostaglandin E2 synthase that is functionally coupled with cyclooxygenase-1 in immediate prostaglandin E2 biosynthesis. J Biol Chem 275:32775–32782

Thomas DW, Mannon RB, Mannon PJ, et al (1998) Coagulation defects and altered hemodynamic responses in mice lacking receptors for thromboxane A2. J Clin Invest 102:1994–2001

Thomas DW, Rocha PN, Nataraj C, et al (2003) Proinflammatory actions of thromboxane receptors to enhance cellular immune responses. J Immunol 171:6389–6395

Tilley SL, Audoly LP, Hicks EH, et al (1999) Reproductive failure and reduced blood pressure in mice lacking the EP2 prostaglandin E2 receptor. J Clin Invest 103:1539–1545

Todaka T, Yokoyama C, Yanamoto H, et al (1999) Gene transfer of human prostacyclin synthase prevents neointimal formation after carotid balloon injury in rats. Stroke 30:419–426

Topper JN, Cai J, Falb D, et al (1996) Identification of vascular endothelial genes differentially responsive to fluid mechanical stimuli: cyclooxygenase-2, manganese superoxide dismutase, and endothelial cell nitric oxide synthase are selectively up-regulated by steady laminar shear stress. Proc Natl Acad Sci U S A 93:10417–10422

Tsuchiya Y, Minami I, Kadotani H, et al (2005) Resetting of peripheral circadian clock by prostaglandin E2. EMBO Rep 6:256–261

Ueno N, Murakami M, Tanioka T, et al (2001) Coupling between cyclooxygenase, terminal prostanoid synthase, and phospholipase A2. J Biol Chem 276:34918–34927

Unoki M, Furuta S, Onouchi Y, et al (2000) Association studies of 33 single nucleotide polymorphisms (SNPs) in 29 candidate genes for bronchial asthma: positive association a T924C polymorphism in the thromboxane A2 receptor gene. Hum Genet 106:440–446

Vezza R, Rokach J, FitzGerald GA (2001) Prostaglandin F2alpha receptor-dependent regulation of prostaglandin transport. Mol Pharmacol 59:1506–1513

Vielhauer GA, Fujino H, Regan JW (2004) Cloning and localization of hFP(S): a six-transmembrane mRNA splice variant of the human FP prostanoid receptor. Arch Biochem Biophys 421:175–185

Wilson SJ, Roche AM, Kostetskaia E, et al (2004) Dimerization of the human receptors for prostacyclin and thromboxane facilitates thromboxane receptor-mediated cAMP generation. J Biol Chem 279:53036–53047

Wong D, Wang M, Cheng Y, et al (2005) Cardiovascular hazard and non-steroidal anti-inflammatory drugs. Curr Opin Pharmacol 5:204–210

Wright DH, Metters KM, Abramovitz M, et al (1998) Characterization of the recombinant human prostanoid DP receptor and identification of L-644,698, a novel selective DP agonist. Br J Pharmacol 123:1317–1324

Xiao CY, Hara A, Yuhki K, et al (2001) Roles of prostaglandin I2 and thromboxane A2 in cardiac ischemia-reperfusion injury: a study using mice lacking their respective receptors. Circulation 104:2210–2215

Yu IS, Lin SR, Huang CC, et al (2004) TXAS-deleted mice exhibit normal thrombopoiesis, defective hemostasis, and resistance to arachidonate-induced death. Blood 104:135–142

Yu Y, Fan J, Chen XS, FitzGerald GA, Funk CD (2006) Genetic model of selective COX-2 inhibition reveals novel heterodimer signalling. Nat Med (published online May 28th)

Zhang Z, Vezza R, Plappert T, et al (2003) COX-2-dependent cardiac failure in Gh/tTG transgenic mice. Circ Res 92:1153–1161

Zou M, Martin C, Ullrich V, et al (1997) Tyrosine nitration as a mechanism of selective inactivation of prostacyclin synthase by peroxynitrite. Biol Chem 378:707–713

Zucker TP, Bonisch D, Hasse A, et al (1998) Tolerance development to antimitogenic actions of prostacyclin but not of prostaglandin E1 in coronary artery smooth muscle cells. Eur J Pharmacol 345:213–220

Nitric Oxide and the Vascular Endothelium

S. Moncada (✉) · E. A. Higgs

The Wolfson Institute for Biomedical Research, University College London,
Gower Street, London WC1E 6BT, UK
s.moncada@ucl.ac.uk

1	Introduction	214
2	Nitric Oxide Synthase	215
2.1	Localisation of eNOS	217
2.2	Up-regulation of eNOS	218
2.3	Activation of eNOS	219
2.4	Inducible Nitric Oxide Synthase	221
3	Actions of Nitric Oxide in the Vascular System	222
3.1	Nitric Oxide and Vascular Tone	222
3.2	Nitric Oxide and Platelets	224
3.3	Nitric Oxide, Vascular Permeability and White Cells	225
3.4	Nitric Oxide and Vascular Smooth Muscle Proliferation	226
3.5	Nitric Oxide and Angiogenesis	227
4	Molecular Targets of the Action of Nitric Oxide	227
4.1	Soluble Guanylate Cyclase	227
4.2	Protein *S*-Nitrosylation	228
4.3	Cytochrome c Oxidase/Mitochondrial Effects	229
5	Nitric Oxide and Pathology	231
5.1	Changes in NO Generation or Activity in Vascular Pathology	231
5.1.1	The Origin of Free Radicals	233
5.1.2	Mechanisms Involved in the Decreased Generation of Nitric Oxide	234
5.1.3	Replacing Nitric Oxide	235
6	Conclusion	236
	References	236

Abstract The vascular endothelium synthesises the vasodilator and anti-aggregatory mediator nitric oxide (NO) from L-arginine. This action is catalysed by the action of NO synthases, of which two forms are present in the endothelium. Endothelial (e)NOS is highly regulated, constitutively active and generates NO in response to shear stress and other physiological stimuli. Inducible (i)NOS is expressed in response to immunological stimuli, is transcriptionally regulated and, once activated, generates large amounts of NO that contribute to pathological conditions. The physiological actions of NO include the regulation of vascular tone and blood pressure, prevention of platelet aggregation and inhibition of vascular smooth muscle proliferation. Many of these actions are a result of the activation by NO of the soluble guanylate cyclase and consequent generation of cyclic guanosine monophosphate (cGMP). An additional target of NO is the cytochrome c oxidase, the terminal enzyme in the electron transport chain, which is inhibited by NO in a manner that is reversible and competitive with oxygen. The consequent reduction of cytochrome c oxidase leads to the release of superoxide anion. This may be an NO-regulated cell signalling system which, under certain circumstances, may lead to the formation of the powerful oxidant species, peroxynitrite, that is associated with a variety of vascular diseases.

Keywords Nitric oxide · eNOS · Guanylate cyclase · Cytochrome c oxidase · Mitochondria · Free radicals

1
Introduction

The release of nitric oxide (NO) by the vascular endothelium was first demonstrated in 1987 (Palmer et al. 1987). Approximately 1 year later it was discovered that endothelial NO was synthesised from the semi-essential amino acid L-arginine (Palmer et al. 1988). These findings established the identity of the so-called endothelium-derived relaxing factor (EDRF) discovered by Furchgott and Zawadzki some 7–8 years earlier (Furchgott and Zawadzki 1980). Furthermore, they threw light on a disparate series of observations, made over more than a decade, suggesting the existence of a widespread metabolic pathway based on the conversion of L-arginine in the central nervous system and in macrophages, and revealed the function of the soluble guanylate cyclase as an intracellular receptor of the newly discovered endogenous ligand (Moncada 1989; Moncada et al. 1991).

Over the following 2 or 3 years the existence of the so-called L-arginine:NO pathway (Moncada et al. 1989) was unequivocally established, playing myriad physiological and pathophysiological roles in the cardiovascular system, the central and peripheral nervous systems and in cellular defence. One of the most extensively investigated areas is that of the role of NO in the vasculature. Nitric oxide generated by the vascular endothelium is a major regulator of vascular homeostasis, and changes in its bioavailability are now known to play a role in the development of a number of clinical conditions in which the function of the vascular system is impaired. Since more than 9,000 papers have been written on NO and the vascular endothelium, it will be impossible to do justice, within the constraints of this book, to all the authors that have made significant contributions to the subject. This chapter will focus on what we consider to be some still unresolved issues. For other aspects we refer the reader to a number of excellent reviews of NO research that have appeared in recent years (Alderton et al. 2001; Von der Leyen and Dzau 2001; Fleming and Busse 2003; Sessa 2004).

2
Nitric Oxide Synthase

Endothelial nitric oxide synthase (eNOS) is one of three isoforms of NO synthase (NOS, EC 1.14.13.39). The isoforms were named after the tissue in which they were first identified, i.e. endothelial, neuronal (nNOS) and inducible (iNOS) for the macrophage enzyme which is induced by activation with lipopolysaccharide plus interferon-γ. The classification of type III, I and II is also used for the three isoforms, respectively. In the last 15 years, important information has been generated about the functioning of NOS. The picture that emerges is that of a highly regulated enzyme, the activity of which can be controlled at different points, including gene expression, phosphorylation at various sites and regulated interactions with other proteins (Alderton et al. 2001; Sessa 2004). The amino acid sequence of the enzyme, as well as its association with other proteins, has been largely elucidated (Janssens et al. 1992; Marsden et al. 1992; Nakane et al. 1993; Geller et al. 1993). The available crystal structure of the oxygen domains of the inducible (Crane et al. 1997; Cubberley et al. 1997) and the endothelial enzymes (Raman et al. 1998; Fischmann et al. 1999) has yielded information about the binding site of the substrate L-arginine, as well as the function of important co-factors such as tetrahydrobiopterin (BH_4; Wei et al. 2002; Werner et al. 2003). Furthermore, increasing knowledge about the molecular action of inhibitors of the enzyme is leading to the "design" of selective compounds for the different isoforms, with characteristics in terms of pharmacokinetics and pharmacodynamics increasingly tailored for their intended use (Hobbs et al. 1999; Mete and Connolly 2003; Alderton et al. 2005). Interestingly, the precise mechanism of the enzymic activity and the nature of the product(s) it generates instead of NO, in addition to NO, or prior to NO has been a matter of controversy. For example, nitroxyl anion (NO^-) has been proposed to be a product of NOS under certain conditions, particularly when the substrate or co-factor concentrations are low (Schmidt et al. 1996; Adak et al. 2000). The elucidation of the exact product has important implications for the understanding of the physiological actions of NO as well as its potential for initiating pathophysiology, as will be discussed later.

eNOS, which is constitutively active in the endothelial cell, was originally identified in 1989 (Palmer and Moncada 1989) and cloned in 1992 (Janssens et al. 1992; Marsden et al. 1992). Human eNOS is encoded by a gene located on chromosome 7, and comprises 1,294 amino acids with a molecular weight of 135 kDa (Lamas et al. 1992; Sessa et al. 1992; Marsden et al. 1993). iNOS, which is induced in the endothelial and other cells following immunological activation, is encoded by a gene located on chromosome 17, comprises 1,153 amino acids and has a molecular weight of 131 kDa (Geller et al. 1993; Sherman et al. 1993; Charles et al. 1993).

The C-terminal portion of the NOS protein closely resembles cytochrome P-450 reductase (Bredt et al. 1991), possesses many of the same co-factor binding sites, and basically performs the same functions. Consequently, this portion is often referred to as the reductase domain (see Fig. 1). At the extreme C terminus is an NADPH (nicotinamide adenine dinucleotide phosphate, reduced)-binding region, which is conserved in all NOS and aligns perfectly with that of cytochrome P-450 reductase. The NADPH binding site is followed, in turn, by flavin adenine dinucleotide (FAD) and flavin mononucleotide (FMN) consensus sequences (Djordjevic et al. 1995). Unlike cytochrome P-450 reductase, NOS is a self-sufficient enzyme in that the oxygenation of its substrate, L-arginine, occurs at a haem-site in the N-terminal portion, termed the oxygenase domain, of the protein. Stoichiometric amounts of haem are present in NOS and are required for catalytic activity (White and Marletta 1992). Close to the haem (catalytic) site is an L-arginine (substrate) binding site. Separation of the reductase and oxygenase domains via limited proteolysis has enabled L-arginine and BH_4 binding sites to be localised to the oxygenase domain. Bridging the reductase and oxygenase domains is a calmodulin-binding site, which acts as a switch to regulate electron flow between the two regions (Abu Soud and Stuehr 1993).

The co-factor requirements of NOS are not only important in aiding catalytic activity, they are also obligatory in permitting the dimerisation of monomers to form active proteins. The active, dimeric proteins possess all the co-factors described above, and dimerisation of the monomeric proteins is promoted by the presence of haem, BH_4 and L-arginine (Baek et al. 1993). Specific to eNOS is a consensus sequence for myristoylation/palmitoylation at its N terminus which contributes to its particulate nature, unlike the cytosolic location of nNOS and iNOS.

The Ca^{2+}/calmodulin dependence of NOS was established early on (Knowles et al. 1989; Bredt and Snyder 1990). Calmodulin is a ubiquitous small Ca^{2+}-binding protein that binds to eNOS—and many other target proteins—and thus transduces the Ca^{2+} signal into a variety of actions (Aoyagi et al. 2003). Ca^{2+}/calmodulin dependence was the basis for an early classification of eNOS and nNOS (Ca^{2+}-dependent) and iNOS (Ca^{2+}-independent). Now it is clear that all three isoforms require Ca^{2+}, with eNOS and nNOS having a much greater requirement due to the presence in their calmodulin/FMN-binding

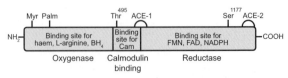

Fig. 1 Diagrammatic representation of eNOS. The oxidase and reductase domains are linked by a calmodulin-binding domain. Myristoylation (*Myr*) and palmitoylation (*Palm*) sites are shown, as well as autoinhibitory control elements (*ACE*)-1 and -2. Thr^{495} and Ser^{1177}, whose dephosphorylation and phosphorylation are likely to be the most significant regulatory steps in eNOS activation, are also indicated

subdomain of an autoinhibitory control element (ACE) of approximately 50 amino acids which hinders the binding of calmodulin to its site (Salerno et al. 1997). This destabilises calmodulin binding at low intracellular Ca^{2+} $[Ca^{2+}]_i$, thus increasing the requirement of the enzyme for Ca^{2+} for activation (Nishida and Ortiz de Montellano 1999; Daff et al. 1999). A second autoinhibitory control element (ACE-2) has been demonstrated at the C-terminus of the so-called Ca^{2+}-dependent isoforms; this has been claimed to act as a barrier to the activation of eNOS by calmodulin binding and to be functionally disabled by phosphorylation of Ser^{1179} (bovine; Ser^{1177}, human) on enzyme activation (Lane and Gross 2002).

2.1
Localisation of eNOS

In the early 1990s it was established that eNOS was mainly localised in the luminal membrane fraction of endothelial cells (Förstermann et al. 1991; Mitchell et al. 1991; Pollock et al. 1991). More specifically, the localisation is in the caveolae, which are specialised plasmalemmal signal-transducing domains (Shaul et al. 1996; Garcia-Cardena et al. 1996). Furthermore, eNOS was shown to interact with caveolin-1 and caveolin-3 (coat proteins of caveolae) via a caveolin-binding motif in the eNOS (Garcia-Cardena et al. 1996), and this interaction has been reported to inhibit the activity of the enzyme and the generation of NO (Bucci et al. 2000).

In addition, there is evidence that eNOS may also be sited in the Golgi apparatus (Garcia-Cardena et al. 1996; Liu et al. 1997). Originally, the localisation to the Golgi was assumed to be an inactive reservoir of immature eNOS on its way to the plasmalemma or simply eNOS bound to internalised caveolae (Govers et al. 2002; Jobin et al. 2003). More recently, however, specific mutagenesis studies targeting eNOS to different cellular locations have established that eNOS located in the plasma membrane differs slightly from that in the Golgi, for example in sensitivity to Ca^{2+} activation. The enzymes from both locations are active and highly regulated (Fulton et al. 2004). The functional significance of these observations, however, remains to be established, since it has been shown that disruption of the Golgi apparatus does not affect NO-dependent relaxation in some coronary arteries (Bauersachs et al. 1997).

The eNOS in its membrane localisation has been shown to be both permanently myristoylated and reversibly palmitoylated (Pollock et al. 1992; Busconi and Michel 1993; Liu et al. 1995). Site-directed mutagenesis studies have demonstrated that myristoylation occurs at Gly^2 and palmitoylation at Cys^{15} and Cys^{26} (Boutin 1997; Dunphy and Linder 1998). Both myristoylation and palmitoylation are required for localisation of eNOS to the membrane (Liu et al. 1995; Robinson and Michel 1995). It is now known that myristoylation is required for targeting the eNOS to the membrane while palmitoylation stabilises membrane association and targets the enzyme to the caveolae (Garcia-Cardena et al. 1996; Sowa et al. 1999; Prabhakar et al. 2000). Myristoylation alone results in a tenfold enhancement in targeting of eNOS to the caveolae, and this can be increased a further tenfold by palmitoylation of the enzyme (Shaul et al. 1996).

The process of palmitoylation/depalmitoylation appears to be necessary not only for the cellular location of eNOS but also for its activity. Indeed, activation of the enzyme seems to lead to its depalmitoylation and its translocation from the caveolae to the cytosol (Michel et al. 1993; Robinson et al. 1995). Once the agonist effect is terminated, the eNOS is re-palmitoylated and relocated to the caveolae (Feron et al. 1998). This dynamic subcellular localisation appears to involve the enzyme acyl-protein thioesterase 1 (APT 1), which regulates eNOS depalmitoylation. Interestingly Ca^{2+}/calmodulin activation of eNOS renders the enzyme more susceptible to APT 1-catalysed depalmitoylation (Yeh et al. 1999). The precise function of eNOS trafficking, however, remains to be fully clarified, since at present it appears to be more involved in terminating rather than initiating the release of NO (Nedvetsky et al. 2002).

2.2
Up-regulation of eNOS

The eNOS gene has been extensively studied and shown to contain a promoter region with multiple regulatory DNA sequences including shear stress response elements (Marsden et al. 1993; Zhang et al. 1995; Karantzoulis-Fegaras et al. 1999). Shear stress induces eNOS messenger RNA (mRNA) expression via a transcriptional pathway (Uematsu et al. 1995), and the detailed mechanism of this action has been described using mutagenesis studies (Ziegler et al. 1998; Silacci et al. 2000; Wedgwood et al. 2003; Davis et al. 2004). Other stimuli that increase eNOS mRNA include chronic exercise (Kojda et al. 2001), vascular endothelial growth factor (VEGF; Bouloumie et al. 1999), transforming growth factor β (Inoue et al. 1995; Saura et al. 2002), lysophosphatidyl choline (Zembowicz et al. 1995), statins (Hernandez-Perera et al. 1998) and oestrogens.

Oestrogens were originally shown to increase eNOS mRNA and activate the enzyme (Weiner et al. 1994). More recently it was demonstrated that ovariectomy in rats results in a decrease in eNOS protein and activity and a simultaneous increase in the abundance of caveolin (Pelligrino et al. 2000). A mechanistic link was thus established between these oestrogen-associated divergent changes in the abundance of caveolin-1 and eNOS protein and eNOS functional activity in cerebral arterioles (Xu et al. 2001). Much of the existing evidence indicates that oestrogens activate eNOS via a Ca^{2+}-dependent mechanism (Caulin-Glaser et al. 1997; Chambliss and Shaul 2002). However, other Ca^{2+}-independent mechanisms have also been claimed, including the promotion by oestrogens of the association between eNOS and heat shock protein 90 (hsp90; Russell et al. 2000) or the reduction by oestrogens of the generation of superoxide anion (O_2^-), thus increasing the availability of NO (Barbacanne et al. 1999).

Free radicals and hydrogen peroxide (H_2O_2) have also been claimed to be involved in the transcriptional regulation of eNOS by cyclosporin A (Lopez-Ongil et al. 1998; Navarro-Antolin et al. 2000). In separate studies, H_2O_2 has been shown to activate eNOS transcription (Cai et al. 2001; Cieslik et al. 2001). Studies on lysophosphatidyl choline have suggested that eNOS is up-regulated by an increased binding of the transcription factor Sp1 to its promoter region via the action of a protein phosphatase 2A (pp2A; Cieslik et al. 2001). Whether this is a mechanism involved in all forms of eNOS up-regulation remains to be investigated.

Interestingly, immunological stimuli such as tumour necrosis factor 1α and lipopolysaccharide decrease eNOS mRNA levels and stability (MacNaul and Hutchinson 1993; Rosenkranz-Weiss et al. 1994; Lu et al. 1996). The effects of oxidised low-density lipoprotein (oxy-LDL) and hypoxia on eNOS mRNA stability remain controversial (Arnet et al. 1996; Govers and Rabelink 2001; Tai et al. 2004); however, it is of interest that statins have been shown to increase eNOS mRNA stability and to prevent the down-regulation of eNOS mRNA induced by oxy-LDL (Hernandez-Perera et al. 1998) and hypoxia (Laufs et al. 1997).

2.3
Activation of eNOS

In resting endothelial cells the scaffolding proteins caveolin-1 and caveolin-3 both bind to and inhibit eNOS (Bucci et al. 2000). Studies in caveolin-1 knockout animals showed a dramatic increase in plasma NO concentration, cell proliferation and enhanced vasodilator responses, indicating that the absence of caveolin-1 leads to increased eNOS activity (Razani et al. 2001; Zhao et al. 2002). In addition, eNOS can interact with calmodulin (Förstermann et al. 1991) and hsp90 (Garcia-Cardena et al. 1998), both of which stimulate NOS activity. Agonists that promote the production of NO, such as bradykinin, histamine and VEGF, are associated with the recruitment of hsp90 to eNOS. hsp90 has been shown to bind to eNOS in a Ca^{2+}-independent manner (Garcia-Cardena et al. 1998), to facilitate its dissociation from caveolin and to form a complex with eNOS and calmodulin in endothelial cells, increasing the activity of the enzyme (Gratton et al. 2000).

It was originally believed that the actions of agonists such as bradykinin and acetylcholine were all dependent on increases in $[Ca^{2+}]_i$. Indeed, chelation of extracellular Ca^{2+} or the presence of an antagonist of calmodulin abolishes NO production in response to these agonists (Luckhoff et al. 1988; Busse and Mulsch 1990). The mechanism of their activation has been largely elucidated, since it is now known that they activate phospholipase C, leading to increases in cytoplasmic Ca^{2+} and diacylglycerol. The increases in $[Ca^{2+}]_i$ cause displacement of the eNOS ACE-1, thus allowing calmodulin access to its binding site on the enzyme. This results in an NADPH-dependent flow of electrons from the reductase domain of one monomer of eNOS to the haem iron in the oxygenase domain of the other monomer (Siddhanta et al. 1996), initiating the synthesis of NO.

Shear stress generated by blood flowing over the endothelial cell surface is likely to be the most important activator of eNOS. Activation of eNOS by shear stress was originally termed "Ca^{2+}-independent"; however it is now clear that when shear stress is applied there is an initial transient increase in $[Ca^{2+}]_i$ (Hoyer et al. 1998). Furthermore, shear stress-induced activation of eNOS can be abolished by chelation of intracellular Ca^{2+}. Thus, it has been proposed that eNOS activation by shear stress actually requires Ca^{2+}, but that phosphorylation of the enzyme at certain sites enables it to be activated at resting Ca^{2+} levels (Dimmeler et al. 1999).

Fluid shear stress has been found to result in stimulation of the phosphatidylinositol 3-kinase (PI3K) pathway, leading to the activation of serine kinase Akt 1 which phosphorylates eNOS on Ser^{1177} (Ayajiki et al. 1996; Go et al. 1998; Dimmeler et al. 1999; Fulton et al. 1999). These original findings led to the uncovering of a series of steps of phosphorylation and dephosphorylation. Phosphorylation of the Ser^{1177} is now thought to remove the steric hindrance caused by ACE-2, resulting in an increase in electron flux through the reductase domain of the enzyme and enhanced NO production (McCabe et al. 2000; Lane and Gross 2002; Sessa 2004). Other sites of eNOS can be phosphorylated and contribute to the regulation of the function of the enzyme; these include Ser^{116} and Ser^{617} (bovine; Bauer et al. 2003; Boo and Jo 2003). While the consequences of phosphorylation of Ser^{116} remain unclear, that of Ser^{617} modulates eNOS activity. However, Ser^{1177} appears to be the main regulatory site, since its mutation prevents Akt-mediated phosphorylation and the release of NO (Dimmeler et al. 1999; Fulton et al. 1999; Luo et al. 2000).

Other protein kinases can also phosphorylate eNOS at Ser^{1177}, for example, in endothelial cells transfected with dominant-negative Akt constructs, shear stress-dependent NO production was found to be dependent on protein kinase A (PKA; Boo et al. 2002). PKA also phosphorylates Ser^{635} which is located in the ACE-1 region of eNOS, rendering it able to produce NO continuously in the absence of any changes in Ca^{2+} (Boo et al. 2003). In addition, the AMP-activated protein kinase (AMPK), which is activated by metabolic stress, phosphorylates Ser^{1177} in the presence of Ca^{2+}/calmodulin, while it phosphorylates Thr^{495} in the absence of Ca^{2+}/calmodulin (Chen et al. 1999). Phosphorylation of Thr^{495} (human; Thr^{497}, bovine) in the calmodulin-binding domain de-activates eNOS by hindering the binding of calmodulin (Fleming et al. 2001). In the presence of stimuli that elevate endothelial $[Ca^{2+}]_i$, Thr^{495} is dephosphorylated, enabling calmodulin to bind to eNOS. Dephosphorylation of Thr^{495} precedes the phosphorylation of Ser^{1177} prior to eNOS activation (Fleming et al. 2001; Harris et al. 2001).

Studies of phosphatases have added significance to the idea that phosphorylation and dephosphorylation of Thr^{495} and Ser^{1177} are likely to be two of the most significant regulatory steps in eNOS activation. Thus, pp1 dephosphorylates Thr^{495}, and inhibition of this phosphatase results in hyperphosphorylation of Thr^{495} and inhibition of eNOS activity (Fleming et al. 2001). On the other hand, pp2A dephosphorylates Ser^{1177}, and inhibitors of this enzyme, such as okadaic acid, increase eNOS activity two- to fourfold (Fisslthaler et al. 2000; Michell et al. 2001).

It is therefore evident that Ca^{2+}-dependent and phosphorylation-dependent mechanisms of activating eNOS are interrelated. Furthermore, different patterns of activation may occur, depending on the stimulus. Several agonists, such as VEGF (Fulton et al. 1999), oestrogen (Simoncini et al. 2000) and bradykinin (Harris et al. 2001), activate eNOS through a Ca^{2+}/calmodulin-dependent mechanism but, at the same time, calmodulin activates CaM kinase II, which may phosphorylate eNOS on Ser^{1177}. Furthermore, binding of these agonists to their receptors can also result in activation of the PI3K/Akt pathway, with consequent phosphorylation of Ser^{1177}. Thus, both Ca^{2+}-dependent and Ca^{2+}-independent activation of eNOS results in phosphorylation at this site.

Other proteins also play a role in the regulation of eNOS activity. These include the C-terminal hsp70-interacting protein (CHIP), which forms part of the eNOS complex and appears to play a role in its intracellular localisation (Jiang et al. 2003), and dynamin-2, a protein whose association with eNOS both affects the localisation of the enzyme and increases its activity (Cao et al. 2003). Two additional proteins, NOS-interacting protein (NOSIP; Dedio et al. 2001) and NOS traffic inducer protein (NOSTRIN; Zimmermann et al. 2002) have been suggested to play a role in eNOS activity and/or subcellular localisation, based on studies in transfected cells. These remain to be confirmed, however, in endothelial cell studies. It has also been shown that eNOS binds directly with porin (a voltage-dependent anion/cation channel), and it has been suggested that this interaction may be important for regulating eNOS activity (Sun and Liao 2002).

eNOS has also been reported to be associated in unstimulated endothelial cells to G protein-coupled receptors of bradykinin, angiotensin II and endothelin. Dissociation is reported to occur on cell stimulation (Marrero et al. 1999). Furthermore, the soluble guanylate cyclase, despite its name, has been found in caveolae in close association with eNOS (Zabel et al. 2002). The nature and implications of this interaction, which are potentially many if confirmed, however, await clarification. One possibility is that this interaction is controlled by the hsp90, which interacts with both enzymes to form a complex (Venema et al. 2003).

2.4
Inducible Nitric Oxide Synthase

iNOS was originally identified in macrophages and recognised as part of the cytostatic and cytotoxic mechanisms that operate in these cells (Hibbs et al. 1990). Unlike eNOS, iNOS is mostly transcriptionally regulated and is not normally produced in most cells (Förstermann et al. 1994; Morris and Billiar

1994). Although the rank order of intrinsic activity of the isomers of NOS per unit time is nNOS>iNOS>eNOS (Santolini et al. 2001), iNOS generates 100- to 1,000-fold more NO than eNOS (Morris and Billiar 1994; Nathan and Xie 1994) since, once it is expressed in response to immunological stimuli, its activity persists for many hours. The murine macrophage iNOS gene was originally cloned in 1992 (Xie et al. 1992) and this was shortly followed by the cloning of the human iNOS in hepatocytes (Geller et al. 1993; Chartrain et al. 1994) and chondrocytes (Charles et al. 1993). While eNOS is more than 90% conserved between species (Lamas et al. 1992; Nishida et al. 1992; Janssens et al. 1992; Marsden et al. 1992), human and murine iNOS show only 80% amino acid sequence identity (Geller et al. 1993). The human and murine iNOS promoters have limited similarity and, while iNOS expression in murine cells is readily observed, its induction in human monocytes and macrophages requires stringent conditions (Albina 1995; Vouldoukis et al. 1995).

In the early 1990s the induction of iNOS was demonstrated in vascular endothelial cells and in the smooth muscle layer of the vasculature (Radomski et al. 1990a; Knowles et al. 1990; Durante et al. 1991), and this was shown to be responsible for the hypotension of septic shock (Kilbourn et al. 1990). iNOS is now known to be expressed in almost every cell type, and its induction is inhibited by glucocorticoids (Radomski et al. 1990a; Knowles et al. 1990). These drugs have been shown to act at multiple levels to regulate iNOS expression and NO generation, including decreased gene transcription, decreased mRNA stability, reduced translation of mRNA and increased degradation of iNOS protein (Walker et al. 1997; Matsumura et al. 2001; Korhonen et al. 2002). Atherosclerosis is associated with increases in iNOS expression, and this has been shown in humans to co-exist with a decrease in eNOS mRNA expression in the endothelial cells overlying advanced atheromatous plaques (Wilcox et al. 1997; Fukuchi and Giaid 1999). This pattern of increased iNOS accompanied by reduced eNOS has been reported in response to ischaemia (Azadzoi et al. 2004), hypercholesterolaemia (Kim et al. 2002) and reactive oxygen species (ROS; Aliev et al. 1998).

3
Actions of Nitric Oxide in the Vascular System

3.1
Nitric Oxide and Vascular Tone

By far the most immediately demonstrable action of NO generated by the vascular endothelium is the provision of a significant vasodilator tone in the cardiovascular system, the absence of which leads to immediate vasoconstriction of all vascular beds or to an increase in blood pressure in all species so far tested. This mechanism was initially discovered and demonstrated using pharmacological inhibitors in animals (Rees et al. 1989) and humans (Vallance et al. 1989). Later experiments in $eNOS^{-/-}$ mice demonstrated that these animals show a hypertensive phenotype (Huang et al. 1995). More recently, the endothelial cell-specific overexpression of eNOS has been shown to reduce blood pressure, further demonstrating the essential role of eNOS in blood pressure regulation (Ohashi et al. 1998).

Although the increase in blood pressure, and especially the vasoconstriction in vascular beds, that follows pharmacological inhibition of NO is dependent on the withdrawal of the NO dilator tone, it has been argued that the hypertension of $eNOS^{-/-}$ mice—which has been demonstrated in all reported experiments in these animals (Stauss et al. 1999, 2000; Wagner et al. 2000b)—is probably not due exclusively to lack of NO dilator tone. Other mechanisms have been proposed related to actions of eNOS in the kidney and in the heart (Ortiz and Garvin 2003). However, to date, the effects of eNOS on, for example, renin release in the kidney (Kurtz and Wagner 1998; Shesely et al. 1996; Beierwaltes et al. 2002) remain controversial and the clear effects that pharmacological inhibition of NO has on medullary blood flow and sodium excretion in wild-type animals (Mattson et al. 1997; Ortiz et al. 2001; Pallone and Mattson 2002) are less clear when investigated in $eNOS^{-/-}$ mice (Ortiz and Garvin 2003). There may be several reasons for these differences; however, a significant one is that in both the kidney and the heart other NOS isoforms, such as nNOS (Mattson and Bellehumeur 1996; Kurihara et al. 1998) and iNOS (Ahn et al. 1994), are present and generate NO that compensates for the lack of that normally produced by eNOS.

In the heart, eNOS is expressed in the vascular endothelium and also in the cardiomyocytes (for review see Massion et al. 2003) and the latter is likely to play a role in cardiac contractility (Paulus and Shah 1999). Indeed, there is evidence that stretch induces phosphorylation of Akt and eNOS, leading to the generation of NO associated with an increase in Ca^{2+}-spark frequency (Petroff et al. 2001). nNOS has also been reported to be present in the mitochondria and sarcoplasmic reticulum of cardiomyocytes (Kanai et al. 2001; Xu et al. 1999), and the interplay between the NO generated by the two isoforms in myocardial physiology remains unknown.

An additional complication in the interpretation of results in eNOS knock-out animals relates to a potential phenotypic adaptation or adaptations which most probably occur during different phases of development. These adaptations may arise in the working of the heart and the kidney, and in other crucial functions such as the modulation of cardiac vagal control and of responses to sympathetic stimulation (Chowdhary and Townend 1999) and β-receptor activation (Balligand et al. 1993; Barouch et al. 2002). All of these have themselves been claimed to be modulated by NO derived either from eNOS or from other isoforms (Massion et al. 2003).

In $eNOS^{-/-}$ mice, while the acetylcholine response in large conductance vessels is completely abolished (Rees et al. 2000; Brandes et al. 2000) and can be restored by gene transfer of eNOS in vitro (Scotland et al. 2002), the vasodilator response is maintained in the mesenteric (Rees et al. 2000) and other resistance vascular beds (Sun et al. 1999; Ding et al. 2000; Scotland et al. 2001; Huang et al. 2001). This "remaining" vasodilator response has been extensively investigated and variously attributed to, among other things, prostaglandins (Sun et al. 1999) or the elusive endothelium-derived hyperpolarising factor (Huang et al. 2001). This is of particular interest since a gender difference has been suggested. Agonist-induced NO-dependent dilations are greater in females than in males (Huang et al. 1998), and the degree of hypertension is greater in male than female $eNOS^{-/-}$ mice (Rees et al. 2000). Recent studies on a double knock-out mouse ($eNOS^{-/-}$ and $COX-1^{-/-}$), unable to generate either NO or prostacyclin, show that in these animals there is indeed a compensatory vasodilator mechanism, especially in female animals (Scotland et al. 2005), which still requires identification (Cohen 2005). It remains to be clarified whether these differences between males and females are also true in humans, in which case they may help to explain, at least in part, the reduced incidence of cardiovascular disease in pre-menopausal women.

In summary, although multiple mechanisms have been ascribed to eNOS-derived NO in the regulation of blood pressure and blood flow, and several mediators have been suggested to play compensatory roles in its absence, some important facts remain; first, the increase in blood pressure following pharmacological inhibition of eNOS is immediate and resembles the response of isolated strips of endothelium-containing vascular tissues in vitro, and second, there do not seem to be compensatory mechanisms that are able to down-regulate the blood pressure of animals treated long-term with NOS inhibitors (Blot et al. 1994; Navarro et al. 1994). All the experiments reported, without exception, concur that in $eNOS^{-/-}$ knock-out animals both males and females are hypertensive, suggesting that any mechanisms that operate to compensate for the lack of eNOS during intra- or extra-uterine development are not sufficient to down-regulate the blood pressure. All these observations single out the unique and crucial role of the continuous vasodilator tone provided by the local generation of NO. It is therefore likely that adaptive or compensatory mechanisms operate in conjunction with the NO dilator tone rather than in its stead. In this respect, there is a great need, especially in in vivo experiments, to differentiate between mechanisms which are modulated directly by NO and those that are the result of the general systemic adaptation to lack of its dilator tone.

3.2
Nitric Oxide and Platelets

Early studies revealed that the vascular endothelial cells possess a non-eicosanoid platelet anti-aggregating and anti-adhesive principle that could be explained by EDRF/NO (Azuma et al. 1986; Radomski et al. 1987a, b). Moreover, it was found that NO strongly synergised with prostacyclin as an inhibitor of platelet aggregation, leading to the suggestion that an interaction between the two compounds explained, at least in part, the non-thrombogenic properties of vascular endothelium (Radomski et al. 1987c).

In the early 1990s, the L-arginine:NO pathway was discovered in platelets and was suggested to act as a negative regulatory mechanism of platelet aggregation (Radomski et al. 1990b, c; Malinski et al. 1993). Since then several groups have described the molecular characteristics of the NOS in platelets (Muruganandam and Mutus 1994; Chen and Mehta 1996; Wallerath et al. 1997; Berkels et al. 1997), leading to the identification of an eNOS mRNA in human and porcine platelets (Wallerath et al. 1997; Berkels et al. 1997).

In vivo, inhibition of NOS has been shown to shorten bleeding time in healthy volunteers (Simon et al. 1995) and to increase platelet accumulation in the vasculature of the rat (Stagliano et al. 1997). A study in which platelets from wild-type or $eNOS^{-/-}$ mice were transfused into thrombocytopaenic eNOS-deficient mice has suggested an independent significant role of platelet-derived eNOS in the modulation of thrombus formation (Freedman et al. 1999).

Interesting developments in the last few years suggest that the eNOS in platelets may be differentially regulated vis-à-vis the vascular endothelial enzyme, in terms of phosphorylation (Fleming et al. 2003), Ca^{2+} sensitivity (Lantoine et al. 1995) and response to certain agonists such as insulin. Although in platelets insulin increases eNOS activity, leading to the attenuation of agonist-induced aggregation (Rao et al. 1990; Trovati et al. 1996), in the vascular endothelium it seems to phosphorylate eNOS without an effect on NO generation or endothelium-dependent relaxation (Fisslthaler et al. 2003; Randiramboavonjy et al. 2004). Studies on the interaction between NO and prostacyclin are also required, especially in relation to the decreased generation of either or both mediators by the vascular endothelium during endothelial dysfunction. Another relevant question is whether or not the mechanisms involved in decreasing endothelial NO in oxidative stress affect NO generation in platelets to a similar extent.

3.3
Nitric Oxide, Vascular Permeability and White Cells

It has long been known that NO modulates leucocyte adhesion to the microcirculation (Kubes et al. 1991). However, the mechanisms responsible for this have not been elucidated and several possibilities remain open. These include modulation of the expression of adhesion molecules such as P-selectin (Gauthier et al. 1994), E-selectin (De Caterina et al. 1995), vascular cell adhesion molecule (Khan et al. 1996) and intercellular adhesion molecule (Biffl et al. 1996)—all of which have been shown to be down-regulated by NO—or the possibility that NO protects cells from oxidative stress by interacting rapidly with and scavenging O_2^- (Gaboury et al. 1993).

Recent results using microvascular endothelial cells from $eNOS^{-/-}$ mice indicate that the role of endothelial NO does not seem to be continuous and tonic, since the simple absence of NO does not in itself lead to endothelial cell activation, measured by the expression of several adhesion proteins. Instead, NO seems to act as a counterbalance for signals that lead to its activation, including the formation of ROS (Kuhlencordt et al. 2004).

Interestingly, NO has been suggested to play a role in maintaining microvascular integrity. Studies have shown that inhibition of eNOS increases microvascular fluid and protein flux (Kubes 1995; Whittle 1997). The increase in vascular permeability due to absence of NO seems to have two distinct phases—an initial one which is white cell-independent and a latter one in which white cells are clearly involved (Kanwar and Kubes 1995). Paradoxically, however, increases in NO generation, even in the small amounts generated by eNOS, have also been claimed to play a role in increasing vascular permeability. This was demonstrated by (1) experiments in which VEGF, which activates eNOS, increases vascular permeability in an NO-dependent manner (Feng et al. 1999), and (2) the way in which inhibition of eNOS by the administration of a chimeric peptide related to caveolin is able to reduce local vascular leakage (Bucci et al. 2000).

Thus, the role of eNOS in maintaining a homeostatic control of vascular permeability, and the way in which it modulates the early white cell-independent and the later cell-dependent changes, remain to be elucidated. Those studies crucially will have to clarify whether or not O_2^- is generated physiologically by the endothelium and, if so, under which circumstances it reduces the bioavailability of NO and when, if at all, it interacts with NO, leading to the generation of peroxynitrite ($ONOO^-$). It also remains to be established when NO generated by other sources, specifically iNOS (Radomski et al. 1990a), comes into play in the process of endothelial activation.

3.4
Nitric Oxide and Vascular Smooth Muscle Proliferation

Nitric oxide inhibits vascular smooth muscle proliferation (Garg and Hassid 1989), and in different models of vascular injury, in both animals and in man, it has been shown that manipulations that increase NO, including transfection of eNOS and administration of NO donors, down-regulate intimal hyperplasia (Lablanche et al. 1997; Janssens et al. 1998; Varenne et al. 1998). Furthermore, in $eNOS^{-/-}$ mice the response to vascular injury leads to intimal hyperplasia which is significantly greater than that observed in wild-type controls (Moroi et al. 1998). Thus, it is clear that exogenous and endogenous NO, including that generated by eNOS, is able to control vascular smooth muscle proliferation once it is activated by injurious stimuli. What remains unclear is whether, under physiological conditions, NO generated by eNOS exerts a tonic control on vascular smooth muscle proliferation, keeping it in a non-proliferative state. Studies in *eNOS* knock-out animals are likely to be complicated by the fact that the hypertensive phenotype per se leads to vascular smooth muscle proliferation.

3.5
Nitric Oxide and Angiogenesis

Nitric oxide derived from eNOS and iNOS has been shown to be involved in angiogenesis (Jenkins et al. 1995; Kroll and Waltenberger 1998; Ziche and Morbidelli 2000) and in capillary organisation (Papapetropoulos et al. 1997). Furthermore, VEGF increases the production of NO via up-regulation of eNOS (van der Zee et al. 1997; Hood et al. 1998), and this NO mediates the migratory and proliferative activity of VEGF (Papapetropoulos et al. 1997; Ziche et al. 1997). The migratory properties of VEGF may be attributable to the activation by NO of podokinesis and its dissolution of the extracellular matrix. Its anti-apoptotic and vasodilator properties may also contribute to the angiogenic actions of NO (Cooke 2003). Angiogenesis induced via VEGF-independent mechanisms is also modulated by NO (Leibovich et al. 1994; Ziche et al. 1994; Vodovotz et al. 1999) and conversely NO is able to induce transforming growth factor (TGF)-β, which is also a potent angiogenic cytokine (Vodovotz et al. 1999).

4
Molecular Targets of the Action of Nitric Oxide

4.1
Soluble Guanylate Cyclase

Activation of the soluble guanylate cyclase is the main mechanism by which NO produces vascular relaxation and inhibition of platelet aggregation (for review see Denninger and Marletta 1999). Activation of the soluble guanylate cyclase leads to an increase in cyclic guanosine monophosphate (cGMP), which in turn decreases $[Ca^{2+}]_i$ flux by inhibiting the flow through voltage-gated Ca^{2+} channels (Blatter and Wier 1994). cGMP also activates cGMP-dependent protein kinases (Schlossmann and Hofmann 2005), in particular, protein kinase GI (PKGI) which is present in vascular smooth muscle (Pfeifer et al. 1998). PKGI phosphorylates proteins in the sarcoplasmic reticulum, including the Ca^{2+}-activated K^+ channels (Sausbier et al. 2000), the 1,4,5 inositol trisphosphate (IP3) receptor-associated cGMP kinase substrate (IRAG; Schlossmann et al. 2000) and phospholamban (Cornwell et al. 1991). Phosphorylation of these proteins leads to the sequestration of Ca^{2+} in the sarcoplasmic reticulum, reduction of cytosolic Ca^{2+} and vascular relaxation (for review see Gewaltig and Kojda 2002). Nitric oxide is also able to prevent Ca^{2+} flux directly by activating Ca^{2+}-dependent K^+ channels through a mechanism independent of cGMP (Bolotina et al. 1994).

In platelets, it has been shown that NO-dependent increases in cGMP also result in a decrease in intracellular Ca^{2+} flux by a mechanism involving PKGI (Massberg et al. 1999). This correlates with inhibition of the association of fibrinogen with glycoprotein IIb/IIIa and with inhibition of platelet activation (see Schwarz et al. 2001). Increases in cGMP can also increase intracellular cyclic adenosine monophosphate (cAMP) indirectly by inhibiting phosphodiesterase III (PDE III; Bowen and Haslam 1991). cAMP, which is the second messenger for the actions of prostacyclin, is also associated with decreases in Ca^{2+} flux (Geiger et al. 1994), thus explaining the synergism between NO and prostacyclin in the platelet.

4.2
Protein S-Nitrosylation

S-Nitrosylation of proteins was identified in 1992 as a post-translational modification potentially involved in NO signalling (Stamler et al. 1992). Early on, a wide variety of proteins, including serum albumin (Stamler et al. 1992), haemoglobin β-subunits (Gow and Stamler 1998), ryanodine-sensitive calcium release channels (Xu et al. 1998), N-methyl-D-aspartate (NMDA) receptors (Choi and Lipton 2000), methionine adenosyl transferase (Perez-Mato et al. 1999) and caspase-3 (Mannick et al. 1999) were identified as targets for S-nitrosylation and a physiological function assigned to it. This list has been extended to many other proteins which have been shown to be susceptible to S-nitrosylation in vitro.

Early in vivo studies in this area were hampered by methodological difficulties; however, the biotin-switch method of Jaffrey et al. (2001) enabled the identification of a number of proteins in the mouse brain that seem to be S-nitrosylated physiologically in vivo, a process which only occurs in some proteins and in specific cysteine groups, and is dependent on the expression of nNOS. More recently, using a similar method, S-nitrosoproteins have been identified in bovine vascular endothelial cells; these were generated not only from exogenously added NO but also from endogenously generated NO, presumably from eNOS (Yang and Loscalzo 2005).

These results have provided evidence that specific S-nitrosylation may occur in vivo and may play a physiological role, leading to a great deal of support for this hypothesis (Hess et al. 2005). However, many questions remain, the most important of which relates to the lack of a clear in vivo correlate of a significant physiological function being modified by this process. Related to this is the lack of understanding of the process of denitrosylation and, more importantly, the precise mechanism by which it actually occurs, since NO itself is a very poor nitrosating species (see Lane et al. 2001). Most of the evidence suggests that S-nitrosylation depends on the formation of higher oxides of nitrogen such as

NO_2, N_2O_3 and $ONOO^-$ (Stamler and Hausladen 1998; Grisham et al. 1999; Viner et al. 1999). In this respect, it is interesting that S-nitrosylated proteins in endothelial cells are either in the mitochondria or localised close to them (Frost et al. 2005; Yang and Loscalzo 2005), a site where NO/O_2 interactions are more likely to occur. This might be suggesting that S-nitrosylation is an early response to oxidative stress (Clementi et al. 1998; Beltran et al. 2000) rather than a physiological mechanism.

4.3
Cytochrome c Oxidase/Mitochondrial Effects

In the mid 1990s it was found that NO modulates the activity of the cytochrome c oxidase, the terminal enzyme in the mitochondrial oxidative phosphorylation chain which catalyses the reduction of O_2 to water (Cleeter et al. 1994; Brown and Cooper 1994; Schweizer and Richter 1994; see Fig. 2a). This effect is reversible, in competition with O_2, and takes place at concentrations of NO likely to occur physiologically. Indeed, the affinity of the cytochrome c oxidase for NO is greater than that for O_2, such that, for example, at 30 µM O_2 the IC_{50} of NO is 30 nM (Brown and Cooper 1994). Later it was demonstrated in vascular endothelial cells that endogenous concentrations of NO modulate cell respiration in an O_2-dependent manner (Clementi et al. 1999) and that exogenous and endogenous NO reduces the consumption of oxygen in isolated canine skeletal and cardiac muscle (Zhao et al. 1999). This led to the suggestion that NO might, on the one hand, modulate cellular bioenergetics by regulating O_2 consumption (Brown 1999; Clementi et al. 1999) and on the other, through inhibition of the cytochrome c oxidase, decrease electron flux through the electron transport chain and favour the generation of O_2^- (Poderoso et al. 1996; Moncada and Erusalimsky 2002; see Fig. 2b). This, as will be discussed later, can lead to the generation of $ONOO^-$ (see Fig. 2c). Furthermore, it is likely that NO, by modulating O_2 consumption in endothelial cell mitochondria, plays a role in diverting O_2 away from the endothelium, thus facilitating the supply of O_2 to the vascular smooth muscle (Poderoso et al. 1996; Hagen et al. 2003).

It has not yet been established whether the NO that inhibits cytochrome c oxidase comes from an eNOS in an extramitochondrial localisation, from eNOS localised to the mitochondria (Bates et al. 1995; Kobzik et al. 1995) or from a different form of NOS present in the mitochondria (Ghafourifar and Richter 1997; Giulivi et al. 1998). Recent evidence from endothelial cells suggests that eNOS is localised to the cytoplasmic face of the outer mitochondrial membrane, where it binds in a manner unrelated to caveolin and therefore unlike the way in which it binds to the outer membrane of the cell (Gao et al. 2004). A different study has demonstrated a protein–protein interaction between mitochondrial nNOS and the cytochrome c oxidase in nervous tissue (Persichini et al. 2005). Both of these studies argue for the co-localisation of NOS with cytochrome c

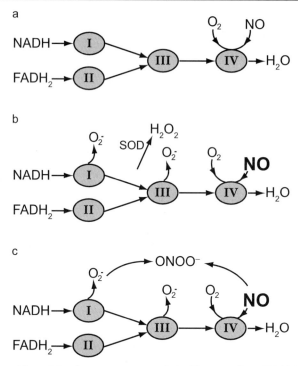

Fig. 2 a–c Nitric oxide and the electron transport chain. Electrons from NADH or FADH$_2$ pass along the electron transport chain. At cytochrome c oxidase (complex IV) they interact with oxygen, and water is produced. **a** This part shows nitric oxide (NO) competing with oxygen at the oxygen-binding site of cytochrome c oxidase. It is a physiological action which occurs under normal conditions. In **b**, the balance between NO and oxygen is shifted in favour of NO, which inhibits cytochrome c oxidase, leading to a reduction of the electron transport chain. This facilitates the generation of superoxide anions (O_2^-) which are subsequently converted to hydrogen peroxide (H_2O_2) by superoxide dismutase (*SOD*). This activates the defence system of the cell. In **c**, a prolonged increase in the generation of O_2^- in the presence of continuous NO results in the formation of peroxynitrite ($ONOO^-$), leading to damage

oxidase, thus favouring the idea of tightly regulated control of this enzyme by NO.

eNOS has also been implicated in mitochondrial biogenesis; interestingly, however, this occurs not through an effect on oxidative phosphorylation, but through an effect on the soluble guanylate cyclase (Nisoli et al. 2003). Calorie restriction has recently been shown to induce eNOS expression and the formation of cGMP in various tissues of the mouse; these effects were attenuated in *eNOS*$^{-/-}$ animals. Thus, NO plays a role in the processes induced by calorie restriction and may be involved in the extension of lifespan in mammals (Nisoli et al. 2005).

5
Nitric Oxide and Pathology

5.1
Changes in NO Generation or Activity in Vascular Pathology

Changes in NO generation have been associated with a number of conditions and disease states in which the vasculature is compromised. These include atherosclerosis, hypercholesterolaemia, hypertension, hyperhomocysteinaemia, pulmonary hypertension, heart failure, smoking, diabetes, Raynaud's syndrome and pre-eclampsia. The objective of this review is not to discuss any of these conditions in detail; there are excellent reviews covering those fields (Maxwell 2002; Barbato and Tzeng 2004) and some aspects are also covered in this book (see J.S. Pober and W. Min; L.E. Spieker et al.; and P. Libby et al., volume II). We will focus instead on the putative mechanisms that might be involved in changes in the generation or actions of NO.

The early stages of a number of the above-mentioned conditions have in common a specific pathophysiological feature, namely endothelial dysfunction (see Stemerman 1981; Luscher et al. 1993). Because of the variety of functions carried out by the vascular endothelium, endothelial dysfunction is likely to include a number of abnormalities, both vascular and haemostatic. However, at present the accepted definition is that of a reduction in endothelial NO, which is measured as a decrease in endothelium-dependent vasodilatation induced either by appropriate agonists (Schachinger et al. 2000) or by flow (Neunteufl et al. 2000). Endothelial dysfunction described in this way occurs prior to any other evidence of cardiovascular disease and can be detected in subjects with a family history of essential hypertension or other risk factors for atherosclerosis (Reddy et al. 1994; Taddei et al. 1996). Furthermore, it has also been associated with smoking (Heitzer et al. 1996) and, in general, its presence is predictive of cardiovascular disease (for review see Asselbergs et al. 2005). Although, as will be described later, decreases in NO formation by the vascular endothelium prior to cardiovascular disease could be due to a variety of reasons, current evidence indicates that the most likely mechanism for this endothelial dysfunction is that of a reduced bioavailability of NO as a result of its interaction with oxygen-derived species, specifically O_2^-. The inactivation of NO by O_2^- is a component of what is called oxidative stress, a term used to describe various deleterious processes resulting from an imbalance between the anti-oxidant defences of tissues and excessive formation of ROS (Turrens 2003).

Although the possibility that free radical formation is involved in vascular damage was considered many years ago (Slater 1972), the discovery of an association between free radical formation and the inactivation of both prostacyclin and NO identified specific biochemical mechanisms responsible for this action (see Moncada 2006). Interactions between NO and O_2^- have been claimed to regulate physiologically the concentrations of NO in the vasculature and therefore to regulate the NO-dependent vasodilator tone. This proposal remains controversial. Generation of O_2^-, however, might occur very early during pathological development in the vascular wall. Consequently, it has been proposed, for example, that generation of O_2^- might be involved in the tolerance to nitroglycerin (Munzel et al. 1996)—a suggestion that also remains controversial (Fung 2004)—as well as in the genesis of angiotensin II-dependent hypertension (Rajagopalan et al. 1996). In this context, it has recently been shown that transgenic mice which generate increased amounts of free radicals from mitochondria have a hypertensive phenotype which can be reversed by anti-oxidants (Bernal-Mizrachi et al. 2005).

The reaction between NO and O_2^- also leads to the formation of ONOO$^-$ (Beckman et al. 1990), a powerful oxidant species that has been implicated in established conditions such as hypercholesterolaemia, diabetes and coronary artery disease (Greenacre and Ischiropoulos 2001). Vascular disease of different origins is, in addition, associated with inflammation (Tracy 2002; Virdis and Schiffrin 2003), which is usually accompanied by the induction of iNOS. Indeed, inflammatory stimuli such as endotoxin lipopolysaccharide and cytokines induce iNOS in many cells and tissues as well as in the vasculature. The excessive production of NO that results from the induction of iNOS in the vasculature is responsible for the profound hypotension and contributes to the tissue damage of septic shock (see Vallance and Moncada 1993). The inducible form of NOS has been identified in macrophages and smooth muscle of animal and human blood vessels in atherosclerosis (Esaki et al. 1997; Buttery et al. 1996; Luoma et al. 1998; De Meyer et al. 2000) and other vascular conditions (Wang et al. 2003; Nagareddy et al. 2005). In advanced atherosclerotic plaques from human blood vessels, iNOS has been found to co-localise with nitrotyrosine, a marker for the formation of ONOO$^-$ (Cromheeke et al. 1999). Interestingly, the use of anti-oxidants improves endothelium-dependent vasodilatation in advanced disease, both in the forearm and coronary arteries of patients with coronary heart disease and diabetes (Levine et al. 1996; Ting et al. 1996; Solzbach et al. 1997).

While the low concentrations of NO generated by the constitutive eNOS protect against atherosclerosis (by, among other things, promoting vasodilatation, preventing leucocyte and platelet activation, preventing the expression of adhesion molecules and inhibiting vascular smooth muscle cell proliferation), it is evident that the higher concentrations generated by iNOS contribute to atherosclerosis through a series of mechanisms which include increased oxidation of low-density lipoprotein (LDL; Cromheeke et al. 1999), and activation of macrophages (De Meyer et al. 2002). These apparent paradoxical actions of NO have been described in other systems (Laszlo et al. 1994) and are supported by recent studies with ApoE knock-out mice in which the concomitant knocking out of eNOS leads to an increase in atherosclerosis (Kuhlencordt et al. 2001a) while knocking out of iNOS reduces atherosclerosis (Kuhlencordt et al. 2001b).

At what stage in the pathophysiological sequence of these conditions does inactivation of NO by O_2^- play a role? This is a particularly pertinent question, since an early intervention may afford the greatest benefit in terms of preventing vascular disease. The switch from the physiological role of NO to its pathological actions seems to be closely related to oxidative stress. The process probably starts by inactivation of NO and reduction of its bioavailability in early disease, and progresses to the formation of pro-oxidant adjuncts, notably $ONOO^-$, which are generated when multiple mechanisms of ROS formation are activated and which overwhelm the anti-oxidant defence of the vascular wall.

5.1.1
The Origin of Free Radicals

There has been a great deal of research investigating the origin of O_2^- in the vasculature. So far, the activation of enzymes such as NADPH oxidases and xanthine oxidase has been implicated, and substantial evidence now exists showing that the activity as well as the expression of these enzymes can be enhanced by pathological stimuli (see Cai and Harrison 2000; Mueller et al. 2005). In addition, vascular cytochrome P-450 enzymes that can generate O_2^- have been described (Fleming 2001), and their inhibition appears to improve endothelium-dependent NO-mediated vasodilatation in patients with coronary artery disease (Fichtlscherer et al. 2004).

Another potential source of O_2^- is what has been called the uncoupled NO synthases. Indeed, eNOS and iNOS have the capacity to generate O_2^- under specific circumstances of low L-arginine or low BH_4 (see Stuehr et al. 2001; Vasquez-Vivar et al. 1998). The uncoupling of eNOS has been demonstrated in several pathological conditions such as diabetes, hypercholesterolaemia and hypertension (Hink et al. 2001; Stroes et al. 1997; Landmesser et al. 2003). Moreover, re-coupling of NOS has been successfully accomplished either by using sepiapterin (Tiefenbacher et al. 1996) or by preventing oxidation of BH_4 (D'Uscio et al. 2003; Landmesser et al. 2003). Uncoupling of eNOS as a result of depletion of both L-arginine and BH_4 is not, however, likely to be an early mechanism of O_2 generation since, if such depletion does occur in pathology, it is likely to result from drastic changes in the vasculature. The uncoupling of eNOS as a result of changes in its association with hsp90 (Pritchard et al. 2001) or in the phosphorylation of Thr^{495} (Lin et al. 2003) might be far more subtle and remains an intriguing possibility. Indeed, the phosphorylation/dephosphorylation of Thr^{495} has been proposed to be an intrinsic switch mechanism that determines whether eNOS generates NO or O_2^- (Lin et al. 2003). Although the relative roles of these mechanisms are at present unknown, increasing emphasis is now placed on the redox balance of the vessel wall, leading to the suggestion that there are a number of vascular diseases in which this balance is disturbed, including heritable deficiency of the anti-oxidant enzymes catalase, haem oxygenase and glutathione peroxidases (Leopold and Loscalzo 2005).

In the last few years, the generation of ROS from mitochondria has become a focus of interest. For many years it has been believed that a small percentage of the O_2 being utilised by these organelles is not completely reduced to water and escapes as O_2^- (Chance et al. 1979). Although it is not clear whether this actually occurs in endothelial cells in vivo, at physiological O_2 concentrations there is the possibility that the redox status of the mitochondrial respiratory chain is determinant in the escape of electrons required to generate O_2^- from O_2. We have recently shown that NO, by favouring the reduction of the cytochrome c oxidase, is able to facilitate the release from mitochondria of O_2^-, which is subsequently converted into H_2O_2 with the resulting signalling consequences (Palacios-Callender et al. 2004; see Fig. 2b). It is likely that such a mechanism, which is an extension of the physiological action of NO on the cytochrome c oxidase, might provide clues to the understanding of the early origins of oxidative stress in the vasculature, specifically in endothelial cells. Recent work has implicated the generation of mitochondrial ROS as initiators of the signalling mechanisms involved in preconditioning (Kimura et al. 2005). Moreover, it has been suggested that the release of H_2O_2 from mitochondria as a signalling molecule occurs under conditions that do not change the redox status of the cells (Go et al. 2004).

It has been known for some time that endothelial cells are highly glycolytic (Mann et al. 2003). We have recently confirmed this observation and demonstrated that the mitochondria of these cells, under the control of NO, seem to act more as signalling organelles, regulating amongst other things the activation of hypoxia-inducible factor-1 and AMP-activated protein kinase, the latter via a ROS-dependent mechanism (Quintero et al. 2006). Thus, the release of mitochondrial ROS may be dependent on a physiologically regulated process, and its primary objective might also be physiological. We have suggested that this may be to maintain a high anti-oxidant potential in these cells (see Moncada and Higgs 2006). Whether an exaggeration of this mechanism may be the initial inactivating mechanism of NO in early disease remains to be established.

5.1.2
Mechanisms Involved in the Decreased Generation of Nitric Oxide

Several polymorphisms have been described in the eNOS promoter, although none is situated within the binding site for known transcription factors (for reviews see Wang and Wang 2000; Wattanapitayakul et al. 2001). The early claimed association of some of these polymorphisms observed in intron 4 or 13 of the eNOS gene with essential hypertension (Nakayama et al. 1997; Uwabo et al. 1998) has not been confirmed in other studies with populations outside the original Japanese cohort (Benjafield and Morris 2000). A different polymorphism, in exon 7 (Glu298Asp) has also been claimed by studies in different populations to be associated with essential hypertension (Miyamoto et al. 1998), coronary artery disease (Hingorani et al. 1999) and myocardial infarction (Shimasaki et al. 1998). However, this polymorphism does not seem to be associated with changes in NO-dependent dilatation (Schneider et al. 2000). Thus, at this stage more research is required to clarify this question and to investigate whether subtle changes in eNOS functioning which are genetically determined become significant only in the presence of genetic polymorphisms affecting the function of other systems, such as angiotensin II (van Geel et al. 1998), or in conjunction with other acquired defects in NO bioavailability or other risk factors such as smoking or advanced age.

Some years ago, it was observed that an endogenous compound, asymmetric dimethylarginine (ADMA), was a competitive inhibitor of the synthesis of NO (Vallance et al. 1992) and it was speculated that this and related compounds may act as endogenous regulators of the L-arginine:NO pathway in health and disease. Since then, increases in plasma concentrations of this compound have been identified in hypercholesterolaemic individuals (Boger et al. 1998) and in other conditions associated with vascular disease such as diabetes (Fard et al. 2000). ADMA has also been identified as an independent risk factor and possible marker in patients with coronary artery disease (Lu et al. 2003). Interestingly, the accumulation of ADMA in blood seems to be the result of a dysfunction of the enzyme responsible for its conversion into L-citrulline. This enzyme, dimethylarginine dimethylaminohydrolase (DDAH), is present in the vascular endothelium (Leiper et al. 1999), and a correlation between oxidised LDL and a decrease in DDAH activity has been described (Ito et al. 1999). This has led to the suggestion that accumulation of ADMA resulting from oxidative stress might play a role in endothelial dysfunction (Fliser 2005).

5.1.3
Replacing Nitric Oxide

It is generally accepted that protection against decreases in eNOS-derived NO in the vasculature may prevent the development of vascular disease or treat it once it is established. In this respect, the most often tried interventions relate to the use of anti-oxidants (see Carr and Frei 2000) and the transfection of eNOS to the vasculature (von der Leyen and Dzau 2001). Each of these interventions has shown some promise in both animal experiments and humans.

There is an unexpected and highly interesting development related to the effect of statins, which in the last few years have been shown to increase production of endothelial NO both in animal and human endothelial cell cultures as well as in animals in vivo. The mechanism(s) by which statins might exert these actions is not clear at present; however, several putative mechanisms have been claimed, including inhibition of the production of LDL cholesterol (Dobrucki et al. 2001) or of mevalonate (Endres et al. 1998), both of which down-regulate the expression of eNOS. Other ways in which statins have been shown to increase eNOS activity include the activation of Akt (Kureishi et al. 2000), increasing the interaction of eNOS with hsp90 (Brouet et al. 2001), increasing the synthesis of BH_4 (Hattori et al. 2002), reducing oxidative stress by decreasing O_2^- generation by NADPH oxidase (Wagner et al. 2000a), and decreasing the abundance of caveolin (Feron et al. 1999).

6
Conclusion

Nitric oxide generated by eNOS has been established as a key regulatory signalling molecule in the vasculature. Its discovery and the elucidation of the myriad roles it plays have contributed greatly to the concept of the vascular endothelium as an active metabolic organ. The details of many of the physiological functions of NO remain to be clarified but, most importantly, its paradoxical role as a pathophysiological agent is only now beginning to be understood. Clarification of this latter role will no doubt throw light on the origin of vascular disease, its prevention and its treatment.

References

Abu Soud HM, Stuehr DJ (1993) Nitric oxide synthases reveal a role for calmodulin in controlling electron transfer. Proc Natl Acad Sci USA 90:10769–10772

Adak S, Wang Q, Stuehr DJ (2000) Arginine conversion to nitroxide by tetrahydrobiopterin-free neuronal nitric oxide synthase. J Biol Chem 275:33554–33561

Ahn KY, Mohaupt MG, Madsen KM, et al. (1994) In situ hybridization localization of mRNA encoding inducible nitric oxide synthase in rat kidney. Am J Physiol 267:F748–F757

Albina JE (1995) On the expression of nitric oxide synthase by human macrophages. Why no NO? J Leukoc Biol 58:643–649

Alderton WK, Cooper CE, Knowles RG (2001) Nitric oxide synthases: structure, function and inhibition. Biochem J 357:593–615

Alderton WK, Angell AD, Craig C, et al. (2005) GW274150 and GW273629 are potent and highly selective inhibitors of inducible nitric oxide synthase in vitro and in vivo. Br J Pharmacol 145:301–312

Aliev G, Bodin P, Burnstock G (1998) Free radical generators cause changes in endothelial and inducible nitric oxide synthases and endothelin-1 immunoreactivity in endothelial cells from hyperlipidemic rabbits. Mol Genet Metab 63:191–197

Aoyagi M, Arvai AS, Tainer JA, et al. (2003) Structural basis for endothelial nitric oxide synthase binding to calmodulin. EMBO J 22:766–775

Arnet UA, McMillan A, Dinerman JL, et al. (1996) Regulation of endothelial nitric oxide synthase during hypoxia. J Biol Chem 271:15069–15073

Asselbergs FW, van der Harst P, Jessurun GAJ, et al. (2005) Clinical impact of vasomotor function assessment and the role of ACE-inhibitors and statins. Vascul Pharmacol 42:125–140

Ayajiki K, Kindermann M, Hecker M, et al. (1996) Intracellular pH and tyrosine phosphorylation but not calcium determine shear stress-induced nitric oxide production in native endothelial cells. Circ Res 78:750–758

Azadzoi KM, Master TA, Siroky MB (2004) Effect of chronic ischemia on constitutive and inducible nitric oxide synthase expression in erectile tissue. J Androl 25:382–388

Azuma H, Ishikawa M, Sekizaki S (1986) Endothelium-dependent inhibition of platelet aggregation. Br J Pharmacol 88:411–415

Baek KJ, Thiel BA, Lucas S, et al. (1993) Macrophage nitric oxide synthase subunits. Purification, characterization, and role of prosthetic groups and substrate in regulating their association into a dimeric enzyme. J Biol Chem 268:21120–21129

Balligand JL, Kelly RA, Marsden PA, et al. (1993) Control of cardiac muscle cell function by an endogenous nitric oxide signaling system. Proc Natl Acad Sci USA 90:347–351

Barbacanne MA, Rami J, Michel JB, et al. (1999) Estradiol increases rat aorta endothelium-derived relaxing factor (EDRF) activity without changes in endothelial NO synthase gene expression: possible role of decreased endothelium-derived superoxide anion production. Cardiovasc Res 41:672–681

Barbato JE, Tzeng E (2004) Nitric oxide and arterial disease. J Vasc Surg 40:187–193

Barouch LA, Harrison RW, Skaf MW, et al. (2002) Nitric oxide regulates the heart by spatial confinement of nitric oxide synthase isoforms. Nature 416:337–339

Bates TE, Loesch A, Burnstock G, et al. (1995) Immunocytochemical evidence for a mitochondrially located nitric oxide synthase in brain and liver. Biochem Biophys Res Commun 213:896–900

Bauer PM, Fulton D, Boo YC (2003) Compensatory phosphorylation and protein-protein interactions revealed by loss of function and gain of function mutants of multiple serine phosphorylation sites in endothelial nitric oxide synthase. J Biol Chem 278:14841–14849

Bauersachs J, Fleming I, Scholz D, et al. (1997) Endothelium-derived hyperpolarizing factor, but not nitric oxide, is reversibly inhibited by brefeldin A. Hypertension 30:1598–1605

Beckman JS, Beckman TW, Chen J, et al. (1990) Apparent hydroxyl radical production by peroxynitrite: implications for endothelial injury from nitric oxide and superoxide. Proc Natl Acad Sci USA 87:1620–1624

Beierwaltes WH, Potter DL, Shesely EG (2002) Renal baroreceptor-stimulated renin in the eNOS knockout mouse. Am J Physiol 282:F59–F64

Beltran B, Orsi A, Clementi E, et al. (2000) Oxidative stress and S-nitrosylation of proteins in cells. Br J Pharmacol 129:953–960

Benjafield AV, Morris BJ (2000) Association analyses of endothelial nitric oxide synthase gene polymorphisms in essential hypertension. Am J Hypertens 13:994–998

Berkels R, Stockklauser K, Rosen P, et al. (1997) Current status of platelet NO synthases. Thromb Res 87:51–55

Bernal-Mizrachi C, Gates AC, Weng S, et al. (2005) Vascular respiratory uncoupling increases blood pressure and atherosclerosis. Nature 435:502–506

Biffl WL, Moore EE, Moore FA, et al. (1996) Nitric oxide reduces endothelial expression of intercellular adhesion molecule (ICAM)-1. J Surg Res 63:328–332

Blatter LA, Wier WG (1994) Nitric oxide decreases $[Ca^{2+}]i$ in vascular smooth muscle by inhibition of the calcium current. Cell Calcium 15:122–131

Blot S, Arnal JF, Xu Y, et al. (1994) Spinal cord infarcts during long-term inhibition of nitric oxide synthase in rats. Stroke 25:1666–1673

Boger RH, Bode-Boger SM, Szuba A, et al. (1998) Asymmetric dimethylarginine (ADMA): a novel risk factor for endothelial dysfunction: its role in hypercholesterolemia. Circulation 98:1842–1847

Bolotina VM, Najibi S, Palacino JJ, et al. (1994) Nitric oxide directly activates calcium-dependent potassium channels in vascular smooth muscle. Nature 368:850–853

Boo YC, Jo H (2003) Flow-dependent regulation of endothelial nitric oxide synthase: role of protein kinases. Am J Physiol 285:C499–C508

Boo YC, Sorescu G, Boyd N, et al. (2002) Shear stress stimulates phosphorylation of endothelial nitric-oxide synthase at Ser1179 by Akt-independent mechanisms: role of protein kinase A. J Biol Chem 277:3388–3396

Boo YC, Sorescu GP, Bauer PM, et al. (2003) Phosphorylation of eNOS at Ser635 stimulates NO production in a Ca^{2+}-independent manner. FASEB J 17:A805

Bouloumie A, Schini-Kerth VB, Busse R (1999) Vascular endothelial growth factor up-regulates nitric oxide synthase expression in endothelial cells. Cardiovasc Res 41:773–780

Boutin JA (1997) Myristoylation. Cell Commun Signal 9:15–35

Bowen R, Haslam RJ (1991) Effects of nitrovasodilators on platelet cyclic nucleotide levels in rabbit blood; role for cyclic AMP in synergistic inhibition of platelet function by SIN-1 and prostaglandin E_1. J Cardiovasc Pharmacol 17:424–433

Brandes RP, Schmitz-Winnenthal FH, Feletou M, et al. (2000) An endothelium-derived hyperpolarizing factor distinct from NO and prostacyclin is a major endothelium-dependent vasodilator in resistance vessels of wild-type and endothelial NO synthase knockout mice. Proc Natl Acad Sci USA 97:9747–9752

Bredt DS, Snyder SH (1990) Isolation of nitric oxide synthetase, a calmodulin-requiring enzyme. Proc Natl Acad Sci USA 87:682–685

Bredt DS, Hwang PM, Glatt CE, et al. (1991) Cloned and expressed nitric oxide synthase structurally resembles cytochrome P-450 reductase. Nature 351:714–718

Brouet A, Sonveaux P, Dessy C, et al. (2001) Hsp90 and caveolin are key targets for the proangiogenic nitric oxide-mediated effects of statins. Circ Res 89:866–873

Brown GC (1999) Nitric oxide and mitochondrial respiration. Biochim Biophys Acta 1411:351–369

Brown GC, Cooper CE (1994) Nanomolar concentrations of nitric oxide reversibly inhibit synaptosomal respiration by competing with oxygen at cytochrome oxidase. FEBS Lett 356:295–298

Bucci M, Gratton JP, Rudic RD, et al. (2000) In vivo delivery of the caveolin-1 scaffolding domain inhibits nitric oxide synthesis and reduces inflammation. Nat Med 6:1362–1367

Busconi L, Michel T (1993) Endothelial nitric oxide synthase: N-terminal myristoylation determines subcellular localization. J Biol Chem 268:8410–8413

Busse R, Mulsch A (1990) Calcium-dependent nitric oxide synthesis in endothelial cytosol is mediated by calmodulin. FEBS Lett 265:133–136

Buttery LD, Springall DR, Chester AH, et al. (1996) Inducible nitric oxide synthase is present within human atherosclerotic lesions and promotes the formation and activity of peroxynitrite. Lab Invest 75:77–85

Cai H, Harrison DG (2000) Endothelial dysfunction in cardiovascular diseases: the role of oxidant stress. Circ Res 87:840–844

Cai H, Davis ME, Drummond GR, et al. (2001) Induction of endothelial NO synthase by hydrogen peroxide via a Ca^{2+}/calmodulin-dependent protein kinase II/janus kinase 2-dependent pathway. Arterioscler Thromb Vasc Biol 21:1571–1576

Cao S, Yao J, Shah V (2003) The proline-rich domain of dynamin-2 is responsible for dynamin-dependent in vitro potentiation of endothelial nitric oxide synthase activity via selective effects on reductase domain function. J Biol Chem 278:5894–5901

Carr A, Frei B (2000) The role of natural antioxidants in preserving the biological activity of endothelium-derived nitric oxide. Free Radic Biol Med 28:1806–1814

Caulin-Glaser T, Garcia-Cardena G, Sarrel P, et al. (1997) 17 β-Estradiol regulation of human endothelial cell basal nitric oxide release, independent of cytosolic Ca^{2+} mobilization. Circ Res 81:885–892

Chambliss KL, Shaul PW (2002) Rapid activation of endothelial NO synthase by estrogen: evidence for a steroid receptor fast-action complex (SRFC) in caveolae. Steroids 67:413–419

Chance B, Sies B, Boveris A (1979) Hydroperoxide metabolism in mammalian organs. Physiol Rev 59:527–605

Charles IG, Palmer RM, Hickery MS, et al. (1993) Cloning, characterization, and expression of a cDNA encoding an inducible nitric oxide synthase from the human chondrocyte. Proc Natl Acad Sci USA 90:11419–11423

Chartrain NA, Geller DA, Koty PP, et al. (1994) Molecular cloning, structure and chromosomal localization of the human inducible nitric oxide synthase gene. J Biol Chem 269:6765–6772

Chen LY, Mehta JL (1996) Further evidence of the presence of constitutive and inducible nitric oxide synthase isoforms in human platelets. J Cardiovasc Pharmacol 27:154–158

Chen ZP, Mitchelhill KI, Michell BJ, et al. (1999) AMP-activated protein kinase phosphorylation of endothelial NO synthase. FEBS Lett 443:285–289

Choi YB, Lipton SA (2000) Redox modulation of the NMDA receptor. Cell Mol Life Sci 57:1535–1541

Chowdhary S, Townend JN (1999) Role of nitric oxide in the regulation of cardiovascular autonomic control. Clin Sci (Lond) 97:5–17

Cieslik K, Abrams CS, Wu KK (2001) Up-regulation of endothelial nitric-oxide synthase promoter by the phosphatidylinositol 3-kinase gamma/Janus kinase 2/MEK-1-dependent pathway. J Biol Chem 276:1211–1219

Cleeter MW, Cooper JM, Darley-Usmar VM, et al. (1994) Reversible inhibition of cytochrome c oxidase, the terminal enzyme of the mitochondrial respiratory chain, by nitric oxide. Implications for neurodegenerative diseases. FEBS Lett 345:50–54

Clementi E, Brown GC, Feelisch M, et al. (1998) Persistent inhibition of cell respiration by nitric oxide: crucial role of S-nitrosylation of mitochondrial complex I and protective action of glutathione. Proc Natl Acad Sci USA 95:7631–7636

Clementi E, Brown GC, Foxwell N, et al. (1999) On the mechanism by which vascular endothelial cells regulate their oxygen consumption. Proc Natl Acad Sci USA 96:1559–1562

Cohen RA (2005) The endothelium-derived hyperpolarizing factor puzzle. A mechanism without a mediator? Circulation 111:724–727

Cooke JP (2003) NO and angiogenesis. Atheroscler Suppl 4:53–60

Cornwell TL, Pryzwansky KB, Wyatt TA, et al. (1991) Regulation of sarcoplasmic reticulum protein phosphorylation by localized cyclic GMP-dependent protein kinase in vascular smooth muscle cells. Mol Pharmacol 40:923–931

Crane BR, Arvai AS, Gachhui R, et al. (1997) The structure of nitric oxide synthase oxygenase domain and inhibitor complexes. Science 278:425–431

Cromheeke KM, Kockx MM, De Meyer GR, et al. (1999) Inducible nitric oxide synthase colocalizes with signs of lipid oxidation/peroxidation in human atherosclerotic plaques. Cardiovasc Res 43:744–754

Cubberley RR, Alderton WK, Boyhan A, et al. (1997) Cysteine-200 of human inducible nitric oxide synthase is essential for dimerization of haem domains and for binding of haem, nitroarginine and tetrahydrobiopterin. Biochem J 323:131–146

D'Uscio LV, Milstien S, Richardson D, et al. (2003) Long-term vitamin C treatment increases vascular tetrahydrobiopterin levels and nitric oxide synthase activity. Circ Res 92:88–95

Daff S, Sagami I, Shimizu T (1999) The 42-amino acid insert in the FMN domain of neuronal nitric-oxide synthase exerts control over Ca$^{(2+)}$/calmodulin-dependent electron transfer. J Biol Chem 274:30589–30595

Davis ME, Grumbach IM, Fukai T, et al. (2004) Shear stress regulates endothelial nitric-oxide synthase promoter activity through nuclear factor kappaB binding. J Biol Chem 279:163–168

De Caterina R, Libby P, Peng HB, et al. (1995) Nitric oxide decreases cytokine-induced endothelial activation. Nitric oxide selectively reduces endothelial expression of adhesion molecules and proinflammatory cytokines. J Clin Invest 96:60–68

Dedio J, Konig P, Wohlfart P, et al. (2001) NOSIP, a novel modulator of endothelial nitric oxide synthase activity. FASEB J 15:79–89

De Meyer GR, Kockx MM, Cromheeke KM, et al. (2000) Periadventitial inducible nitric oxide synthase expression and intimal thickening. Arterioscler Thromb Vasc Biol 20:1896–1902

De Meyer GR, De Cleen DM, Cooper S, et al. (2002) Platelet phagocytosis and processing of beta-amyloid precursor protein as a mechanism of macrophage activation in atherosclerosis. Circ Res 90:1197–1204

Denninger JW, Marletta MA (1999) Guanylate cyclase and the NO/cGMP signaling pathway. Biochim Biophys Acta 1411:334–350

Dimmeler S, Fleming I, Fisslthaler B, et al. (1999) Activation of nitric oxide synthase in endothelial cells by Akt-dependent phosphorylation. Nature 399:601–605

Ding H, Kubes P, Triggle C (2000) Potassium- and acetylcholine-induced vasorelaxation in mice lacking endothelial nitric oxide synthase. Br J Pharmacol 129:1194–1200

Djordjevic S, Roberts DL, Wang M (1995) Crystallization and preliminary X-ray studies of NADPH-cytochrome P450 reductase. Proc Natl Acad Sci USA 92:3214–3218

Dobrucki LW, Kalinowski L, Dobrucki IT, et al. (2001) Statin-stimulated nitric oxide release from endothelium. Med Sci Monit 7:622–627

Dunphy JT, Linder ME (1998) Signalling functions of protein palmitoylation. Biochem Biophys Acta 1436:245–261

Durante W, Schini VB, Scott-Burden T, et al. (1991) Platelet inhibition by an L-arginine-derived substance released by IL-1 beta-treated vascular smooth muscle cells. Am J Physiol 261:H2024–H2030

Endres M, Laufs U, Huang Z, et al. (1998) Stroke protection by 3-hydroxy-3-methylglutaryl (HMG)-CoA reductase inhibitors mediated by endothelial nitric oxide synthase. Proc Natl Acad Sci USA 95:8880–8885

Esaki T, Hayashi T, Muto E, et al. (1997) Expression of inducible nitric oxide synthase in T lymphocytes and macrophages of cholesterol-fed rabbits. Atherosclerosis 128:39–46

Fard A, Tuck CH, Donis JA, et al. (2000) Acute elevations of plasma asymmetric dimethylarginine and impaired endothelial function in response to a high-fat meal in patients with type 2 diabetes. Arterioscler Thromb Vasc Biol 20:2039–2044

Feng Y, Venema VJ, Venema RC, et al. (1999) VEGF-induced permeability increase is mediated by caveolae. Invest Ophthalmol Vis Sci 40:157–167

Feron O, Saldana F, Michel JB, et al. (1998) The endothelial nitric oxide synthase-caveolin regulatory cycle. J Biol Chem 273:3125–3128

Feron O, Dessy C, Moniotte S, et al. (1999) Hypercholesterolemia decreases nitric oxide production by promoting the interaction of caveolin and endothelial nitric oxide synthase. J Clin Invest 103:897–905

Fichtlscherer S, Dimmeler S, Breuer S, et al. (2004) Inhibition of cytochrome P450 2C9 improves endothelium-dependent, nitric oxide-mediated vasodilatation in patients with coronary artery disease. Circulation 109:178–183

Fischmann TO, Hruza A, Niu XD, et al. (1999) Structural characterization of nitric oxide synthase isoforms reveals striking active-site conservation. Nat Struct Biol 6:233–242

Fisslthaler B, Dimmeler S, Hermann C, et al. (2000) Phosphorylation and activation of the endothelial nitric oxide synthase by fluid shear stress. Acta Physiol Scand 168:81–88

Fisslthaler B, Benzing T, Busse R, et al. (2003) Insulin enhances the expression of the endothelial nitric oxide synthase in native endothelial cells: a dual role for Akt and AP-1. Nitric Oxide 8:253–261

Fleming I (2001) Cytochrome p450 and vascular homeostasis. Circ Res 89:753–762

Fleming I, Busse R (2003) Molecular mechanisms involved in the regulation of the endothelial nitric oxide synthase. Am J Physiol 284:R1–R12

Fleming I, Fisslthaler B, Dimmeler S, et al. (2001) Phosphorylation of Thr^{495} regulates Ca^{2+}/calmodulin-dependent endothelial nitric oxide synthase activity. Circ Res 88:e68–e75

Fleming I, Schulz C, Fichtlscherer B, et al. (2003) AMP-activated protein kinase (AMPK) regulates the insulin-induced activation of the nitric oxide synthase in human platelets. Thromb Haemost 90:863–871

Fliser D (2005) Asymmetric dimethylarginine (ADMA): the silent transition from an 'uraemic toxin' to a global cardiovascular risk molecule. Eur J Clin Invest 35:71–79

Förstermann U, Pollock JS, Schmidt HH, et al. (1991) Calmodulin-dependent endothelium-derived relaxing factor/nitric oxide synthase activity is present in the particulate and cytosolic fractions of bovine aortic endothelial cells. Proc Natl Acad Sci USA 88:1788–1792

Förstermann U, Closs EI, Pollock JS, et al. (1994) Nitric oxide synthase isozymes. Characterization, purification, molecular cloning and functions. Hypertension 23:1121–1131

Freedman JE, Sauter R, Battinelli EM, et al. (1999) Deficient platelet-derived nitric oxide and enhanced hemostasis in mice lacking the NOSIII gene. Circ Res 84:1416–1421

Frost MT, Wang Q, Moncada S, et al. (2005) Hypoxia accelerates nitric oxide-dependent inhibition of mitochondrial complex I in activated macrophages. Am J Physiol 288:R394–R400

Fukuchi M, Giaid A (1999) Endothelial expression of endothelial nitric oxide synthase and endothelin-1 in human coronary artery disease. Specific reference to underlying lesion. Lab Invest 79:659–670

Fulton D, Gratton JP, McCabe TJ, et al. (1999) Regulation of endothelium-derived nitric oxide production by the protein kinase Akt. Nature 399:597–601

Fulton D, Babbitt R, Zoellner S, et al. (2004) Targeting of endothelial nitric oxide synthase to the cytoplasmic face of the golgi complex or plasma membrane regulates Akt- versus calcium-dependent mechanisms for nitric oxide release. J Biol Chem 279:30349–30357

Fung HL (2004) Biochemical mechanism of nitroglycerin action and tolerance: is this old mystery solved? Annu Rev Pharmacol Toxicol 44:67–85

Furchgott RF, Zawadzki JV (1980) The obligatory role of endothelial cells in the relaxation of arterial smooth muscle by acetylcholine. Nature 288:373–376

Gaboury J, Woodman RC, Granger DN, et al. (1993) Nitric oxide prevents leukocyte adherence: role of superoxide. Am J Physiol 265:H862–H867

Gao S, Chen J, Brodsky SV, et al. (2004) Docking of endothelial nitric oxide synthase (eNOS) to the mitochondrial outer membrane: a pentabasic amino acid sequence in the autoinhibitory domain of eNOS targets a proteinase K-cleavable peptide on the cytoplasmic face of mitochondria. J Biol Chem 279:15968–15974

Garcia-Cardena G, Oh P, Liu J, et al. (1996) Targeting of nitric oxide synthase to endothelial cell caveolae via palmitoylation: implications for nitric oxide signaling. Proc Natl Acad Sci USA 93:6448–6453

Garcia-Cardena G, Fan R, Shah V, et al. (1998) Dynamic activation of endothelial nitric oxide synthase by Hsp90. Nature 392:821–824

Garg UC, Hassid A (1989) Nitric oxide-generating vasodilators and 8-bromo-cyclic guanosine monophosphate inhibit mitogenesis and proliferation of cultured rat vascular smooth muscle cells. J Clin Invest 83:1774–1777

Gauthier TW, Davenpeck KL, Lefer AM (1994) Nitric oxide attenuates leukocyte-endothelial interaction via P-selectin in splanchnic ischemia-reperfusion. Am J Physiol 267:G562–G568

Geiger J, Nolte C, Walter U (1994) Regulation of calcium mobilization and entry in human platelets by endothelium-derived factors. Am J Physiol 267:C236–C244

Geller DA, Lowenstein CJ, Shapiro RA, et al. (1993) Molecular cloning and expression of inducible nitric oxide synthase from human hepatocytes. Proc Natl Acad Sci USA 90:3491–3495

Gewaltig MT, Kojda G (2002) Vasoprotection by nitric oxide: mechanisms and therapeutic potential. Cardiovasc Res 55:250–260

Ghafourifar P, Richter C (1997) Nitric oxide synthase activity in mitochondria. FEBS Lett 418:291–296

Giulivi C, Poderoso JJ, Boveris A (1998) Production of nitric oxide by mitochondria. J Biol Chem 273:11038–11043

Go YM, Park H, Maland MC, et al. (1998) Phosphatidylinositol 3-kinase γ mediates shear stress-dependent activation of JNK in endothelial cells. Am J Physiol 275:H1898–H1904

Go YM, Gipp JJ, Mulcahy RT, et al. (2004) H_2O_2-dependent activation of GCLC-ARE4 reporter occurs by mitogen-activated protein kinase pathways without oxidation of cellular glutathione or thioredoxin-1. J Biol Chem 279:5837–5845

Govers R, Rabelink TJ (2001) Cellular regulation of endothelial nitric oxide synthase. Am J Physiol Renal Physiol 280:F193–F206

Govers R, van der Sluijs P, van Donselaar E, et al. (2002) Endothelial nitric oxide synthase and its negative regulator caveolin-1 localize to distinct perinuclear organelles. J Histochem Cytochem 50:779–788

Gow AJ, Stamler JS (1998) Reactions between nitric oxide and haemoglobin under physiological conditions. Nature 391:169–173

Gratton JP, Fontana J, O'Connor DS, et al. (2000) Reconstitution of an endothelial nitric oxide synthase (eNOS), hsp90, and caveolin-1 complex in vivo. J Biol Chem 275:22268–22272

Greenacre SA, Ischiropoulos H (2001) Tyrosine nitration: localization, quantification, consequences for protein function and signal transduction. Free Radic Res 34:541–581

Grisham MB, Jourd'Heuil D, Wink DA (1999) Nitric oxide. I. Physiological chemistry of nitric oxide and its metabolites: implications in inflammation. Am J Physiol 276:G315–G321

Hagen T, Taylor CT, Lam F, et al. (2003) Redistribution of intracellular oxygen in hypoxia by nitric oxide: effect on HIF1α. Science 302:1975–1978

Harris MB, Ju H, Venema VJ, et al. (2001) Reciprocal phosphorylation and regulation of endothelial nitric oxide synthase in response to bradykinin stimulation. J Biol Chem 276:16587–16591

Hattori Y, Nakanishi N, Kasai K (2002) Statin enhances cytokine-mediated induction of nitric oxide synthesis in vascular smooth muscle cells. Cardiovasc Res 54:649–658

Heitzer T, Yla-Herttuala S, Luoma J, et al. (1996) Cigarette smoking potentiates endothelial dysfunction of forearm resistance vessels in patients with hypercholesterolemia. Role of oxidized LDL. Circulation 93:1346–1353

Hernandez-Perera O, Perez-Sala D, Navarro-Antolin J, et al. (1998) Effects of the 3-hydroxy-3-methylglutaryl-CoA reductase inhibitors, atorvastatin and simvastatin, on the expression of endothelin-1 and endothelial nitric oxide synthase in vascular endothelial cells. J Clin Invest 101:2711–2719

Hess DT, Matsumoto A, Kim SO, et al. (2005) Protein S-nitrosylation: purview and parameters. Nat Rev Mol Cell Biol 6:150–166

Hibbs JB, Taintor RR, Vavrin Z, et al. (1990) Synthesis of nitric oxide from a terminal guanidine nitrogen atom of L-arginine: a molecular mechanism regulating cellular proliferation that targets intracellular iron. In: Moncada S, Higgs EA (eds) Nitric oxide from L-arginine: a bioregulatory system. Elsevier, Amsterdam, pp 189–223

Hingorani AD, Liang CF, Fatibene J, et al. (1999) A common variant of the endothelial nitric oxide synthase (Glu298Asp) is a major risk factor for coronary artery disease in the UK. Circulation 100:1515–1520

Hink U, Li H, Mollnau H, et al. (2001) Mechanisms underlying endothelial dysfunction in diabetes mellitus. Circ Res 88:E14–E22

Hobbs AJ, Higgs A, Moncada S (1999) Inhibition of nitric oxide synthase as a potential therapeutic target. Annu Rev Pharmacol Toxicol 39:191–220

Hood JD, Meininger CJ, Ziche M, et al. (1998) VEGF upregulates ecNOS message, protein, and NO production in human endothelial cells. Am J Physiol 274:H1054–H1058

Hoyer J, Köhler R, Distler A (1998) Mechanosensitive Ca^{2+} oscillations and STOC activation in endothelial cells. FASEB J 12:359–366

Huang A, Sun D, Koller A, et al. (1998) Gender difference in flow-induced dilation and regulation of shear stress: role of estrogen and nitric oxide. Am J Physiol 275:R1571–R1577

Huang A, Sun D, Carroll MA, et al. (2001) EDHF mediates flow-induced dilation in skeletal muscle arterioles of female eNOS-KO mice. Am J Physiol 280:H2462–H2469

Huang PL, Huang Z, Mashimo H, et al. (1995) Hypertension in mice lacking the gene for endothelial nitric oxide synthase. Nature 377:239–242

Inoue N, Venema RC, Sayegh HS, et al. (1995) Molecular regulation of the bovine endothelial cell nitric oxide synthase by transforming growth factor-beta 1. Arterioscler Thromb Vasc Biol 15:1255–1261

Ito A, Tsao PS, Adimoolam S, et al. (1999) Novel mechanism for endothelial dysfunction: dysregulation of dimethylarginine dimethylaminohydrolase. Circulation 99:3092–3095

Jaffrey SR, Erdjument-Bromage H, Ferris CD, et al. (2001) Protein S-nitrosylation: a physiological signal for neuronal nitric oxide. Nat Cell Biol 3:193–197

Janssens S, Flaherty D, Nong Z, et al. (1998) Human endothelial nitric oxide synthase gene transfer inhibits vascular smooth muscle cell proliferation and neointima formation after balloon injury in rats. Circulation 97:1274–1281

Janssens SP, Shimouchi A, Quertermous T, et al. (1992) Cloning and expression of a cDNA encoding human endothelium-derived relaxing factor/nitric oxide synthase. J Biol Chem 267:14519–14522

Jenkins DC, Charles IG, Thomsen LL, et al. (1995) Roles of nitric oxide in tumor growth. Proc Natl Acad Sci USA 92:4392–4396

Jiang J, Cyr D, Babbitt RW, et al. (2003) Chaperone-dependent regulation of endothelial nitric oxide synthase intracellular trafficking by the co-chaperone/ubiquitin ligase CHIP. J Biol Chem 278:49332–49341

Jobin CM, Chen H, Lin AJ, et al. (2003) Receptor-regulated dynamic interaction between endothelial nitric oxide synthase and calmodulin revealed by fluorescence resonance energy transfer in living cells. Biochemistry 42:11716–11725

Kanai AJ, Pearce LL, Clemens PR, et al. (2001) Identification of a neuronal nitric oxide synthase in isolated cardiac mitochondria using electrochemical detection. Proc Natl Acad Sci USA 98:14126–14131

Kanwar S, Kubes P (1995) Nitric oxide is an antiadhesive molecule for leukocytes. New Horiz 3:93–104

Karantzoulis-Fegaras F, Antoniou H, Lai SL, et al. (1999) Characterization of the human endothelial nitric-oxide synthase promoter. J Biol Chem 274:3076–3093

Khan BV, Harrison DG, Olbrych MT, et al. (1996) Nitric oxide regulates vascular cell adhesion molecule 1 gene expression and redox-sensitive transcriptional events in human vascular endothelial cells. Proc Natl Acad Sci USA 93:9114–9119

Kilbourn RG, Jubran A, Gross SS, et al. (1990) Reversal of endotoxin-mediated shock by N^G-methyl-L-arginine, an inhibitor of nitric oxide synthesis. Biochem Biophys Res Commun 172:1132–1138

Kim JW, Kang KW, Oh GT, et al. (2002) Induction of hepatic inducible nitric oxide synthase by cholesterol in vivo and in vitro. Exp Mol Med 34:137–144

Kimura S, Zhang GX, Nishiyama A (2005) Mitochondria-derived reactive oxygen species and vascular MAP kinases: comparison of angiotensin II and diazoxide. Hypertension 45:438–444

Knowles RG, Palacios M, Palmer RMJ, et al. (1989) Formation of nitric oxide from L-arginine in the central nervous system: a transduction mechanism for stimulation of the soluble guanylate cyclase. Proc Natl Acad Sci USA 86:5159–5162

Knowles RG, Salter M, Brooks SL, et al. (1990) Anti-inflammatory glucocorticoids inhibit the induction by endotoxin of nitric oxide synthase in the lung, liver and aorta of the rat. Biochem Biophys Res Commun 172:1042–1048

Kobzik L, Stringer B, Balligand JL, et al. (1995) Endothelial type nitric oxide synthase in skeletal muscle fibers: mitochondrial relationships. Biochem Biophys Res Commun 211:375–381

Kojda G, Cheng YC, Burchfield J, et al. (2001) Dysfunctional regulation of endothelial nitric oxide synthase (eNOS) expression in response to exercise in mice lacking one eNOS gene. Circulation 103:2839–2844

Korhonen R, Lahti A, Hamalainen M, et al. (2002) Dexamethasone inhibits inducible nitric oxide synthase expression and nitric oxide production by destabilizing mRNA in lipopolysaccharide-treated macrophages. Mol Pharmacol 62:698–704

Kroll J, Waltenberger J (1998) VEGF-A induces expression of eNOS and iNOS in endothelial cells via VEGF receptor-2 (KDR). Biochem Biophys Res Commun 252:743–746

Kubes P (1995) Nitric oxide affects microvascular permeability in the intact and inflamed vasculature. Microcirculation 2:235–244

Kubes P, Suzuki M, Granger DN (1991) Nitric oxide: an endogenous modulator of leukocyte adhesion. Proc Natl Acad Sci USA 88:4651–4655

Kuhlencordt PJ, Gyurko R, Han F, et al. (2001a) Accelerated atherosclerosis, aortic aneurysm formation, and ischemic heart disease in apolipoprotein E/endothelial nitric oxide synthase double-knockout mice. Circulation 104:448–454

Kuhlencordt PJ, Chen J, Han F, et al. (2001b) Genetic deficiency of inducible nitric oxide synthase reduces atherosclerosis and lowers plasma lipid peroxides in apolipoprotein E-knockout mice. Circulation 103:3099–3104

Kuhlencordt PJ, Rosel E, Gerszten RE, et al. (2004) Role of endothelial nitric oxide synthase in endothelial activation: insights from eNOS knockout endothelial cells. Am J Physiol 286:C1195–C1202

Kureishi Y, Luo Z, Shiojima I, et al. (2000) The HMG-CoA reductase inhibitor simvastatin activates the protein kinase Akt and promotes angiogenesis in normocholesterolemic animals. Nat Med 6:1004–1010

Kurihara N, Alfie ME, Sigmon DH, et al. (1998) Role of nNOS in blood pressure regulation in eNOS null mutant mice. Hypertension 32:856–861

Kurtz A, Wagner C (1998) Role of nitric oxide in the control of renin secretion. Am J Physiol 275:F849–F862

Lablanche JM, Grollier G, Lusson JR, et al. (1997) Effect of the direct nitric oxide donors linsidomine and molsidomine on angiographic restenosis after coronary balloon angioplasty. The ACCORD Study. Angioplastic Coronaire Corvasal Diltiazem. Circulation 95:83–89

Lamas S, Marsden PA, Li GK, et al. (1992) Endothelial nitric oxide synthase: molecular cloning and characterization of a distinct constitutive enzyme isoform. Proc Natl Acad Sci USA 89:6348–6352

Landmesser U, Dikalov S, Price SR, et al. (2003) Oxidation of tetrahydrobiopterin leads to uncoupling of endothelial cell nitric oxide synthase in hypertension. J Clin Invest 111:1201–1209

Lane P, Gross SS (2002) Disabling a C-terminal autoinhibitory control element in endothelial nitric-oxide synthase by phosphorylation provides a molecular explanation for activation of vascular NO synthesis by diverse physiological stimuli. J Biol Chem 277:19087–19094

Lane P, Hao G, Gross SS (2001) S-nitrosylation is emerging as a specific and fundamental post-translational protein modification: head-to-head comparison with O-phosphorylation. Sci STKE 86:RE1

Lantoine F, Brunet A, Bedioui F, et al. (1995) Direct measurement of nitric oxide production in platelets: relationship with cytosolic Ca^{2+} concentration. Biochem Biophys Res Commun 215:842–848

Laszlo F, Whittle BJR, Moncada S (1994) Time-dependent enhancement or inhibition of endotoxin-induced vascular injury in rat intestine by nitric oxide synthase inhibitors. Br J Pharmacol 111:1309–1315

Laufs U, Fata VL, Liao JK (1997) Inhibition of 3-hydroxy-methylglutaryl (HMG)-CoA reductase blocks hypoxia-mediated down-regulation of endothelial nitric oxide synthase. J Biol Chem 272:31725–31729

Leibovich SJ, Polverini PJ, Fong TW, et al. (1994) Production of angiogenic activity by human monocytes requires an L-arginine/nitric oxide-synthase-dependent effector mechanism. Proc Natl Acad Sci USA 91:4190–4194

Leiper JM, Santa Maria J, Chubb A, et al. (1999) Identification of two human dimethylarginine dimethylaminohydrolases with distinct tissue distributions and homology with microbial arginine deiminases. Biochem J 343:209–214

Leopold JA, Loscalzo J (2005) Oxidative enzymopathies and vascular disease. Arterioscler Thromb Vasc Biol 25:1332–1340

Levine GN, Frei B, Koulouris SN, et al. (1996) Ascorbic acid reverses endothelial vasomotor dysfunction in patients with coronary artery disease. Circulation 93:1107–1113

Lin MI, Fulton D, Babbitt R, et al. (2003) Phosphorylation of threonine 497 in endothelial nitric oxide synthase coordinates the coupling of L-arginine metabolism to efficient nitric oxide production. J Biol Chem 278:44719–44726

Liu J, Garcia Cardena G, Sessa WC (1995) Biosynthesis and palmitoylation of endothelial nitric oxide synthase: mutagenesis of palmitoylation sites, cysteines-15 and/or -26, argues against depalmitoylation-induced translocation of the enzyme. Biochemistry 34:12333–12340

Liu J, Hughes TE, Sessa WC (1997) The first 35 amino acids and fatty acylation sites determine the molecular targeting of endothelial nitric oxide synthase into the golgi region of cells: a green fluorescent protein study. J Cell Biol 137:1525–1535

Lopez-Ongil S, Hernandez-Perera O, Navarro-Antolin J, et al. (1998) Role of reactive oxygen species in the signalling cascade of cyclosporine A-mediated up-regulation of eNOS in vascular endothelial cells. Br J Pharmacol 124:447–454

Lu JL, Schmiege LM 3rd, Kuo L, et al. (1996) Downregulation of endothelial constitutive nitric oxide synthase expression by lipopolysaccharide. Biochem Biophys Res Commun 225:1–5

Lu TM, Ding YA, Charng MJ, et al. (2003) Asymmetrical dimethylarginine: a novel risk factor for coronary artery disease. Clin Cardiol 26:458–464

Luckhoff A, Pohl U, Mulsch A, et al. (1988) Differential role of extra- and intracellular calcium in the release of EDRF and prostacyclin from cultured endothelial cells. Br J Pharmacol 95:189–196

Luo Z, Fujio Y, Kureishi Y, et al. (2000) Acute modulation of endothelial Akt/PKB activity alters nitric oxide-dependent vasomotor activity in vivo. J Clin Invest 106:493–499

Luoma JS, Stralin P, Marklund SL, et al. (1998) Expression of extracellular SOD and iNOS in macrophages and smooth muscle cells in human and rabbit atherosclerotic lesions: colocalization with epitopes characteristic of oxidized LDL and peroxynitrite-modified proteins. Arterioscler Thromb Vasc Biol 18:157–167

Luscher TF, Tanner FC, Tschudi MR, et al. (1993) Endothelial dysfunction in coronary artery disease. Annu Rev Med 44:395–418

MacNaul KL, Hutchinson NI (1993) Differential expression of iNOS and cNOS mRNA in human vascular smooth muscle cells and endothelial cells under normal and inflammatory conditions. Biochem Biophys Res Commun 196:1330–1334

Malinski T, Radomski MW, Taha Z, et al. (1993) Direct electrochemical measurement of nitric oxide released from human platelets. Biochem Biophys Res Commun 194:960–965

Mann GE, Yudilevich DL, Sobrevia L (2003) Regulation of amino acid and glucose transporters in endothelial and smooth muscle cells. Physiol Rev 83:183–252

Mannick JB, Hausladen A, Liu L, et al. (1999) Fas-induced caspase denitrosylation. Science 284:651–654

Marrero MB, Venema VJ, Ju H, et al. (1999) Endothelial nitric oxide synthase interactions with G-protein-coupled receptors. Biochem J 343:335–340

Marsden PA, Shappert KT, Chen HS, et al. (1992) Molecular cloning and characterization of human endothelial nitric oxide synthase. FEBS Lett 307:287–293

Marsden PA, Heng HH, Scherer SW, et al. (1993) Structure and chromosomal localization of the human constitutive endothelial nitric oxide synthase gene. J Biol Chem 268:17478–17488

Massberg S, Sausbier M, Klatt P, et al. (1999) Increased adhesion and aggregation of platelets lacking cyclic guanosine $3',5'$-monophosphate kinase I. J Exp Med 189:1255–1264

Massion PB, Feron O, Dessy C, et al. (2003) Nitric oxide and cardiac function: ten years after, and continuing. Circ Res 93:388–398

Matsumura M, Kakishita H, Suzuki M, et al. (2001) Dexamethasone suppresses iNOS gene expression by inhibiting NF-κB in vascular smooth muscle cells. Life Sci 69:1067–1077

Mattson DL, Bellehumeur TG (1996) Neural nitric oxide synthase in the renal medulla and blood pressure regulation. Hypertension 28:297–303

Mattson DL, Lu S, Cowley AW Jr (1997) Role of nitric oxide in the control of the renal medullary circulation. Clin Exp Pharmacol Physiol 24:587–590

Maxwell AJ (2002) Mechanisms of dysfunction of the nitric oxide pathway in vascular diseases. Nitric Oxide 6:101–124

McCabe TJ, Fulton D, Roman LJ, et al. (2000) Enhanced electron flux and reduced calmodulin dissociation may explain "calcium-independent" eNOS activation by phosphorylation. J Biol Chem 275:6123–6128

Mete A, Connolly S (2003) Inhibitors of the NOS enzymes: a patent review. IDrugs 6:57–65

Michel T, Gordon K, Busconi L (1993) Phosphorylation and subcellular translocation of endothelial nitric oxide synthase. Proc Natl Acad Sci USA 90:6252–6256

Michell BJ, Chen Z, Tiganis T, et al. (2001) Coordinated control of endothelial NO synthase phosphorylation by protein kinase C and the cAMP-dependent protein kinase. J Biol Chem 276:17625–17628

Mitchell JA, Förstermann U, Warner TD, et al. (1991) Endothelial cells have a particulate enzyme system responsible for EDRF formation: measurement by vascular relaxation. Biochem Biophys Res Commun 176:1417–1423

Miyamoto Y, Saito Y, Kajiyama N, et al. (1998) Endothelial nitric oxide synthase gene is positively associated with essential hypertension. Hypertension 32:3–8

Moncada S (1989) Introduction. In: Moncada S, Higgs EA (eds) Nitric oxide from L-arginine: a bioregulatory system. Elsevier, Amsterdam, pp 1–4

Moncada S (2006) Adventures in vascular biology: a tale of two mediators. Phil Trans Roy Soc B 361:735–759

Moncada S, Erusalimsky JD (2002) Does nitric oxide modulate mitochondrial energy generation and apoptosis? Nat Rev Mol Cell Biol 3:214–220

Moncada S, Higgs EA (2006) The discovery of nitric oxide and its role in vascular biology. Br J Pharmacol 147:S193–S201

Moncada S, Palmer RM, Higgs EA (1989) Biosynthesis of nitric oxide from L-arginine. A pathway for the regulation of cell function and communication. Biochem Pharmacol 38:1709–1715

Moncada S, Palmer RMJ, Higgs EA (1991) Nitric oxide: physiology, pathophysiology and pharmacology. Pharmacol Rev 43:109–142

Moroi M, Zhang L, Yasuda T, et al. (1998) Interaction of genetic deficiency of endothelial nitric oxide, gender, and pregnancy in vascular response to injury in mice. J Clin Invest 101:1225–1232

Morris SM Jr, Billiar TR (1994) New insights into the regulation of inducible nitric oxide synthesis. Am J Physiol 266:E829–E839

Mueller CF, Laude K, McNally JS, et al. (2005) Redox mechanisms in blood vessels. Arterioscler Thromb Vasc Biol 25:274–278

Munzel T, Kurz S, Heitzer T, et al. (1996) New insights into mechanisms underlying nitrate tolerance. Am J Cardiol 77:24C–30C

Muruganandam A, Mutus B (1994) Isolation of nitric oxide synthase from human platelets. Biochim Biophys Acta 1200:1–6

Nagareddy PR, Xia Z, McNeill JH, et al. (2005) Increased expression of iNOS is associated with endothelial dysfunction and impaired pressor responsiveness in streptozotocin-induced diabetes. Am J Physiol Heart Circ Physiol 289:H2144–H2152

Nakane M, Schmidt HH, Pollock JS, et al. (1993) Cloned human brain nitric oxide synthase is highly expressed in skeletal muscle. FEBS Lett 316:175–180

Nakayama T, Soma M, Takahashi Y, et al. (1997) Association analysis of CA repeat polymorphism of the endothelial nitric oxide synthase gene with essential hypertension in Japanese. Clin Genet 51:26–30

Nathan C, Xie QW (1994) Nitric oxide synthases: roles, tolls and controls. Cell 78:915–918

Navarro J, Sanchez A, Saiz J, et al. (1994) Hormonal, renal and metabolic alterations during hypertension induced by chronic inhibition of NO in rats. Am J Physiol 267:R1516–R1521

Navarro-Antolin J, Rey-Campos J, Lamas S (2000) Transcriptional induction of endothelial nitric oxide gene by cyclosporine A. A role for activator protein-1. J Biol Chem 275:3075–3080

Nedvetsky PI, Sessa WC, Schmidt HH (2002) There's NO binding like NOS binding: protein–protein interactions in NO/cGMP signaling. Proc Natl Acad Sci USA 99:16510–16512

Neunteufl T, Heher S, Katzenschlager R, et al. (2000) Late prognostic value of flow-mediated dilation in the brachial artery of patients with chest pain. Am J Cardiol 86:207–210

Nishida CR, Ortiz de Montellano PR (1999) Autoinhibition of endothelial nitric-oxide synthase. Identification of an electron transfer control element. J Biol Chem 274:14692–14698

Nishida K, Harrison DG, Navas JP, et al. (1992) Molecular cloning and characterization of the constitutive bovine aortic endothelial cell nitric oxide synthase. J Clin Invest 90:2092–2096

Nisoli E, Clementi E, Paolucci C, et al. (2003) Mitochondrial biogenesis in mammals: the role of endogenous nitric oxide. Science 299:896–899

Nisoli E, Tonello C, Cardile A, et al. (2005) Calorie restriction promotes mitochondrial biogenesis by inducing the expression of eNOS. Science 310:314–317

Ohashi Y, Kawashima S, Hirata K, et al. (1998) Hypotension and reduced nitric oxide-elicited vasorelaxation in transgenic mice overexpressing endothelial nitric oxide synthase. J Clin Invest 102:2061–2071

Ortiz PA, Garvin JL (2003) Cardiovascular and renal control in NOS-deficient mouse models. Am J Physiol 284:R628–R638

Ortiz PA, Hong NJ, Garvin JL (2001) NO decreases thick ascending limb chloride absorption by reducing Na^+-K^+-$2Cl^-$ cotransporter activity. Am J Physiol 281:F819–F825

Palacios-Callender M, Quintero M, Hollis VS, et al. (2004) Endogenous NO regulates superoxide production at low oxygen concentrations by modifying the redox state of cytochrome c oxidase. Proc Natl Acad Sci USA 101:7630–7635

Pallone TL, Mattson DL (2002) Role of nitric oxide in regulation of the renal medulla in normal and hypertensive kidneys. Curr Opin Nephrol Hypertens 11:93–98

Palmer RM, Ferrige AG, Moncada S (1987) Nitric oxide release accounts for the biological activity of endothelium-derived relaxing factor. Nature 327:524–526

Palmer RM, Ashton DS, Moncada S (1988) Vascular endothelial cells synthesize nitric oxide from L-arginine. Nature 333:664–666

Palmer RMJ, Moncada S (1989) A novel citrulline-forming enzyme implicated in the formation of nitric oxide by vascular endothelial cells. Biochem Biophys Res Commun 158:348–352

Papapetropoulos A, Garcia-Cardena G, Madri JA, et al. (1997) Nitric oxide production contributes to the angiogenic properties of vascular endothelial growth factor in human endothelial cells. J Clin Invest 100:3131–3139

Paulus WJ, Shah AM (1999) NO and cardiac diastolic function. Cardiovasc Res 43:595–606

Pelligrino DA, Ye S, Tan F, et al. (2000) Nitric oxide-dependent pial arteriolar dilation in the female rat: effects of chronic estrogen depletion and repletion. Biochem Biophys Res Commun 269:165–171

Perez-Mato I, Castro C, Ruiz FA, et al. (1999) Methionine adenosyltransferase S-nitrosylation is regulated by the basic and acidic amino acids surrounding the target thiol. J Biol Chem 274:17075–17079

Persichini T, Mazzone V, Polticelli F, et al. (2005) Mitochondrial type I nitric oxide synthase physically interacts with cytochrome c oxidase. Neurosci Lett 384:254–259

Petroff MG, Kim SH, Pepe S, et al. (2001) Endogenous nitric oxide mechanisms mediate the stretch dependence of Ca^{2+} release in cardiomyocytes. Nat Cell Biol 3:867–873

Pfeifer A, Klatt P, Massberg S, et al. (1998) Defective smooth muscle regulation in cGMP kinase I-deficient mice. EMBO J 17:3045–3051

Poderoso JJ, Carreras MC, Lisdero C, et al. (1996) Nitric oxide inhibits electron transfer and increases superoxide radical production in rat heart mitochondria and submitochondrial particles. Arch Biochem Biophys 328:85–92

Pollock JS, Förstermann U, Mitchell JA, et al. (1991) Purification and characterization of particulate endothelium-derived relaxing factor synthase from cultured and native bovine aortic endothelial cells. Proc Natl Acad Sci USA 88:10480–10484

Pollock JS, Klinghofer V, Förstermann U, et al. (1992) Endothelial nitric oxide synthase is myristoylated. FEBS Lett 309:402–404

Prabhakar P, Cheng V, Michel T (2000) A chimeric transmembrane domain directs endothelial nitric oxide synthase palmitoylation and targeting to plasmalemmal caveolae. J Biol Chem 275:19416–19421

Pritchard KA Jr, Ackerman AW, Gross ER, et al. (2001) Heat shock protein 90 mediates the balance of nitric oxide and superoxide anion from endothelial nitric oxide synthase. J Biol Chem 276:17621–17624

Quintero M, Colombo S, Godfrey A, Moncada S (2006) Mitochondria as signaling organelles in the vascular endothelium. Proc Natl Acad Sci USA 103:5379–5384

Radomski MW, Palmer RMJ, Moncada S (1987a) Comparative pharmacology of endothelium-derived relaxing factor, nitric oxide and prostacyclin in platelets. Br J Pharmacol 92:181–187

Radomski MW, Palmer RMJ, Moncada S (1987b) The role of nitric oxide and cGMP in platelet adhesion to vascular endothelium. Biochem Biophys Res Commun 148:1482–1489

Radomski MW, Palmer RMJ, Moncada S (1987c) The anti-aggregating properties of vascular endothelium: interactions between prostacyclin and nitric oxide. Br J Pharmacol 92:639–646

Radomski MW, Palmer RMJ, Moncada S (1990a) Glucocorticoids inhibit the expression of an inducible, but not the constitutive, nitric oxide synthase in vascular endothelial cells. Proc Natl Acad Sci USA 87:10043–10047

Radomski MW, Palmer RMJ, Moncada S (1990b) An L-arginine/nitric oxide pathway present in human platelets regulates aggregation. Proc Natl Acad Sci USA 87:5193–5197

Radomski MW, Palmer RMJ, Moncada S (1990c) Characterization of the L-arginine:nitric oxide pathway in human platelets. Br J Pharmacol 101:325–328

Rajagopalan S, Kurz S, Munzel T, et al. (1996) Angiotensin II-mediated hypertension in the rat increases vascular superoxide production via membrane NADH/NADPH oxidase activation. Contribution to alterations of vasomotor tone. J Clin Invest 97:1916–1923

Raman CS, Li H, Martasek P, Kral V, et al. (1998) Crystal structure of constitutive endothelial nitric oxide synthase: a paradigm for pterin function involving a novel metal center. Cell 95:939–950

Randiramboavonjy V, Schrader J, Busse R, et al. (2004) Insulin induces the release of vasodilator compounds from platelets by a nitric oxide-G-kinase-VAMP-3-dependent pathway. J Exp Med 199:347–356

Rao GH, Krishnamurthi S, Raij L, et al. (1990) Influence of nitric oxide on agonist-mediated calcium mobilization in platelets. Biochem Med Metab Biol 43:271–275

Razani B, Engelman JA, Wang XB, et al. (2001) Caveolin-1 null mice are viable but show evidence of hyperproliferative and vascular abnormalities. J Biol Chem 276:38121–38138

Reddy KG, Nair RN, Sheehan HM, et al. (1994) Evidence that selective endothelial dysfunction may occur in the absence of angiographic or ultrasound atherosclerosis in patients with risk factors for atherosclerosis. J Am Coll Cardiol 23:833–843

Rees DD, Palmer RMJ, Moncada S (1989) Role of endothelium-derived nitric oxide in the regulation of blood pressure. Proc Natl Acad Sci USA 86:3375–3378

Rees DD, Higgs EA, Moncada S (2000) Nitric oxide and the vessel wall. In: Colman RW, Hirsch J, Marder VJ, Clowes AW, George NJ (eds) Hemostasis and thrombosis. Lippincott Williams and Wilkins, Philadelphia, pp 673–682

Robinson LJ, Michel T (1995) Mutagenesis of palmitoylation sites in endothelial nitric oxide synthase identifies a novel motif for dual acylation and subcellular targeting. Proc Natl Acad Sci USA 92:11776–11780

Robinson LJ, Busconi L, Michel T (1995) Agonist-modulated palmitoylation of endothelial nitric oxide synthase. J Biol Chem 270:995–998

Rosenkranz-Weiss P, Sessa WC, Milstien S, et al. (1994) Regulation of nitric oxide synthesis by proinflammatory cytokines in human umbilical vein endothelial cells. Elevations in tetrahydrobiopterin levels enhance endothelial nitric oxide synthase specific activity. J Clin Invest 93:2236–2243

Russell KS, Haynes MP, Caulin-Glaser T, et al. (2000) Estrogen stimulates heat shock protein 90 binding to endothelial nitric oxide synthase in human vascular endothelial cells. Effects on calcium sensitivity and NO release. J Biol Chem 275:5026–5030

Salerno JC, Harris DE, Irizarry K, et al. (1997) An autoinhibitory control element defines calcium-regulated isoforms of nitric oxide synthase. J Biol Chem 272:29769–29777

Santolini J, Meade AL, Stuehr DJ (2001) Differences in three kinetic parameters underpin the unique catalytic profiles of nitric-oxide synthases I, II, and III. J Biol Chem 276:48887–48898

Saura M, Zaragoza C, Cao W, et al. (2002) Smad2 mediates transforming growth factor-beta induction of endothelial nitric oxide synthase expression. Circ Res 91:806–813

Sausbier M, Schubert R, Voigt V, et al. (2000) Mechanisms of NO/cGMP-dependent vasorelaxation. Circ Res 87:825–830

Schachinger V, Britten MB, Zeiher AM (2000) Prognostic impact of coronary vasodilator dysfunction on adverse long-term outcome of coronary heart disease. Circulation 101:1899–1906

Schlossmann J, Hofmann F (2005) cGMP-dependent protein kinases in drug discovery. Drug Discov Today 10:627–634

Schlossmann J, Ammendola A, Ashman K, et al. (2000) Regulation of intracellular calcium by a signalling complex of IRAG, IP3 receptor and cGMP kinase Iβ. Nature 404:197–201

Schmidt H, Hofmann H, Schindler U, et al. (1996) No NO from NO synthase. Proc Natl Acad Sci USA 93:14492–14497

Schneider MP, Erdmann J, Delles C, et al. (2000) Functional gene testing of the Glu298Asp polymorphism of the endothelial NO synthase. J Hypertens 18:1767–1773

Schwarz UR, Walter U, Eigenthaler M (2001) Taming platelets with cyclic nucleotides. Biochem Pharmacol 62:1153–1161

Schweizer M, Richter C (1994) Nitric oxide potently and reversibly deenergizes mitochondria at low oxygen tension. Biochem Biophys Res Commun 204:169–175

Scotland RS, Chauhan S, Vallance PJ, et al. (2001) An endothelium-derived hyperpolarizing factor-like factor moderates myogenic constriction of mesenteric resistance arteries in the absence of endothelial nitric oxide synthase-derived nitric oxide. Hypertension 38:833–839

Scotland RS, Morales-Ruiz M, Chen Y, et al. (2002) Functional reconstitution of endothelial nitric oxide synthase reveals the importance of serine 1179 in endothelium-dependent vasomotion. Circ Res 90:904–910

Scotland RS, Madhani M, Chauhan S, et al. (2005) Investigation of vascular responses in endothelial nitric oxide synthase/cyclooxygenase-1 double-knockout mice. Key role for endothelium-derived hyperpolarizing factor in the regulation of blood pressure in vivo. Circulation 111:796–803

Sessa WC (2004) eNOS at a glance. J Cell Sci 117:2427–2429

Sessa WC, Harrison JK, Barber CM, et al. (1992) Molecular cloning and expression of a cDNA encoding endothelial cell nitric oxide synthase. J Biol Chem 267:15274–15276

Shaul PA, Smart EJ, Robinson LJ, et al. (1996) Acylation targets endothelial nitric-oxide synthase to plasmalemmal caveolae. J Biol Chem 271:6518–6522

Sherman PA, Laubach VE, Reep BR, et al. (1993) Purification and cDNA sequence of an inducible nitric oxide synthase from a human tumor cell line. Biochemistry 32:11600–11605

Shesely EG, Maeda N, Kim HS, et al. (1996) Elevated blood pressures in mice lacking endothelial nitric oxide synthase. Proc Natl Acad Sci USA 93:13176–13181

Shimasaki Y, Yasue H, Yoshimura M, et al. (1998) Association of the missense Glu298Asp variant of the endothelial nitric oxide synthase gene with myocardial infarction. J Am Coll Cardiol 31:1506–1510

Siddhanta U, Wu C, Abu-Soud HM, et al. (1996) Heme iron reduction and catalysis by a nitric oxide synthase heterodimer containing one reductase and two oxygenase domains. J Biol Chem 271:7309–7312

Silacci P, Formentin K, Bouzourene K, et al. (2000) Unidirectional and oscillatory shear stress differentially modulate NOS III gene expression. Nitric Oxide 4:47–56

Simon DI, Stamler JS, Loh E, et al. (1995) Effect of nitric oxide synthase inhibition on bleeding time in humans. J Cardiovasc Pharmacol 26:339–342

Simoncini T, Hafezi-Moghadam A, Brazil DP, et al. (2000) Interaction of oestrogen receptor with the regulatory subunit of phosphatidylinositol-3-OH kinase. Nature 407:538–541

Slater TF (1972) Free radical mechanisms in tissue injury. Pion, London

Solzbach U, Hornig B, Jeserich M, et al. (1997) Vitamin C improves endothelial dysfunction of epicardial coronary arteries in hypertensive patients. Circulation 96:1513–1519

Sowa G, Liu J, Papapetropoulos A, et al. (1999) Trafficking of endothelial nitric oxide synthase in living cells. J Biol Chem 274:22524–22531

Stagliano NE, Zhao W, Prado R, et al. (1997) The effect of nitric oxide synthase inhibition on acute platelet accumulation and hemodynamic depression in a rat model of thromboembolic stroke. J Cereb Blood Flow Metab 17:1182–1190

Stamler JS, Hausladen A (1998) Oxidative modifications in nitrosative stress. Nat Struct Biol 5:247–249

Stamler JS, Simon DI, Osborne JA, et al. (1992) S-nitrosylation of proteins with nitric oxide: synthesis and characterization of biologically active compounds. Proc Natl Acad Sci USA 89:444–448

Stauss HM, Godecke A, Mrowka R, et al. (1999) Enhanced blood pressure variability in eNOS knock-out mice. Hypertension 33:1359–1363

Stauss HM, Nafz B, Mrowka R, et al. (2000) Blood pressure control in eNOS knock-out mice: comparison with other species under NO blockade. Acta Physiol Scand 168:155–160

Stemerman MB (1981) Vascular injury: platelets and smooth muscle cell response. Philos Trans R Soc Lond B Biol Sci 294:217–224

Stroes E, Kastelein J, Cosentino F, et al. (1997) Tetrahydrobiopterin restores endothelial function in hypercholesterolemia. J Clin Invest 99:41–46

Stuehr D, Pou S, Rosen GM (2001) Oxygen reduction by nitric oxide synthases. J Biol Chem 276:14533–14536

Sun D, Huang A, Smith CJ, et al. (1999) Enhanced release of prostaglandins contributes to flow-induced arteriolar dilation in eNOS knockout mice. Circ Res 85:288–293

Sun J, Liao JK (2002) Functional interaction of endothelial nitric oxide synthase with a voltage-dependent anion channel. Proc Natl Acad Sci USA 99:13108–13113

Taddei S, Virdis A, Mattei P, et al. (1996) Defective L-arginine-nitric oxide pathway in offspring of essential hypertensive patients. Circulation 94:1298–1303

Tai SC, Robb GB, Marsden PA (2004) Endothelial nitric oxide synthase: a new paradigm for gene regulation in the injured blood vessel. Arterioscler Thromb Vasc Biol 24:405–412

Tiefenbacher CP, Chilian WM, Mitchell M, et al. (1996) Restoration of endothelium-dependent vasodilation after reperfusion injury by tetrahydrobiopterin. Circulation 94:1423–1429

Ting HH, Timimi FK, Boles KS, et al. (1996) Vitamin C improves endothelium-dependent vasodilation in patients with non-insulin-dependent diabetes mellitus. J Clin Invest 97:22–28

Tracy RP (2002) Diabetes and atherothrombotic disease: linked through inflammation? Semin Vasc Med 2:67–73

Trovati M, Massucco P, Mattiello L, et al. (1996) The insulin-induced increase of guanosine-3′,5′-cyclic monophosphate in human platelets is mediated by nitric oxide. Diabetes 45:768–770

Turrens JF (2003) Mitochondrial formation of reactive oxygen species. J Physiol 552:335–344

Uematsu M, Ohara Y, Navas JP, et al. (1995) Regulation of endothelial cell nitric oxide synthase mRNA expression by shear stress. Am J Physiol 269:C1371–C1378

Uwabo J, Soma M, Nakayama T, et al. (1998) Association of a variable number of tandem repeats in the endothelial constitutive nitric oxide synthase gene with essential hypertension in Japanese. Am J Hypertens 11:125–128

Vallance P, Moncada S (1993) Role of endogenous nitric oxide in septic shock. New Horiz 1:77–86

Vallance P, Collier J, Moncada S (1989) Nitric oxide synthesized from L-arginine mediates endothelium-dependent dilatation in human veins in vivo. Cardiovasc Res 23:1053–1057

Vallance P, Leone A, Calver A, et al. (1992) Accumulation of an endogenous inhibitor of nitric oxide synthesis in chronic renal failure. Lancet 339:572–575

van der Zee R, Murohara T, Luo Z, et al. (1997) Vascular endothelial growth factor/vascular permeability factor augments nitric oxide release from quiescent rabbit and human vascular endothelium. Circulation 95:1030–1037

van Geel PP, Pinto YM, Buikema H, et al. (1998) Is the A1166C polymorphism of the angiotensin II type 1 receptor involved in cardiovascular disease? Eur Heart J 19:G13–G17

Varenne O, Pislaru S, Gillijns H, et al. (1998) Local adenovirus-mediated transfer of human endothelial nitric oxide synthase reduces luminal narrowing after coronary angioplasty in pigs. Circulation 98:919–926

Vasquez-Vivar J, Kalyanaraman B, Martasek P, et al. (1998) Superoxide generation by endothelial nitric oxide synthase: the influence of cofactors. Proc Natl Acad Sci USA 95:9220–9225

Venema RC, Venema VJ, Ju H, et al. (2003) Novel complexes of guanylate cyclase with heat shock protein 90 and nitric oxide synthase. Am J Physiol 285:H669–H678

Viner RI, Williams TD, Schoneich C (1999) Peroxynitrite modification of protein thiols: oxidation, nitrosylation, and S-glutathiolation of functionally important cysteine residue(s) in the sarcoplasmic reticulum Ca-ATPase. Biochemistry 38:12408–12415

Virdis A, Schiffrin EL (2003) Vascular inflammation: a role in vascular disease in hypertension? Curr Opin Nephrol Hypertens 12:181–187

Vodovotz Y, Chesler L, Chong H, et al. (1999) Regulation of transforming growth factor β1 by nitric oxide. Cancer Res 59:2142-2149

von der Leyen HE, Dzau VJ (2001) Therapeutic potential of nitric oxide synthase gene manipulation. Circulation 103:2760-2765

Vouldoukis I, Riveros-Moreno V, Dugas B, et al. (1995) The killing of Leishmania major by human macrophages is mediated by nitric oxide induced after ligation of the Fc epsilon RII/CD23 surface antigen. Proc Natl Acad Sci USA 92:7804-7808

Wagner AH, Kohler T, Ruckschloss U, et al. (2000a) Improvement of nitric oxide-dependent vasodilatation by HMG-CoA reductase inhibitors through attenuation of endothelial superoxide anion formation. Arterioscler Thromb Vasc Biol 20:61-69

Wagner C, Godecke A, Ford M, et al. (2000b) Regulation of renin gene expression in kidneys of eNOS- and nNOS-deficient mice. Pflugers Arch 439:567-572

Walker G, Pfeilshifter J, Kunz D (1997) Mechanisms of suppression of inducible nitric oxide synthase (iNOS) expression in interferon (IFN)-γ-stimulated RAW 264.7 cells by dexamethasone. Evidence for glucocorticoid-induced degradation of iNOS protein by calpain as a key step in post-transcriptional regulation. J Biol Chem 272:16679-16687

Wallerath T, Gath I, Aulitzky WE, et al. (1997) Identification of the NO synthase isoforms expressed in human neutrophil granulocytes, megakaryocytes and platelets. Thromb Haemost 77:163-167

Wang X, Wang J, Trudinger B (2003) Gene expression of nitric oxide synthase by human umbilical vein endothelial cells: the effect of fetal plasma from pregnancy with umbilical placental vascular disease. Br J Obstet Gynaecol 110:53-58

Wang XL, Wang J (2000) Endothelial nitric oxide synthase gene sequence variations and vascular disease. Mol Genet Metab 70:241-251

Wattanapitayakul SK, Mihm MJ, Young AP, et al. (2001) Therapeutic implications of human endothelial nitric oxide synthase gene polymorphism. Trends Pharmacol Sci 22:361-368

Wedgwood S, Mitchell CJ, Fineman JR, et al. (2003) Developmental differences in the shear stress-induced expression of endothelial NO synthase: changing role of AP-1. Am J Physiol Lung Cell Mol Physiol 284:L650-L662

Wei CC, Wang ZQ, Meade AL, et al. (2002) Why do nitric oxide synthases use tetrahydrobiopterin? J Inorg Biochem 91:6618-6624

Weiner CP, Lizasoain I, Baylis SA, et al. (1994) Induction of calcium-dependent nitric oxide synthases by sex hormones. Proc Natl Acad Sci USA 91:5212-5216

Werner ER, Gorren AC, Heller R, et al. (2003) Tetrahydrobiopterin and nitric oxide: mechanistic and pharmacological aspects. Exp Biol Med 228:1291-1302

White KA, Marletta MA (1992) Nitric oxide synthase is a cytochrome P-450 type hemoprotein. Biochemistry 31:6627-6631

Whittle BJR (1997) Nitric oxide—a mediator of inflammation or mucosal defence. Eur J Gastroenterol Hepatol 9:1026-1032

Wilcox JN, Subramanian RR, Sundell CL, et al. (1997) Expression of multiple isoforms of nitric oxide synthase in normal and atherosclerotic vessels. Arterioscler Thromb Vasc Biol 17:2479-2488

Xie QW, Cho HJ, Calaycay J, et al. (1992) Cloning and characterization of inducible nitric oxide synthase from mouse macrophages. Science 256:225-228

Xu HL, Galea E, Santizo RA, et al. (2001) The key role of caveolin-1 in estrogen-mediated regulation of endothelial nitric oxide synthase function in cerebral arterioles in vivo. J Cereb Blood Flow Metab 21:907-913

Xu KY, Huso DL, Dawson TM, et al. (1999) Nitric oxide synthase in cardiac sarcoplasmic reticulum. Proc Natl Acad Sci USA 96:657-662

Xu L, Eu JP, Meissner G, et al. (1998) Activation of the cardiac calcium release channel (ryanodine receptor) by poly-S-nitrosylation. Science 279:234–237

Yang Y, Loscalzo J (2005) S-nitrosoprotein formation and localization in endothelial cells. Proc Natl Acad Sci USA 102:117–122

Yeh DC, Duncan JA, Yamashita S, et al. (1999) Depalmitoylation of endothelial nitric oxide synthase by acyl-protein thioesterase 1 is potentiated by Ca^{2+}-calmodulin. J Biol Chem 274:33148–33154

Zabel U, Kleinschnitz C, Oh P, et al. (2002) Calcium-dependent membrane association sensitizes soluble guanylyl cyclase to nitric oxide. Nat Cell Biol 4:307–311

Zembowicz A, Tang JL, Wu KK (1995) Transcriptional induction of endothelial nitric oxide synthase type III by lysophosphatidylcholine. J Biol Chem 270:17006–17010

Zhang R, Min W, Sessa WC (1995) Functional analysis of the human endothelial nitric oxide synthase promoter. Sp1 and GATA factors are necessary for basal transcription in endothelial cells. J Biol Chem 270:15320–15326

Zhao G, Bernstein RD, Hintze TH (1999) Nitric oxide and oxygen utilization: exercise, heart failure and diabetes. Coron Artery Dis 10:315–320

Zhao YY, Liu Y, Stan RV, et al. (2002) Defects in caveolin-1 cause dilated cardiomyopathy and pulmonary hypertension in knockout mice. Proc Natl Acad Sci USA 99:11375–11380

Ziche M, Morbidelli L (2000) Nitric oxide and angiogenesis. J Neurooncol 50:139–148

Ziche M, Morbidelli L, Masini E, et al. (1994) Nitric oxide mediates angiogenesis in vivo and endothelial cell growth and migration in vitro promoted by substance P. J Clin Invest 94:2036–2044

Ziche M, Morbidelli L, Choudhuri R, et al. (1997) Nitric oxide synthase lies downstream from vascular endothelial growth factor-induced but not basic fibroblast growth factor-induced angiogenesis. J Clin Invest 99:2625–2634

Ziegler T, Silacci P, Harrison VJ, et al. (1998) Nitric oxide synthase expression in endothelial cells exposed to mechanical forces. Hypertension 32:351–355

Zimmermann K, Optiz N, Dedio J, et al. (2002) NOSTRIN: a protein modulating nitric oxide release and subcellular distribution of endothelial nitric oxide synthase. Proc Natl Acad Sci USA 99:17167–17172

Angiotensin, Bradykinin and the Endothelium

C. Dimitropoulou · A. Chatterjee · L. McCloud · G. Yetik-Anacak · J. D. Catravas (✉)

Vascular Biology Center and Department of Pharmacology and Toxicology, Medical College of Georgia, Augusta GA, 30912-2500, USA
jcatrava@mcg.edu

1	Angiotensin	256
1.1	Synthesis	257
1.2	Receptors	257
1.3	Receptor Antagonists	258
1.4	Signalling	259
1.5	Effects	260
1.5.1	Haemostasis and Fibrinolysis	260
1.5.2	Apoptosis and Neovascularisation	260
1.5.3	Fibrosis	261
1.5.4	Hypertrophy	261
1.5.5	Hypertension	261
1.5.6	Inflammation	262
1.5.7	Atherosclerosis	263
2	Bradykinin	264
2.1	Synthesis	264
2.2	Receptors	264
2.3	Receptor Antagonists	265
2.4	Signalling	265
2.5	Effects	266
2.5.1	Neovascularisation	266
2.5.2	Hypertension	266
2.5.3	Inflammation	266
2.5.4	Diabetes	267
2.6	Interactions with eNOS	268
3	Angiotensin–BK Interactions	268
3.1	Biosynthesis and Degradation	269
3.2	BK Interactions with Angiotensin (1–7)	270
3.2.1	Stimulation of BK Release by ANG-(1–7)	270
3.2.2	Potentiation of the Effects of BK by ANG-(1–7)	270
3.2.3	Resensitisation of BK Receptors by ANG-(1–7)	271
3.3	BK Interactions with Angiotensin Type 2 Receptors	271
3.3.1	BK and AT2R: Mediated NO Release	271
3.3.2	BK and AT2R: Mediated Flow-Dependent Vasodilatation	272
3.3.3	BK and AT2R: Mediated Effects on the Myocardium	272
3.4	BK Interactions with Angiotensin Type 1 Receptors	273
3.4.1	Modulation of BK Levels by Angiotensin Peptides	273

3.4.2　Upregulation of BK Type 2 Receptors 273
3.4.3　Angiotensin–BK Receptor Heterodimerisation 273

4　Angiotensin Converting Enzyme (ACE) 274
4.1　Expression 274
4.2　Molecular Regulation 275
4.3　Inhibitors 277
4.4　ACE as an Index of Endothelial Function 278

References 278

Abstract Angiotensins and kinins are endogenous peptides with diverse biological actions; as such, they represent current and future targets of therapeutic intervention. The field of angiotensin biology has changed significantly over the last 50 years. Our original understanding of the crucial role of angiotensin II in the regulation of vascular tone and electrolyte homeostasis has been expanded to include the discovery of new angiotensins, their important role in cardiovascular inflammation and the development of clinically useful synthesis inhibitors and receptor antagonists. While less applied progress has been achieved in the kinin field, there are continuous discoveries in bradykinin physiology and in the complexity of kinin interactions with other proteins. The present review focuses on mechanisms and interactions of angiotensins and kinins that deal specifically with vascular endothelium.

Keywords Angiotensin receptors · Bradykinin receptors · Angiotensin-converting enzyme · Angiotensin receptor blockers · Angiotensin-converting enzyme inhibitors

1
Angiotensin

The octapeptide angiotensin (ANG) II (Asp^1-Arg^2-Val^3-Tyr^4-Ile^5-His^6-Pro^7-Phe^8) stimulates the release of catecholamines from the adrenal medulla and sympathetic nerve endings, increases sympathetic nervous system activity, stimulates thirst and appetite, and regulates sodium and water homeostasis by stimulating aldosterone release from the adrenal cortex (Luft et al. 1989; Mitchell and Navar 1989; Ferrario and Flack 1996). It regulates endothelial function and stimulates inflammatory, proliferative, fibrotic and thrombotic processes in the vasculature. It has potent effects on vascular tone, constricts smooth muscle cells, regulates vascular cell growth, apoptosis, fibrosis, matrix metalloproteinase production and extracellular matrix degradation (Griendling et al. 1997; Tomita et al. 1998; Yoo et al. 1998). ANG IV, the (3–8) hexapeptide fragment of ANG II (Swanson et al. 1992), and ANG-(1–7) can be formed metabolically by peptidase or protease cleavage from either ANG II or ANG I (Wright and Harding 1995). ANG IV interacts specifically with the AT4R subtype (Harding et al. 1992).

1.1
Synthesis

The biologically inactive decapeptide, ANG I, is the metabolic product of the liver-synthesised angiotensinogen and the enzyme renin. ANG I has a half-life of a few seconds, as it is quickly converted—mostly by angiotensin-converting enzyme (ACE)—to the biologically active octapeptide, ANG II. Alternatively, ANG I can be converted to the biologically active ANG-(1-7) by plasma or neutral endopeptidases or to ANG-(1-9) by ACE2 and subsequently to ANG-(1-7). ANG II can also be converted to ANG-(1-7) by plasma and neutral endopeptidases or by prolyl carboxypeptidase or, more frequently, degraded to inactive metabolites by various angiotensinases (Fig. 1).

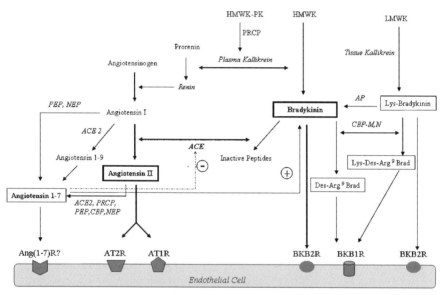

Fig. 1 Synthesis and degradation of the major components of the renin-angiotensin (RAS) and kallikrein-kinin (KKS) systems. Enzymes are in *italics*. *ACE*, angiotensin converting enzyme; *AP*, aminopeptidase P; *AT(1–7)R*, putative angiotensin (1–7) receptor; *AT1R*, angiotensin type 1 receptor; *AT2R*, angiotensin type 2 receptor; *BKB1R*, bradykinin B1 receptor; *BKB2R*, bradykinin B2 type receptor; *Brad*, bradykinin; *CBP-M*, carboxypeptidase M; *CBP-N*, carboxypeptidase N; *HMWK*, high molecular weight kininogen; *LMWK*, low molecular weight kininogen; *NEP*, neutral endopeptidase; *PEP*, plasma endopeptidase; *PK*, pre-kallikrein; *PRCP*, prolyl carboxypeptidase

1.2
Receptors

There are three subtypes of ANG receptors: type 1 (AT1R), type 2 (AT2R) and type 4 (AT4R). AT1R are localised on cardiomyocytes, vascular smooth muscle and endothelial cells, nerve endings and conductive tissues. AT1R are present and functionally active in fetal systemic arteries; the umbilical circulation displays a greater responsiveness to ANG II than the systemic vasculature (Segar et al. 2001). Rodents express two AT1R (AT1R$_A$ and AT1R$_B$) receptor genes, whereas humans express only a single AT1R protein. AT2R are present in endothelial and vascular smooth muscle cells (Nora et al. 1998) and in fibrous tissue of the heart (Regitz-Zagrosek et al. 1998). In brain, AT2R have regenerative capabilities and are upregulated after global cerebral ischaemia (Makino et al. 1996) and during tissue wound healing (Viswanathan and Saavedra 1992). AT4R exhibit a broad distribution, including in the adrenal gland, kidney, lung and heart. In the kidney, ANG IV increases renal cortical blood flow and decreases Na^+ transport in isolated renal proximal tubules. In high concentrations, ANG IV activates AT1R and evokes cardiovascular effects that can be inhibited with AT1R antagonists (Li et al. 1997a).

1.3
Receptor Antagonists

AT1R are selectively antagonised by biphenylimidazoles, such as losartan, whereas tetrahydroimidazopyridines specifically inhibit AT2R (Ardaillou 1999). The AT2R is the first identified example of a G protein-coupled receptor which also acts as a receptor-specific antagonist. AT2R bind directly to AT1R and thereby antagonise AT1R function (AbdAlla et al. 2001a). The AT1R antagonists approved for use in hypertension by the U.S. Food and Drug Administration (FDA) include losartan, valsartan, irbesartan, candesartan and telmisartan. ACE inhibitors and AT1R blockers share a number of common properties, including their ability to lower blood pressure. However, they have different effects on the renin–angiotensinsystem (RAS), the fibrinolytic system and the actions of bradykinin (BK). In animal models of atherosclerosis, ACE inhibition is associated with a significant reduction in the surface area of lesions, while no similar effect is evident following AT1R blockade. In the fibrinolytic system, both ACE inhibition and AT1R blockade are associated with reduced aldosterone levels, although the effect is greater with ACE inhibition; only ACE inhibition is associated with a reduction in plasminogen activation inhibitor-1. By blocking the degradation of BK, ACE inhibitors potentiate the ability of BK to reduce blood pressure and stimulate the release of tissue-type plasminogen activator from the vasculature, an effect not seen with AT1R blockers (Vaughan 2000). AT1R antagonists are as effective as ACE inhibitors in improving the age-related decline in endothelium-derived hyperpolarising factor (EDHF)-mediated hyperpolarisation and relaxation; both AT1R and ACE inhibitors may be useful in preventing endothelial dysfunction associated with ageing (Kansui et al. 2002). Unlike ACE inhibitors, AT1R blockers (ARBs) are not significantly associated with cough.

1.4
Signalling

ANG II has an important role in cardiovascular regulation and electrolyte balance. Endothelial AT1R modulate Na^+/K^+ ATPase activity; this and the ANG II effect on the Na^+/H^+ exchanger are believed to be responsible for the increased transendothelial Na^+ flux (Muscella et al. 1999). ANG II modulates the production of nitric oxide (NO) in the endothelium (Yan et al. 2003). It stimulates NO release by activating endothelial NO synthase (eNOS) via endothelial AT1R (Saito et al. 1996) and, occasionally (e.g. in porcine pulmonary arterial endothelial cells) through AT4R (Hill-Kapturczak et al. 1999). This stimulation of NO may be beneficial in counterbalancing the direct vasoconstrictor effect of ANG II on the underlying smooth muscle cells (Luscher et al. 1992; Bayraktutan and Ulker 2003). ANG II suppresses endothelial $[Ca^{2+}]_i$, but stimulates pericyte $[Ca^{2+}]_i$ via AT1R. Conversely, acting through AT2R, ANG II antagonises the AT1R-mediated endothelial $[Ca^{2+}]_i$ suppression and vasoconstriction (Rhinehart et al. 2003). Additionally, by stimulating the production of reactive oxygen species (ROS) (Griendling et al. 1994), ANG II induces NO degradation and inactivation (Sowers 2002). Endothelial AT1R are linked to phospholipase C and phospholipase A2 activation (Pueyo et al. 1996). AT1R blockers enhance endothelium-dependent relaxation in coronary artery disease (CAD). By a mechanism involving both BK and NO, candesartan improves flow-dependent, endothelium-mediated vasodilatation in patients with CAD (Hornig et al. 2003). Endothelial AT2R increase with age (Batenburg et al. 2004a) and exert an antiproliferative action (Stoll et al. 1995a). In the human heart, AT2R stimulation dilates coronary arterioles via NO release. Endothelial AT4R are G protein-coupled receptors (Riva and Galzin 1996) that induce vasodilatation by activating the NO–cyclic guanosine monophosphate (cGMP) pathway (Patel et al. 1998). In the lung, AT4 activates eNOS to produce pulmonary arterial vasorelaxation (Patel et al. 1998) through a Ca^{2+} release via phospholipase C-phosphoinositol (PI)3-kinase signalling mechanisms (Chen et al. 2000).

ANG II upregulates vascular endothelial growth factor (VEGF), which plays a significant role in ANG II-induced hyperpermeability (Chua et al. 1998). AT1R mediate the stimulatory effects of ANG II on E-selectin expression and leucocyte adhesion on endothelial cells (Grafe et al. 1997) and regulate endothelin-1 release by endothelial cells (Imai et al. 1992; Chua et al. 1993), without influencing circulating endothelin-1 levels (Ferri et al. 1999). Corticosteroids upregulate ANG II receptors by synthesis of new receptor protein rather than by alterations in receptor trafficking (Ullian et al. 1996).

1.5 Effects

1.5.1 Haemostasis and Fibrinolysis

ACE inhibitors and AT1R antagonists exert antithrombotic actions by enhancing NO and prostacyclin release and attenuating ANG II actions (Buczko et al. 1999). ANG II modulates haemostasis and fibrinolysis by inducing the expression of plasminogen activator inhibitor-1 (PAI-1), via AT1R and a pathway involving Rho/Rho kinase, cyclic adenosine monophosphate (cAMP) and ROS (Kramer et al. 2002; Mehta et al. 2002). ANG II also upregulates tissue-type plasminogen activator (t-PA) gene activity, but this may reflect autoregulation in response to PAI-1 release. ANG IV also upregulates PAI-1 expression in endothelial cells (Kerins et al. 1995; Mehta et al. 2002). AT1R blockers also exert AT1R-independent anticoagulant effects by inhibiting cyclooxygenase (COX)-2 and consequently inhibiting thromboxane-induced platelet aggregation (Li et al. 2000).

1.5.2 Apoptosis and Neovascularisation

The role of ANG II in endothelial cell apoptosis remains unclear (Ohashi et al. 2004). ANG II induces endothelial cell apoptosis via activation of the caspase cascade, an effect completely blocked by NO (Dimmeler et al. 1997). ANG II, via AT1R, also activates protein kinase (PK)C, increases Fas (Li et al. 1999a), increases intracellular concentration of ceramide (Lehtonen et al. 1999) and decreases bcl-2 protein expression via extracellular signal-regulated kinase (ERK) phosphorylation (Dimmeler and Zeiher 2000), all of which may promote the development of apoptosis (Li et al. 1999b). On the other hand, ANG II exerts antiapoptotic effects in endothelial cells by a mechanism involving PI3-kinase/Akt activation, subsequent upregulation of survivin and suppression of caspase-3 activity. ARBs also exhibit AT1R-independent anti-apoptotic effects via Akt/eNOS phosphorylation (Watanabe et al. 2005). ANG II potentiates VEGF-induced endothelial cell proliferation and network formation by upregulating the kinase insert domain (KDR, Flk-1 or VEGFR2) receptor (Imanishi et al. 2004). The growth modulating actions of ANG II depend on the type of ANG receptor present on a given cell. Stimulation of AT2R may counterbalance the effects of AT1R stimulation, and initiate tissue regenerative events or apoptosis. The antiproliferative actions of the AT2R offset the growth-promoting effects mediated by the AT1R (Stoll et al. 1995b). Stimulation of AT2R inhibits VEGF-induced endothelial cell migration and tube formation via activation of a pertussis toxin (PTX)-sensitive G protein (Benndorf et al. 2003). It also increases tyrosine phosphatase activity and functionally antagonises the AT1R-induced superoxide formation (Sohn et al. 2000). Changes in AT2R expression may occur during treatment with ARBs, suggesting the existence of cross-talk between AT1R and AT2R (De Paolis et al. 1999). In addition to ANG II, ANG IV modulates the actions of basic fibroblast growth factor (bFGF) on endothelial cells (Hall et al. 1995). Balloon injury increases AT4R binding in the media, large neointima and re-endothelialised cell layer, suggesting a role for ANG IV in the adaptive response and remodelling of the vascular wall following damage (Moeller et al. 1999).

1.5.3
Fibrosis

ANG II stimulates transforming growth factor (TGF)-β1 production via PKC and upregulates tissue inhibitor of metalloproteinase-1 (TIMP-1) gene expression in endothelial cells. The release of TGF-β1 or TIMP-1 by endothelial cells may provide the initial trigger leading to cardiac fibrosis in angiotensin-renin-dependent hypertension (Chua et al. 1994, 1996). Upregulation of TSP-1 by ANGII also leads to perivascular fibrosis in the heart (Chua et al. 1997).

1.5.4
Hypertrophy

The selective AT1R antagonist losartan, even at doses that reduce blood pressure, only moderately induces regression of cardiovascular hypertrophy and endothelial dysfunction in genetically hypertensive rats (Li et al. 1997b). AT1R mediate myocyte hypertrophy, fibroblast proliferation, collagen synthesis, smooth muscle cell growth, endothelial adhesion molecule expression and catecholamine synthesis. AT1R are downregulated in cardiac failure as well as in the hypertrophied transplanted heart, indicating that a 50% loss of AT1R does not impede cardiac hypertrophy. In heart failure therapy, ARBs differ from ACE inhibitors in that they lack the ability to inhibit the degradation of BK (Regitz-Zagrosek et al. 1998).

1.5.5
Hypertension

ANG II—acting through AT1R—has been implicated in the pathophysiology of hypertension and chronic renal failure (Dalmay et al. 2001; Delles et al. 2004). Endothelial dysfunction occurs in large or smaller vessels, especially in the presence of risk factors such as diabetes, smoking, dyslipidaemia and advanced atherosclerosis. Treatment with ACE inhibitors, AT1R antagonists and calcium channel blockers corrects small artery structure and endothelial dysfunction in hypertensive patients (Schiffrin 2001). For example, endothelium-dependent relaxation and the media/lumen ratio of resistance arteries of hypertensive patients are normalised after 1 year of treatment with losartan but not with atenolol (Schiffrin et al. 2000, 2002). Endothelial function of the retinal vasculature is also impaired in early essential hypertension and is improved by ARBs. Following myocardial infarction, AT1R blockade reduces ROS generation and protects the coronary arteries from endothelial dysfunction (Kuno et al. 2002; Liu et al. 2002). Since both polymorphonuclear leucocytes (PMN) and endothelial cells express AT1R, it is believed that AT1R blockers ameliorate endothelial injury, in part by inhibiting PMN adhesion to endothelial cells (Ito et al. 2001). ANG II increases systemic blood pressure not only via direct vasoconstriction, but also via release of aldosterone, leading to water and salt retention. Kidney damage is also caused by elevations in intraglomerular pressure, leading to mechanical damage of glomerular capillaries (Eiskjaer et al. 1992). In kidneys, AT2R activation causes endothelium-dependent vasodilatation via a cytochrome P450 pathway, possibly by epoxyeicosatrienoic acids (EETs) (Arima et al. 1997; Takeuchi 1999), thus modulating the AT1R-mediated vasoconstriction. Impaired function of renovascular AT2R may contribute to the pathophysiology of hypertension (Arima 2003). Old spontaneously hypertensive rats (SHR) exhibit reduced acetylcholine-induced relaxation, probably due to diminished EDHF availability; losartan corrects this defect by increasing NO availability (Maeso et al. 1998). Oestrogen exerts a vasoprotective effect by upregulating AT2R expression in the kidney, resulting in increased prostaglandin E_2 and cGMP concentrations in the renal medulla, and eNOS expression in cortical arteries (Baiardi et al. 2005).

1.5.6
Inflammation

ANG II is a potent proinflammatory agent that causes activation, chemotaxis and proliferation of mononuclear cells and upregulation of proinflammatory mediators, including cytokines and adhesion molecules. The proinflammatory AT1R is found on endothelial cells and circulating blood cells, including PMN, monocytes, T lymphocytes and platelets. The pro-oxidative effect of ANG II is due to AT1R-mediated activation of NAD(P)H oxidase and is blocked by ARBs. The expression of NAD(P)H oxidase subunit gp91-phox is critical for ANG II-induced superoxide formation in endothelial cells (Griendling et al. 1994). Products include not only superoxide but also peroxynitrite, and counteract the beneficial effect of ANG II-stimulated NO release (Pueyo et al. 1998). Endothelial cell migration is pivotal for the maintenance of vessel wall integrity and is stimulated by NO. ANG II inhibits endothelial cell motility by reducing NO availability via an AT1R- and ROS-dependent effect (Desideri et al. 2003). Conversely, as a compensatory signalling mechanism, small amounts of hydrogen peroxide, also derived from NAD(P)H oxidase, elicit endothelial NO production in response to ANG II (Cai et al. 2002). ANG II-derived ROS induce P-selectin expression (Tayeh and Scicli 1998) and mobilisation on the endothelial cell surface (Alvarez and Sanz 2001), as well as activation of nuclear factor (NF)-κB and induction of redox-sensitive genes for endothelial adhesion molecules, cytokines and chemokines (Phillips and Kagiyama 2002; Costanzo et al. 2003). ANG II also downregulates nox4, while it markedly upregulates the nox-1 isoform in smooth muscle cells (Lassegue et al. 2001) and regulates xanthine oxidase-mediated superoxide production (Mervaala et al. 2001).

AT1R mediate the ANG II-induced increase in VEGF in endothelial cells via induction of hypoxia-inducible factor-1, resulting in vascular remodelling, increased permeability and oedema formation (Tamarat et al. 2002). ARBs also possess AT1R-independent anti-inflammatory effects (Kramer et al. 2002). Oxidised low-density lipoprotein receptor (LOX-1) expression is also stimulated by ANG II. Human coronary arterial endothelial cells possess abundant LOX-1 receptors, which appear to mediate uptake of oxidised low-density lipoprotein (ox-LDL) via AT1R activation, thus enhancing ox-LDL-mediated injury (Li et al. 1999c). Conversely, peroxisome proliferator-activated receptor-γ (PPARγ) activators (insulin sensitisers, e.g. the glitazones pioglitazone and rosiglitazone) and peroxisome proliferator-activated receptor-α (PPARα) activators (fibrates, e.g. fenofibrate) exhibit cardiovascular anti-inflammatory and antioxidant properties and correct endothelial dysfunction induced by ANG II (Diep et al. 2002).

1.5.7
Atherosclerosis

Disruption of the NO–ROS balance contributes to endothelial dysfunction and leads to vascular injury and atherosclerosis. Endothelial, leucocyte and platelet AT1R contribute to the development of hypercholesterolaemia and atherosclerosis (Papademetriou 2002; Strawn and Ferrario 2002). AT1R blockers are anti-atherosclerotic and reduce oxidative stress in the vessel wall (Rueckschloss et al. 2002). Hypercholesterolaemia is associated with AT1R upregulation, endothelial dysfunction and increased NAD(P)H oxidase-dependent superoxide production (Warnholtz et al. 1999; Nickenig and Harrison 2002), which is prevented by statin treatment through a mechanism that is independent of the lipid-lowering effect of the drugs. Thus, ARBs may represent a novel approach for the prevention of vascular dysfunction associated with hypercholesterolaemia, independent of lipid-lowering and blood pressure-lowering interventions (Wassmann et al. 2002). The recent observation of insulin resistance-induced upregulation of AT1R expression could further explain the association of insulin resistance with endothelial dysfunction and hypertension (Shinozaki et al. 2004). Conversely, evidence from AT2R knockout mice suggests that AT2R protect both heart and brain tissue from ischaemia (Iwai et al. 2004). Treatment with an AT1R antagonist before vascular injury decreases neointima formation in wild-type but not AT2R knockout mice, whereas treatment with an AT2R antagonist before injury has no effect. These results suggest that AT2R-mediated

ANG II signalling is not essential for the development of neointimal formation, although it may modify it (Harada et al. 1999).

2
Bradykinin

Components of the kallikrein–kinin system (KKS) have been under investigation since 1909, when a hypotensive factor was found in the urine and was later identified as kallikrein. In 1949, Rocha e Silva and collaborators discovered that blood containing the venom of *Bothrops jararaca* (South American pit viper) caused slow contractions in an isolated preparation of guinea-pig ileum, which was made refractory to the venom itself. They coined a name derived from Greek for this active factor, using the word *kinin* (indicating movement) with the prefix *brady* (indicating slow) to describe the slow effect of the substance on the guinea-pig ileum (Beraldo and Rocha e Silva 1949).

2.1
Synthesis

There are two pathways that generate BK (Fig. 1). The simpler of the two has two components: the enzyme, tissue kallikrein (Margolius 1998), secreted by many cells (especially salivary glands, pancreatic exocrine gland, lung, kidney, intestine, brain) and the substrate, low molecular weight kininogen (LMWK), an α_2-globulin synthesised in the liver (Muller-Esterl et al. 1985). Tissue kallikrein digests LMWK to yield the decapeptide, lysyl-BK (kallidin). The second pathway of BK formation is part of the intrinsic coagulation pathway (Kaplan et al. 1998). BK is formed when plasma kallikrein acts on high molecular weight kininogen (HMWK), which is synthesised and secreted from the liver by alternative splicing of the same gene that encodes LMWK (Nakanishi 1987). Tissue kallikrein prefers LMWK but is also capable of cleaving HMWK, whereas plasma kallikrein cleaves HMWK exclusively. BK is subject to rapid enzymatic degradation and has a plasma half-life of 10–50 s (Decarie et al. 1996). It is metabolised by several peptidases (collectively known as kininases). A plasma amino-peptidase named carboxypeptidase-N converts BK to [des-Arg^9]-BK and kallidin to [des-Arg^{10}]-lysyl-BK. ACE is probably the most important enzyme for degrading BK in the circulation (Erdos 1990a), while neutral endopeptidase (NEP) (EC 3.4.24.11) appears to be the most important enzyme for the degradation of BK in the airways (Frossard et al. 1990).

2.2
Receptors

At least two BK receptor subtypes (B1 or BKB1R and B2 or BKB2R) are recognised, based on the rank order of potency of kinin agonists (Regoli and

Barabe 1980). BKB1R demonstrate decreasing affinity for [des-Arg^{10}]-lysyl-BK>[des-Arg^9]-BK = lysyl BK >> BK; BKB2R demonstrate decreasing affinity for BK = lysyl-BK >> [des-Arg^{10}]-lysyl-BK > [des-Arg^9]-BK. BK and lysyl-BK (kallidin) stimulate constitutively-produced BKB2R (Vavrek and Stewart 1985), whereas [des-Arg^9]-BK or [des-Arg^{10}]-lysyl-BK stimulate BKB1R (Regoli and Barabe 1980), induced as a result of inflammation (Marceau et al. 1980). Both BKB1R and BKB2R are G protein-coupled receptors primarily linked to phospholipase C activation, and cause intracellular calcium mobilisation by inositol 1,4,5-trisphosphate. Unlike the BKB2R, BKB1R are resistant to desensitisation and are not phosphorylated or internalised after agonist stimulation (Blaukat et al. 1999).

2.3
Receptor Antagonists

There is significant interest in developing BKB1R antagonists as possible interventions in chronic inflammation. There have been limited clinical trials of a few BKB1R antagonists. Deltibant had some efficacy in closed head trauma, but was not effective in septic shock (Fein et al. 1997). B-9340 was found to be effective against vasodilatation in patients with heart failure (Witherow et al. 2001). A third antagonist, B-9870, is in the pre-clinical stage for the potential treatment of lung cancer (Chan et al. 2002). One BKB1R agonist has been used in patients with brain tumours to increase permeability of the blood–brain barrier in order to increase penetration of chemotherapeutic drugs (Bartus et al. 1996).

2.4
Signalling

Application of exogenous BK on human or animal tissues reproduces the four classic signs of inflammation: redness, local heat, swelling and pain. Redness and local heat are caused by local endothelium-dependent vasodilatation. The stimulation of endothelial cells also results in increased microvascular permeability, which contributes to accumulation of protein-rich fluid from the circulation (swelling). BK produces pain through stimulation of its receptors in the sensory endings of non-myelinated afferent neurons and causes contraction of several types of smooth muscle preparations, including human bronchi, colon and bladder. BK releases NO, prostaglandin (PG)I$_2$ and PGE$_2$ from endothelial cells in a number of tissues (Jose et al. 1981), via the breakdown of inositol lipids to inositol 1,4,5-trisphosphate (Derian and Moskowitz 1986). Endothelium-dependent hyperpolarisation of smooth muscle cells appears to be the principal mechanism involved in BK-induced relaxation of isolated human coronary arterioles (Batenburg et al. 2004b).

2.5
Effects

2.5.1
Neovascularisation

Daily administration of BK into sponge implants enhances basal sponge-induced neovascularisation (Hu and Fan 1993). This effect is significantly potentiated by interleukin (IL)-1α. The BK/IL-1α-induced neovascularisation is abolished by the BKB1R antagonist [Leu^8] des-Arg^9-BK, but not by the BKB2R antagonist Ac-D-Arg-[Hyp^3, D-Phe^7, Leu^8]-BK, suggesting that blockade of BKB1R may provide effective treatment for chronic inflammatory diseases. BK promotes growth of endothelial cells from postcapillary venules (Morbidelli et al. 1998) by upregulating c-Fos expression and potentiating the growth promoting effect of FGF-2 via activation of the NOS pathway. Only the BKB1R appear to be responsible for BK-induced proliferation, suggesting that these receptors might be implicated in promoting angiogenesis (Parenti et al. 2001). On the other hand, BKB2R-mediated angiogenesis occurs via recruitment of inflammatory mediators, requires higher tissue levels of BK, does not involve endothelial cell proliferation and is linked to phospholipase C activation. Like VEGF, BK also induces angiogenesis via BKB2R-mediated transactivation of KDR/Flk-1 accompanied by eNOS activation (Miura et al. 2003). The proangiogenic effect of ACE inhibitors is mediated through BKB2R activation and increased eNOS protein levels (Silvestre et al. 2001). BK antagonists stimulate apoptosis in cancer by blocking intracellular increase of calcium and stimulating the mitogen-activated protein (MAP) kinase pathway to produce caspase activation (Stewart 2003).

2.5.2
Hypertension

BK interacts with the RAS to stimulate renin gene expression (Yosipiv et al. 2001). BKB2R knockout mice overloaded with a high salt diet develop malignant hypertension (Alfie et al. 1996), suppression of the RAS, abnormal kidney development (El-Dahr et al. 2000) and cardiac impairment (Emanueli et al. 1999). The vasodilator response to BK is absent in BKB2R-null mice, suggesting the importance of BKB2R in this action of BK (Berthiaume et al. 1997). The damaging effects of salt overload in the heart implicate the AT1R, and it is believed that the lack of BKB2R is responsible for failing to counterbalance the AT1R action in BKB2R-null mice (Madeddu et al. 2000).

2.5.3
Inflammation

Inhibition of BKB2R with the non-peptide FR174657 or with the peptide icatibant attenuates exudate formation in various models of cutaneous inflammation (Griesbacher and Legat 2000). BK stimulates leucocyte–endothelial cell interactions via a BKB2R-initiated, cytochrome P450 epoxygenase-, oxidant- and PKC-mediated upregulation of cell adhesion molecule (e.g. P-selectin and ICAM-1) expression (Tayeh and Scicli 1998; Shigematsu et al. 2002). BK also produces venular protein leakage, an effect that is initiated by stimulation of BKB2R and involves cytochrome P450E and PKC activation, oxidant generation and cytoskeletal reorganisation. BK, acting through BKB2R, induces activation of the Ras/Raf-1/ERK pathway, which initiates inhibitor of κB kinase (IKK)-α and NF-κB activation, and ultimately induces COX-2 expression in a human airway epithelial cell line (Chen et al. 2004). Synthesis of BK from HMWK also results in the formation of a two-chain peptide (HKa, cleaved high molecular weight kininogen) which has been reported to bind the β_2-integrin Mac-1 on PMN in a Zn^{2+}-dependent manner and to exert anti-adhesive properties through inhibition of ICAM-1 and Mac-1 binding (Chavakis et al. 2001). Locally generated HKa can balance the BK-induced recruitment of leucocytes, thereby providing a physiological feedback mechanism. Bacterial lipopolysaccharide-induced BKB1R expression in the rat paw sensitises the rat paw to the oedema-forming effect of [des-Arg^9]-BK in a manner dependent on neutrophil influx, local NF-κB activation and local formation of tumour necrosis factor (TNF)-α and IL-1β (Passos et al. 2004). The KKS can be massively activated in bacterial sepsis, with increased expression of peripheral BKB1R expression (Marceau et al. 1998). BK antagonists might thus be able to antagonise the circulatory and systemic components of sepsis. Deltibant, a BKB2R antagonist, has reached clinical trials for sepsis but has not shown decisive benefits (Fein et al. 1997). BKB1R have been implicated in nociception (Rupniak et al. 1997) and accumulation of leucocytes in inflamed tissues (Perron et al. 1999). BKB1R-null mice develop normally but show a drastic reduction in PMN infiltration at sites of inflammation (Pesquero et al. 2000).

2.5.4
Diabetes

Infusion of BK and ACE inhibitors reduces the hyperglycaemia associated with streptozotocin-induced insulin-dependent diabetes mellitus in rodents (Rett et al. 1986). Chronic treatment of animals with a BKB1R antagonist prevents streptozotocin-induced diabetes and reduces β cell damage (Zuccollo et al. 1999). ACE inhibitors also improve insulin sensitivity in non-insulin-dependent diabetes mellitus (Gans et al. 1991; Torlone et al. 1991), whereas BKB2R antagonists reduce insulin sensitivity in normotensive rats (Kohlman et al. 1995). ACE inhibitors improve sensitivity to insulin and other metabolic end-points in animal models of type II diabetes (insulin resistant Zucker rats and diabetic mice KK-Aγ); this is inhibited by icatibant (Wang et al. 2003), supporting the idea that these effects are mediated by endogenous BK and *BKB*2R (Shiuchi et al. 2002). BK not only increases glucose uptake but also increases in sulin secretion. Both in vitro (Yang and Hsu 1995) and in vivo (Mikrut et al.

2001) studies have confirmed that BK stimulates insulin release and a reduction in blood glucose levels, both of which were inhibited by HOE140, a selective BKB2R antagonist. BK directly triggers GLUT4 and GLUT1 translocation to increase the rate of glucose uptake in various cell types (Isami et al. 1996; Rett et al. 1996; Kishi et al. 1998). The insulin receptor is a protein tyrosine kinase that, when activated by insulin binding, undergoes rapid autophosphorylation and phosphorylates intracellular protein substrates such as insulin receptor substrate-1 (IRS-1). Following tyrosine phosphorylation, IRS-1 acts as docking protein for several molecules including PI3-kinase. BK and the ACE inhibitor captopril increase insulin-stimulated tyrosine phosphorylation of the insulin-receptor and IRS-1 in the liver and muscle of rats (Carvalho et al. 1997). Aprotinin, an inhibitor of kallikrein, antagonises the exercise- or hypoxia-induced increase in blood flow and glucose uptake in skeletal muscle (Dietze et al. 1980).

2.6
Interactions with eNOS

Endothelial BKB2R co-immunoprecipitate with eNOS (Ju et al. 1998). BK stimulation causes a transient rise in endothelial $[Ca^{2+}]_i$ levels, followed by dephosphorylation of eNOS at Thr^{497}, dissociation of eNOS from BKB2R and subsequent eNOS activation accompanied by phosphorylation of Ser^{617}, Ser^{635} and Ser^{1179} (Venema 2002). Additionally, BK stimulation of bovine aortic endothelial cells causes dissociation of eNOS-Raf-1-ERK-Akt heterotrimeric complex, leading to activation of ERK and phosphorylation of eNOS (Bernier et al. 2000). Sustained activation of eNOS by BK results in downregulation of eNOS synthesis, whereas sustained inhibition of BK receptors results in an upregulation of eNOS synthesis (Vaziri et al. 2005), suggesting an adaptive physiologic response of eNOS expression mediated by BK-derived NO.

3
Angiotensin–BK Interactions

There are multiple levels of interaction between ANG II and BK. Both ANG II and BK stimulate phosphoinositide turnover and pathways that generate Ca^{2+} (Ogino and Costa 1992). In sympathetically innervated tissues, ANG II and BK facilitate the release of noradrenaline evoked by electrical stimulation (Starke and Schumann 1972; Guimaraes et al. 1998; Boehm and Kubista 2002). The two peptides also share common binding sites on ACE. Additionally, while AT2R stimulate the production of BK in smooth muscle cells (Tsutsumi et al. 1999), BK stimulates renin gene activity (Yosipiv et al. 2001). Furthermore, AT1R and BKB2R form heterodimers. ANG-(1–7) interacts with both BK and ACE. Endothelial cells contain BKB2R, which potently stimulate production of NO. ANG II is a potent stimulus for vasoconstriction and vascular smooth muscle hypertrophy, whereas NO has a vasodepressor effect and has been shown to be an antiproliferative agent. In rats, AT2R stimulation induces a systemic vasodilator response mediated by BK and NO that counterbalances the vasoconstrictor action of ANG II via the AT1R (Carey et al. 2001). ANG II infusion in BKB2R-null mice produces much stronger hypertension than in wild-type, suggesting that the KKS selectively buffers the vasoconstrictor activity of ANG II (Maly et al. 2001). In isolated rat hearts, inhibition of BKB2R with HOE140 increases myocardial ischaemia/reperfusion injury, whereas inhibition of AT1R with losartan reduces it (Sato et al. 2000). AT1R knockout mice exhibit activated KKS that ameliorates the severity of renal vascular disease (Tsuchida et al. 1999). In the developing kidney, there is much cross-talk between the RAS and KKS. High salt load during gestation suppresses fetal RAS and provokes abnormal renal development in the BKB2R knockout mouse (El-Dahr et al. 2000). These interactions, along with the roles of ACE, cross-talk between BK and ANG-(1–7) and the opposite effects of AT1R and AT2R activation, support the hypothesis of a counterbalance between the KKS and the RAS.

3.1
Biosynthesis and Degradation

The first recognised important link between ANG II and BK was the discovery that kininase II, a major BK-degrading enzyme, was indeed ACE, the enzyme that catalyzes the formation of ANG II (Erdos and Yang 1967; Yang and Erdos 1967; Yang et al. 1971). ACE inhibitors exert their beneficial cardiovascular effects via the inhibition of both ANG II formation and BK breakdown. Recently, a homologue of ACE, ACE2, has been recognised (Tipnis et al. 2000; Bernstein 2002; Crackower et al. 2002). ACE2 degrades ANG I by removing the carboxy terminal lysine, making the peptide ANG-(1–9), which enhances arachidonic acid release by BK and resensitises the BKB2R (Marcic et al. 1999; Bernstein 2002). The ACE2 product, ANG-(1–7), also acts as an ACE inhibitor (Tom et al. 2001), and may stimulate BK release via AT2R. Recent studies indicate that the enzyme prolylcarboxypeptidase, an ANG II-inactivating enzyme, is a prekallikrein activator. The ability of prolylcarboxypeptidase to act in the KKS and the RAS indicates a novel interaction between these two systems. There is also evidence that the BK-potentiating effects of ACE inhibitors may include a mechanism independent of BK hydrolysis, i.e. there may be ACE-BKB2R cross-talk, resulting in BKB2R upregulation as well as direct activation of BKB1R by ACE inhibitors (Busse and Fleming 1996; Minshall et al. 1997; Benzing et al. 1999). The mechanism behind this phenomenon may require ACE-BKB2R co-localisation on the endothelial cell membrane (Erdos et al. 1999; Marcic et al. 1999; Tom et al. 2003).

3.2
BK Interactions with Angiotensin (1–7)

3.2.1
Stimulation of BK Release by ANG-(1–7)

ANG-(1–7) was originally considered to be an inactive product of ANG II metabolism because of its inability to mimic the vasoconstrictor or aldosterone-secreting actions of ANG II (Ferrario et al. 1991). It is now known that ANG-(1–7) is a biologically active peptide with distinct and often opposite effects from those of ANG II (Ferrario et al. 1997). ANG-(1–7) elicits prostaglandin production from astrocytes, smooth muscle and endothelial cells (Jaiswal et al. 1992). In contrast to the vasoconstrictive effects of ANG II, ANG-(1–7) is a vasodilator (Benter et al. 1995), relaxes coronary arterial rings (Porsti et al. 1994), pial arterioles (Meng and Busija 1993) and mesenteric arteries (Osei et al. 1993), and reduces blood pressure in SHR (Benter et al. 1995) and renovascular hypertensive dogs (Nakamoto et al. 1995). These effects are blocked by removal of endothelium or pretreatment with an NO synthase inhibitor (Porsti et al. 1994; Brosnihan et al. 1996). Moreover ANG-(1–7)-induced relaxation is not affected by AT1R or AT2R blockade, but is attenuated by the BKB2R antagonist HOE140, or prior exposure to the competitive nonselective ATR antagonist [Sar^1,Thr^8]-ANGII (saralasin). These results suggest that the biological activity of ANG-(1–7) is mediated through activation of another AT receptor and that it involves release of vasoactive kinins (Porsti et al. 1994; Brosnihan et al. 1996). In the presence of NO synthesis inhibitors, ANG-(1–7) elicits an endothelium-dependent antagonism of ANG II, via activation of AT2R and ANG-(1–7) receptors, in rats under normal or high sodium intake, which is abolished by low sodium intake, suggesting that it may also serve as a negative feedback towards ANG II in response to altered sodium intake (Roks et al. 2004).

3.2.2
Potentiation of the Effects of BK by ANG-(1–7)

The potentiating effect of ANG-(1–7) on BK was first described in conscious rats (Paula et al. 1995); intravenous ANG-(1–7) potentiated—by two- to tenfold—the vasodepressor response to BK. Similar results were obtained in normotensive and hypertensive rats and isolated rat heart (Lima et al. 1997; Almeida et al. 2000). This response is specific to ANG-(1–7), since neither acetylcholine nor sodium nitroprusside—or prostaglandins—augment BK-induced relaxation (Li et al. 1997c), and involves BKB2R and a novel ATR (i.e. not AT1R or AT2R) in an endothelium-dependent manner (Tallant et al. 1997). The ACE inhibitor lisinopril enhances BK-induced vasodilatation, but abolishes the synergistic action of ANG-(1–7) on BK. ANG-(1–7) also reduces the degradation of ^{125}I-[Tyr]-BK and the appearance of the BK-(1–7) and BK-(1–5) metabolites by inhibiting ACE activity with an IC_{50} of 650 nM, supporting the idea that ANG-(1–7) acts as a local synergistic modulator of kinin-induced vasodilatation by inhibiting ACE and releasing NO (Li et al. 1997c). The ANG-(1–7)-dependent release of NO from endothelial cells is attenuated by NO synthase inhibition or the BKB2R antagonist icatibant (HOE140) and is associated with very low concomitant production of superoxide (Heitsch et al. 2001). This potentiating effect, which is present in Wistar rats and SHR (Paula et al. 1995; Lima et al. 1997; Almeida et al. 2000), has been shown to disappear in arterioles of the mesenteric arteriolar bed of diabetic rats and is restored by chronic but not acute insulin treatment (Oliveira et al. 2002, 2003). Infusion of the ANG-(1–7) antagonist A-779 does not modify the ANG II pressor effect or the inhibition of ANG I metabolism by captopril. However, A-779 reduces the potentiating effect of captopril on the hypotensive effect of BK (Maia et al. 2004), demonstrating that endogenous ANG-(1–7), or an ANG-(1–7)-related peptide (or both) plays an important role in the BK potentiation by ACE inhibitors through a mechanism not dependent upon inhibition of the ACE hydrolytic activity. The mechanisms behind the BK potentiating activity of ANG-(1–7) appear complex and involve receptor-mediated facilitation of NO (Li et al. 1997c; Almeida et al. 2000; Heitsch et al. 2001) and prostaglandin release (Paula et al. 1995; Aparecida Oliveira et al. 1999; Almeida et al. 2000; Fernandes et al. 2001), endothelium derived hyperpolarising factor (Fernandes et al. 2001), ACE inhibition (Li et al. 1997c; Tom et al. 2001) and binding of ANG-(1–7) to ACE to facilitate the cross-talk between ACE and BKB2R (Deddish et al. 2002; Tsutsumi et al. 1999).

3.2.3
Resensitisation of BK Receptors by ANG-(1–7)

ANG-(1–7) indirectly resensitises B_2 receptors via induction of a cross-talk between the BKB2R and ACE on plasma membranes without having a direct effect on the BKB2R and BK hydrolysis (Deddish et al. 2002).

3.3
BK Interactions with Angiotensin Type 2 Receptors

It has been suggested that some of the beneficial effects of AT2R stimulation may be mediated through the BK/NO cascade.

3.3.1
BK and AT2R: Mediated NO Release

Evidence for the interaction between ANG II and BK at the level of the ATR was recognised by the finding that formation of nitrite in response to angiotensin peptides is due to the activation of kinin production (Seyedi et al. 1995). The RAS stimulates renal BK production and cGMP formation through the

AT2R; inhibition of renin, not of AT1R, decreases renal BK levels during salt depletion (Siragy et al. 1996). Furthermore, in stroke-prone SHR, infusion of ANG II increases aortic cGMP content, an effect inhibited by either AT2R blockade, NO-synthesis inhibition or BKB2R blockade (Gohlke et al. 1998) and suggesting that stimulation of AT2R releases BK and NO (Seyedi et al. 1995; Liu et al. 1997; Gohlke et al. 1998; Henrion et al. 2001). Mice overexpressing AT2R exhibit an attenuated pressor response to ANG II infusion; pretreatment with an AT2R antagonist, a BKB2R antagonist or an NO synthase inhibitor restored the pressor response to ANG II. ANG II produces a paradoxical decrease in blood pressure after AT1R blockade, suggesting that selective AT2R stimulation has a vasodepressor effect, which is associated with an endothelium-dependent increase in cGMP and activation of the KKS (Tsutsumi et al. 1999).

3.3.2
BK and AT2R: Mediated Flow-Dependent Vasodilatation

BK is thought to be a primary mediator of ANG II-induced flow-dependent vasodilatation, since blockade of BKB2R reduces the dilator response to flow (Bergaya et al. 2001; Katada and Majima 2002). Inhibition of AT2R with PD123319 reduces flow-induced dilatation in wild-type ($TK^{+/+}$) mice, but not in tissue kallikrein-deficient mice ($TK^{-/-}$). Combining PD123319 with the BKB2R antagonist HOE140 has no additional effect on AT2R blockade alone in $TK^{+/+}$ arteries (Bergaya et al. 2004). Furthermore, HOE140 reduces the response to flow in $AT2R^{+/+}$, but not in $AT2R^{-/-}$ mice. AT2R also stimulate NO production by two alternative pathways: through the BKB2R and by direct stimulation of NO and cGMP, as demonstrated in BKB2R-null mice (Abadir et al. 2003).

3.3.3
BK and AT2R: Mediated Effects on the Myocardium

BK exerts cardioprotective actions which are mediated via BKB2R (Dendorfer et al. 1999). In a rat model of chronic heart failure, left ventricular remodelling and cardiac function were improved by blockade of AT1R (Liu et al. 1997). This effect was inhibited by treatment with an AT2R antagonist and also, in part, by treatment with a BKB2R antagonist. Following regional myocardial ischaemia in pigs, infarct size was reduced by AT1R blockade, and this reduction was abolished by pretreatment with the AT2R antagonist PD123319 and by BKB2R blockade (Jalowy et al. 1998). Reduction of perivascular fibrosis by overexpressing cardiac AT2R after pressure overload was abolished after BKB2R blockade or NO synthase inhibition, suggesting that the inhibition of perivascular fibrosis by stimulation of myocyte AT2R was BK/NO-dependent (Kurisu et al. 2003).

3.4
BK Interactions with Angiotensin Type 1 Receptors

Although ANG II stimulates AT2R to release NO, and indirectly BK, there is additional evidence that there also is an interaction between AT1R and the BKB2R (Schmaier 2003). Following myocardial infarction in rats, either ACE inhibitors or AT1R antagonists prevent remodelling of the left ventricle (Li et al. 1997c), and this effect is blocked by BKB2R inhibition.

3.4.1
Modulation of BK Levels by Angiotensin Peptides

Canine cardiac interstitial fluid (ISF) BK levels increase during ANG I and ANG-(1–7), but not ANG II, infusions. ANG I binding to the active site of ACE and neutral endopeptidases, combined with the formation of large amounts of ANG-(1–7) with its inhibitory effects on ACE, could provide a mechanism for the increase of ISF BK (Wei et al. 2002).

3.4.2
Upregulation of BK Type 2 Receptors

Infusion of ANG II results in the upregulation of BKB2R messenger RNA (mRNA) levels (Kintsurashvili et al. 2001). Targeted disruption of $AT1_AR$ results in decreased expression of BKB2R, thus implicating a role for the $ATA1_AR$ in modulating the expression of BKB2R. ANG II stimulates BKB2R expression at the transcriptional level via activation of the p42/p44MAPK pathway, since selective inhibition of the p42/p44MAPK blocks the ANG II-induced increase in BKB2R expression, whereas inhibition of the p38MAPK pathway does not (Tan et al. 2004).

3.4.3
Angiotensin–BK Receptor Heterodimerisation

AT1R communicate with BKB2R and form stable heterodimers, which activate $G\alpha_q$ and $G\alpha_i$ proteins in response to ANG II stimulation. Heterodimerisation also results in a change in the endocytic pathways of both receptors (AbdAlla et al. 2000). Heterodimerisation between AT1R and BKB2R occurs in platelets and omental vessels in pre-eclamptic women (AbdAlla et al. 2001b). This interaction results in a four- to five-fold increase in protein levels of the BKB2R. AT2R also bind AT1R to form additional heterodimers that antagonise AT1R function. BKB2R, BKB1R and AT1R are linked to $G\alpha_i$ and $G\alpha_q$, but with different physiological functions, suggesting that signalling may occur outside the classic G protein interactions. EP24.15, a widely distributed cytosolic enzyme, which can degrade ANG I and II and BK and which is identified as a putative soluble ANG II binding protein (Kiron and Soffer 1989), associates with AT1R and BKB2R both at the plasma membrane and after receptor internalisation. This association suggests a possible mechanism for endosomal disposition of ligand that may facilitate receptor recycling (Shivakumar et al. 2004).

4
Angiotensin Converting Enzyme (ACE)

4.1
Expression

ACE, an ectoenzyme anchored to the plasma membrane with the bulk of its mass exposed to the extracellular surface of the cell (Corvol et al. 1995), is a key enzyme of the RAS. ACE was originally identified (Skeggs et al. 1956) as a "hypertensin-converting enzyme". A soluble form of ACE is also present in serum and other body fluids; however, it is the tissue-bound form of ACE that is proposed to control both blood pressure and renal function (Esther et al. 1997). Through its actions on ANG I and BK, ACE regulates the balance between the RAS and the KKS and has an important role in vascular tone and blood pressure regulation. The primary specificity of ACE is to cleave carboxyterminal dipeptides from oligopeptide substrates with a free C terminus in the absence of a penultimate proline residue. It is via this action that ACE hydrolyses both ANG I and BK (Skeggs et al. 1956; Yang et al. 1970; Corvol et al. 1995). In addition to acting on ANG II and BK, ACE is also able to act as an endopeptidase on certain substrates which are amidated at the C termini by cleaving a C terminal dipeptide amide. ACE can also cleave a C terminal tripeptide amide from substance P and luteinising hormone-releasing hormone (LHRH). ACE exists in two distinct forms. The somatic form of ACE is present on the endothelial surface of all vessels examined to date and on the brush-border membranes of the kidney, intestine, placenta and choroid plexus. The germinal form, found exclusively in testis, plays a crucial role in fertility (Turner and Hooper 2002). Somatic ACE (M_r 180,000) is composed of two homologous domains, the (NH_2) N-domain and the (COOH) C-domain, each of which contains an active site (Soubrier et al. 1988). Each domain contains the typical zinc-binding motif (His-Glu-X-X-His) found in many zinc peptidases. In this motif, the two histidines represent two of the zinc ligands, with the third being a glutamate residue on the C-terminal side of the motif. Thus, ACE is classified as a member of the M2 gluzincin family. The two domains of somatic ACE differ in substrate specificity; for example, the N-domain hydrolyses the Trp^3-Ser^4 bond of LHRH much faster than does the C-domain. The haemoregulatory peptide N-acetyl-Ser-Asp-Lys-Pro is the most specific substrate identified to date for the N-domain (Rousseau et al. 1995), but a substrate specific for the C-domain has not yet been found. The two domains hydrolyse ANG I and BK at a comparable rate, although the C-domain requires high concentrations of Cl⁻ for optimal activity, a property that seems to be conferred by a single arginine residue (Arg^{1098}) in this domain (Liu et al. 2001). Germinal ACE (M_r 100,000) contains a single catalytic site corresponding to the C-domain of somatic ACE (Ehlers et al. 1989). The somatic and germinal forms of ACE mRNA are transcribed from the same gene using alternative promoters (Hubert et al. 1991).

Recently, two groups (Donoghue et al. 2000; Tipnis et al. 2000) reported data on the first known homologue of ACE, which they termed ACE2 and ACEH, respectively. This enzyme, now commonly referred to as ACE2, has many similarities to ACE. ACE2 is a type I integral membrane peptidase showing 40% identity and 61% similarity with ACE and conserving the critical active site residues. Like germinal ACE, ACE2 contains a single catalytic domain. Also similar to ACE, ACE2 is expressed in endothelial cells; however, its basal expression is restricted to heart, kidney and testis (Donoghue et al. 2000; Tipnis et al. 2000). ACE2 does display some differences from ACE; it functions exclusively as a carboxypeptidase, hydrolysing either aromatic or basic residues from the C-terminus and preferring a prolyl residue in the P_1 position (Turner and Hooper 2002; Vickers et al. 2002). ACE2 hydrolyses both ANG I and ANG II but not BK. ACE2 cleaves ANG I to a nonapeptide ANG-(1–9) and directly converts ANG II to ANG-(1–7) (Iyer et al. 2000; Lemos et al. 2002; Ren et al. 2002; Turner and Hooper 2002). Kinetically, ACE2 is a 100-fold faster degrading enzyme of ANG II to ANG-(1–7) than prolylcarboxypeptidase (Odya et al. 1978; Vickers et al. 2002). Although it does not degrade BK, it degrades [des-Arg^9]-BK at its carboxy terminal amino acid (Donoghue et al. 2000). To date, ACE2 has proved to be insensitive to all ACE inhibitors. ACE2 is thought to remove the C-terminal residue from three other vasoactive peptides, neurotensin, kinetensin (a neurotensin-related peptide) and [des-Arg^9]-BK. ACE2 also acts on apelin-13 and apelin-36, peptides with high catalytic efficiency (Vickers et al. 2002). Although the role of the apelins is not fully elucidated, systemic administration of apelin-13 promotes hypotension in rats (Tatemoto et al. 2001). Despite their homologous catalytic domains, ACE2 and ACE are biochemically and pharmacologically distinct. It has been suggested that both ACE and ACE2 are involved in blood pressure regulation (Danilczyk et al. 2003).

4.2
Molecular Regulation

The primary structure of ACE was revealed by protein sequencing of human kidney ACE followed by complementary DNA (cDNA) cloning in endothelial cell libraries (Soubrier et al. 1988). The mouse ACE enzyme has a high overall homology with human ACE (Bernstein et al. 1989). In human endothelial cells, ACE is encoded by a 4.3-kb mRNA species. The coding sequence comprises 1,306 residues, including a signal peptide of 29 amino acids (Costerousse et al. 1992). In addition to the membrane-bound form, ACE exists as a soluble protein. A membrane-associated protease, the ACE secretase, acts on both ACE isozymes to liberate the soluble forms that circulate in the serum and other body fluids (Ramchandran and Sen 1995). The soluble form of ACE circulates in plasma at the relatively high concentration of 10^{-9} M, although the plasma enzyme is considered physiologically less important for the processing of peptides in the circulation than the membrane-bound endothelial enzyme (Alhenc-Gelas et al. 1983; Erdos 1990b). Plasma ACE levels vary widely between individuals; however, when measured repeatedly in a given subject, levels remain remarkably constant (Alhenc-Gelas et al. 1983, 1991). A study of plasma ACE levels in nuclear families revealed intrafamilial correlations between genetically related members, with the genetic analysis suggesting that a major gene effect was responsible for a large part of the inter-individual variability in plasma ACE levels (Cambien et al. 1988). This has been confirmed after the cloning of ACE DNA where an insertion-deletion polymorphism, located in an intron of the ACE gene, was discovered, and it was recognised that this polymorphism was associated with differences in the concentration of ACE in plasma (Rigat et al. 1990). Homozygotes for the insertion (II) have lower serum ACE levels than those homozygotes for the deletion (DD); heterozygotes (ID) have intermediate levels. The molecular mechanisms involved in the genetic control of ACE expression as well as the physiological consequences of this regulation are still being investigated.

To determine whether local vascular production of ANG II is necessary for the normal regulation of blood pressure, a line of genetically altered mice lacking endothelial ACE was developed using targeted homologous recombination to separate the transcriptional control of somatic ACE from its endogenous promoter, by substituting control to the albumin promoter (Cole et al. 2002). These mice, termed ACE.3, express ACE in the liver but not in the lung, the aorta or any vascular structure. Liver ACE appeared to compensate for the lack of endothelial ACE expression, so that ACE.3$^{-/-}$ mice have normal levels of plasma ANG II, normal blood pressure levels, normal response to ACE inhibitors and normal renal function. Conversely, mice lacking all ACE presented a phenotype of approximately 35 mmHg lower blood pressure than control animals (Krege et al. 1995; Esther et al. 1996). Despite having all other compensatory systems intact, these mice cannot effectively compensate and maintain their blood pressure. Similarly, decreased blood pressure was observed in mice lacking angiotensinogen (Tanimoto et al. 1994), renin (Yanai et al. 2000) or both isoforms of the AT1R (Oliverio et al. 1998; Tsuchida et al. 1998). In all these animals, the RAS proved to be vital for blood pressure regulation.

ACE-like proteins have been identified in lower organisms, such as *Drosophila* (Brakebusch et al. 1994; Williams et al. 1996), indicating that it is an evolutionary-conserved protein that also shows an overall sequence homology of 80%–90% across mammalian species (Santhamma et al. 2004). This evolutionary conservation of ACE, along with its widespread distribution in many organs within a species, suggests that it plays a bigger role than just its role in the RAS (Santhamma et al. 2004). The abnormal phenotype of ACE knockout mice further supports this notion.

4.3
Inhibitors

Inhibition of tissue ACE decreases ANG II, oxidative stress and ANG II-induced inflammation. BK formation is also increased, resulting in increased NO and prostacyclin, which have anti-inflammatory, antithrombotic and vasorelaxant actions (Dzau 2001). Tissue ACE inhibition has therefore emerged as an important therapeutic target for treating cardiovascular disease. In addition to hypertension and congestive heart failure, ACE inhibitors are effective in the treatment of coronary heart disease (Dzau et al. 2002) and myocardial infarction (Mukae et al. 2000). ACE inhibitors interfere with the metabolism of both ANG I and BK.

The first ACE inhibitors were developed from the venom on the South American pit viper, *B. jararaca* (Ferreira et al. 1970). Ferreira and colleagues described a mixture of peptides extracted from this venom as BK potentiating factor (BPF) (Ferreira et al. 1970). Further observations revealed that these peptides inhibited a converting enzyme responsible for cleaving ANG I and catalysing the degradation and inactivation of BK, and reduced blood pressure (Davis and Freeman 1982). The first marketed ACE inhibitor was the nonapeptide BPF_{9a}, or teprotide, an effective, parenterally administered, competitive inhibitor of ACE, with a short half-life in vivo (Antonaccio and Cushman 1981). The first orally active ACE inhibitor, captopril, was designed based on the hypothesis that ACE and carboxypeptidase A were structurally similar and functioned via comparable mechanisms (Cushman et al. 1980). Captopril is a very potent ACE inhibitor, partly due to the sulphydryl moiety present in its structure, which binds tightly to the zinc ion of ACE (Cushman et al. 1978). However, the sulphydryl moiety was also responsible for many of the side-effects associated with the use of captopril, such as skin rash and loss of taste (Cushman et al. 1978; Todd and Heel 1986).

Non-thiol ACE inhibitors were subsequently developed (the first one being enalapril) with the additional expectation that the removal of the sulphydryl group would result in a drug with longer duration of action, since the sulphur of captopril easily undergoes oxidation and disulphide exchange reaction (Patchett 1984). Enalapril is an orally active precursor that is rapidly metabolised to the active compound, enalaprilat, a very potent ACE inhibitor (Sweet 1983). Although many ACE inhibitors are now available, there is continuing uncertainty about the mechanism of their therapeutic benefit and the effect of ACE inhibition on ANG II levels (Campbell et al. 2004). Some patients on ACE inhibitors fail to show reduced ANG II levels, leading to the proposal that alternate enzymes such as chymase may convert ANG I to ANG II (Dell'Italia and Husain 2002). However, work in both mice with reduced ACE gene expression and in lisinopril-treated mice indicates that ACE is the predominant pathway of ANG II formation (Campbell et al. 2004). The persistence of measurable levels of ANG II in mice with reduced ACE gene expression or ACE inhibition indicates that non-ACE enzymes contribute to ANG II formation in the absence of ACE. Better understanding of the role of non-ACE enzymes in ANG II formation will help clarify the mechanism of the therapeutic effects of ACE inhibition.

4.4
ACE as an Index of Endothelial Function

Pulmonary vascular endothelial enzymatic processes may be altered as a prequel to morphological or clinical signs of lung dysfunction (Orfanos et al. 1999). In the lung, ACE is uniformly distributed along the luminal surface of the endothelial cells and thus could serve as index of tissue integrity (Orfanos et al. 1994). There are several reports of estimates of ACE activity in vitro using endogenous or synthetic substrates (Soffer et al. 1974; Ryan et al. 1977, 1978; Cushman et al. 1978). Synthetic substrates for ACE show low affinity for compounds other than ACE and yield products that are easily separated from the parent compound when hydrolysed by ACE. The synthetic substrate, benzoyl-Phe-Ala-Pro (BPAP) is a specific substrate for blood, lung and urine ACE (Ryan et al. 1978). In the presence of ACE, BPAP is converted to benzoyl-phenylalanine and alanyl-proline. BPAP is extensively metabolised during a single transpulmonary passage in various animal models (Catravas and Gillis 1981; Dobuler et al. 1982; Pitt and Lister 1983). Pulmonary ACE activity thus measured decreases in various forms of lung injury, and these changes occur before changes in other structural or clinical parameters (Dobuler et al. 1982; McCormick et al. 1987; Orfanos et al. 2000a, b; McCloud et al. 2004a, b). Several studies from our laboratory and others suggest a complex role of endothelium-bound ACE in the pathogenesis of acute lung injury (Dobuler et al. 1982; Hilgenfeldt et al. 1987; McCormick et al. 1987; Orfanos et al. 2000a). Downregulation of endothelium-bound ACE activity may be a response mediated by overproduction of peroxynitrite, hydroxyl radicals and other reactive oxygen and reactive nitrogen species, aimed at reducing oxidant stress to the tissue. This decrease in ACE would allow time for the re-establishment of an anti-inflammatory environment and promote vascular protection and lung repair.

References

Abadir PM, Carey RM, Siragy HM (2003) Angiotensin AT2 receptors directly stimulate renal nitric oxide in bradykinin B2-receptor-null mice. Hypertension 42:600–604

AbdAlla S, Lother H, Quitterer U (2000) AT1-receptor heterodimers show enhanced G-protein activation and altered receptor sequestration. Nature 407:94–98

AbdAlla S, Lother H, Abdel-tawab AM, Quitterer U (2001a) The angiotensin II AT2 receptor is an AT1 receptor antagonist. J Biol Chem 276:39721–39726

AbdAlla S, Lother H, el Massiery A, Quitterer U (2001b) Increased AT(1) receptor heterodimers in preeclampsia mediate enhanced angiotensin II responsiveness. Nat Med 7:1003–1009

Alfie ME, Yang XP, Hess F, Carretero OA (1996) Salt-sensitive hypertension in bradykinin B2 receptor knockout mice. Biochem Biophys Res Commun 224:625–630

Alhenc-Gelas F, Weare JA, Johnson RL Jr, Erdos EG (1983) Measurement of human converting enzyme level by direct radioimmunoassay. J Lab Clin Med 101:83–96

Alhenc-Gelas F, Richard J, Courbon D, Warnet JM, Corvol P (1991) Distribution of plasma angiotensin I-converting enzyme levels in healthy men: relationship to environmental and hormonal parameters. J Lab Clin Med 117:33–39

Almeida AP, Frabregas BC, Madureira MM, Santos RJ, Campagnole-Santos MJ, Santos RA (2000) Angiotensin-(1–7) potentiates the coronary vasodilatory effect of bradykinin in the isolated rat heart. Braz J Med Biol Res 33:709–713

Alvarez A, Sanz MJ (2001) Reactive oxygen species mediate angiotensin II-induced leukocyte-endothelial cell interactions in vivo. J Leukoc Biol 70:199–206

Antonaccio MJ, Cushman DW (1981) Drugs inhibiting the renin—angiotensin system. Fed Proc 40:2275–2284

Aparecida Oliveira M, Bruno Fortes Z, Santos RAS, Kosla MC, De Carvalho MHC (1999) Synergistic effect of angiotensin-(1–7) on bradykinin arteriolar dilation in vivo. Peptides 20:1195–1201

Ardaillou R (1999) Angiotensin II receptors. J Am Soc Nephrol 10 Suppl 11:S30–S39

Arima S (2003) Role of angiotensin II and endogenous vasodilators in the control of glomerular hemodynamics. Clin Exp Nephrol 7:172–178

Arima S, Endo Y, Yaoita H, Omata K, Ogawa S, Tsunoda K, Abe M, Takeuchi K, Abe K, Ito S (1997) Possible role of P-450 metabolite of arachidonic acid in vasodilator mechanism of angiotensin II type 2 receptor in the isolated microperfused rabbit afferent arteriole. J Clin Invest 100:2816–2823

Baiardi G, Macova M, Armando I, Ando H, Tyurmin D, Saavedra JM (2005) Estrogen upregulates renal angiotensin II AT1 and AT2 receptors in the rat. Regul Pept 124:7–17

Bartus RT, Elliott PJ, Dean RL, Hayward NJ, Nagle TL, Huff MR, Snodgrass PA, Blunt DG (1996) Controlled modulation of BBB permeability using the bradykinin agonist, RMP-7. Exp Neurol 142:14–28

Batenburg WW, Garrelds IM, Bernasconi CC, Juillerat-Jeanneret L, van Kats JP, Saxena PR, Danser AH (2004a) Angiotensin II type 2 receptor-mediated vasodilation in human coronary microarteries. Circulation 109:2296–2301

Batenburg WW, Garrelds IM, van Kats JP, Saxena PR, Danser AH (2004b) Mediators of bradykinin-induced vasorelaxation in human coronary microarteries. Hypertension 43:488–492

Bayraktutan U, Ulker S (2003) Effects of angiotensin II on nitric oxide generation in proliferating and quiescent rat coronary microvascular endothelial cells. Hypertens Res 26:749–757

Benndorf R, Boger RH, Ergun S, Steenpass A, Wieland T (2003) Angiotensin II type 2 receptor inhibits vascular endothelial growth factor-induced migration and in vitro tube formation of human endothelial cells. Circ Res 93:438–447

Benter IF, Ferrario CM, Morris M, Diz DI (1995) Antihypertensive actions of angiotensin-(1–7) in spontaneously hypertensive rats. Am J Physiol Heart Circ Physiol 269:H313–319

Benzing T, Fleming I, Blaukat A, Muller-Esterl W, Busse R (1999) Angiotensin-converting enzyme inhibitor ramiprilat interferes with the sequestration of the B2 kinin receptor within the plasma membrane of native endothelial cells. Circulation 99:2034–2040

Beraldo WT, Rocha e Silva M (1949) Biological assay of antihistaminics, atropine and antispasmodics upon the guinea pig gut. J Pharmacol Exp Ther 97:388–398

Bergaya S, Meneton P, Bloch-Faure M, Mathieu E, Alhenc-Gelas F, Levy BI, Boulanger CM (2001) Decreased flow-dependent dilation in carotid arteries of tissue kallikrein-knockout mice. Circ Res 88:593–599

Bergaya S, Hilgers RHP, Meneton P, Dong Y, Bloch-Faure M, Inagami T, Alhenc-Gelas F, Levy BI, Boulanger CM (2004) Flow-dependent dilation mediated by endogenous kinins requires angiotensin AT2 receptors. Circ Res 94:1623–1629

Bernier SG, Haldar S, Michel T (2000) Bradykinin-regulated interactions of the mitogen-activated protein kinase pathway with the endothelial nitric-oxide synthase. J Biol Chem 275:30707–30715

Bernstein KE (2002) Two ACEs and a heart. Nature 417:799–802

Bernstein KE, Martin BM, Edwards AS, Bernstein EA (1989) Mouse angiotensin-converting enzyme is a protein composed of two homologous domains. J Biol Chem 264:11945–11951

Berthiaume N, Hess F, Chen A, Regoli D, D'Orleans-Juste P (1997) Pharmacology of kinins in the arterial and venous mesenteric bed of normal and B2 knockout transgenic mice. Eur J Pharmacol 333:55–61

Blaukat A, Herzer K, Schroeder C, Bachmann M, Nash N, Muller-Esterl W (1999) Overexpression and functional characterization of kinin receptors reveal subtype-specific phosphorylation. Biochemistry 38:1300–1309

Boehm S, Kubista H (2002) Fine tuning of sympathetic transmitter release via ionotropic and metabotropic presynaptic receptors. Pharmacol Rev 54:43–99

Brakebusch C, Varfolomeev EE, Batkin M, Wallach D (1994) Structural requirements for inducible shedding of the p55 tumor necrosis factor receptor. J Biol Chem 269:32488–32496

Brosnihan KB, Li P, Ferrario CM (1996) Angiotensin-(1–7) dilates canine coronary arteries through kinins and nitric oxide. Hypertension 27:523–528

Buczko W, Matys T, Kucharewicz I, Chabielska E (1999) The role of endothelium in antithrombotic effect of the renin-angiotensin system blockade. J Physiol Pharmacol 50:499–507

Busse R, Fleming I (1996) Molecular responses of endothelial tissue to kinins. Diabetes 45 Suppl 1:S8–13

Cai H, Li Z, Dikalov S, Holland SM, Hwang J, Jo H, Dudley SC Jr, Harrison DG (2002) NAD(P)H oxidase-derived hydrogen peroxide mediates endothelial nitric oxide production in response to angiotensin II. J Biol Chem 277:48311–48317

Cambien F, Alhenc-Gelas F, Herbeth B, Andre JL, Rakotovao R, Gonzales MF, Allegrini J, Bloch C (1988) Familial resemblance of plasma angiotensin-converting enzyme level: the Nancy Study. Am J Hum Genet 43:774–780

Campbell DJ, Alexiou T, Xiao HD, Fuchs S, McKinley MJ, Corvol P, Bernstein KE (2004) Effect of reduced angiotensin-converting enzyme gene expression and angiotensin-converting enzyme inhibition on angiotensin and bradykinin peptide levels in mice. Hypertension 43:854–859

Carey RM, Howell NL, Jin XH, Siragy HM (2001) Angiotensin type 2 receptor-mediated hypotension in angiotensin type-1 receptor-Blocked rats. Hypertension 38:1272–1277

Carvalho CR, Thirone AC, Gontijo JA, Velloso LA, Saad MJ (1997) Effect of captopril, losartan, and bradykinin on early steps of insulin action. Diabetes 46:1950–1957

Catravas JD, Gillis CN (1981) Metabolism of [3H]benzoyl-phenylalanyl-alanyl-proline by pulmonary angiotensin converting enzyme in vivo: effects of bradykinin, SQ 14225 or acute hypoxia. J Pharmacol Exp Ther 217:263–270

Chan DC, Gera L, Stewart JM, Helfrich B, Zhao TL, Feng WY, Chan KK, Covey JM, Bunn PA Jr (2002) Bradykinin antagonist dimer, CU201, inhibits the growth of human lung cancer cell lines in vitro and in vivo and produces synergistic growth inhibition in combination with other antitumor agents. Clin Cancer Res 8:1280–1287

Chavakis T, Kanse SM, Pixley RA, May AE, Isordia-Salas I, Colman RW, Preissner KT (2001) Regulation of leukocyte recruitment by polypeptides derived from high molecular weight kininogen. FASEB J 15:2365–2376

Chen BC, Yu CC, Lei HC, Chang MS, Hsu MJ, Huang CL, Chen MC, Sheu JR, Chen TF, Chen TL, Inoue H, Lin CH (2004) Bradykinin B2 receptor mediates NF-kappaB activation and cyclooxygenase-2 expression via the Ras/Raf-1/ERK pathway in human airway epithelial cells. J Immunol 173:5219–5228

Chen S, Patel JM, Block ER (2000) Angiotensin IV-mediated pulmonary artery vasorelaxation is due to endothelial intracellular calcium release. Am J Physiol Lung Cell Mol Physiol 279:L849–856

Chua BH, Chua CC, Diglio CA, Siu BB (1993) Regulation of endothelin-1 mRNA by angiotensin II in rat heart endothelial cells. Biochim Biophys Acta 1178:201–206

Chua CC, Diglio CA, Siu BB, Chua BH (1994) Angiotensin II induces TGF-beta 1 production in rat heart endothelial cells. Biochim Biophys Acta 1223:141–147

Chua CC, Hamdy RC, Chua BH (1996) Angiotensin II induces TIMP-1 production in rat heart endothelial cells. Biochim Biophys Acta 1311:175–180

Chua CC, Hamdy RC, Chua BH (1997) Regulation of thrombospondin-1 production by angiotensin II in rat heart endothelial cells. Biochim Biophys Acta 1357:209–214

Chua CC, Hamdy RC, Chua BH (1998) Upregulation of vascular endothelial growth factor by H_2O_2 in rat heart endothelial cells. Free Radic Biol Med 25:891–897

Cole J, Quach du L, Sundaram K, Corvol P, Capecchi MR, Bernstein KE (2002) Mice lacking endothelial angiotensin-converting enzyme have a normal blood pressure. Circ Res 90:87–92

Corvol P, Williams TA, Soubrier F (1995) Peptidyl dipeptidase A: angiotensin I-converting enzyme. Methods Enzymol 248:283–305

Costanzo A, Moretti F, Burgio VL, Bravi C, Guido F, Levrero M, Puri PL (2003) Endothelial activation by angiotensin II through NFkappaB and p38 pathways: involvement of NFkappaB-inducible kinase (NIK), free oxygen radicals, and selective inhibition by aspirin. J Cell Physiol 195:402–410

Costerousse O, Jaspard E, Wei L, Corvol P, Alhenc-Gelas F (1992) The angiotensin I-converting enzyme (kininase II): molecular organization and regulation of its expression in humans. J Cardiovasc Pharmacol 20 Suppl 9:S10–15

Crackower MA, Sarao R, Oudit GY, Yagil C, Kozieradzki I, Scanga SE, Oliveira-dos-Santos AJ, da Costa J, Zhang L, Pei Y, Scholey J, Ferrario CM, Manoukian AS, Chappell MC, Backx PH, Yagil Y, Penninger JM (2002) Angiotensin-converting enzyme 2 is an essential regulator of heart function. Nature 417:822–828

Cushman DW, Cheung HS, Sabo EF, Ondetti MA (1978) Design of new antihypertensive drugs: potent and specific inhibitors of angiotensin-converting enzyme. Prog Cardiovasc Dis 21:176–182

Cushman DW, Ondetti MA, Cheung HS, Antonaccio MJ, Murthy VS, Rubin B (1980) Inhibitors of angiotensin-converting enzyme. Adv Exp Med Biol 130:199–225

Dalmay F, Mazouz H, Allard J, Pesteil F, Achard JM, Fournier A (2001) Non-AT(1)-receptor-mediated protective effect of angiotensin against acute ischaemic stroke in the gerbil. J Renin Angiotensin Aldosterone Syst 2:103–106

Danilczyk U, Eriksson U, Crackower MA, Penninger JM (2003) A story of two ACEs. J Mol Med 81:227–234

Davis JO, Freeman RH (1982) Historical perspectives on the renin-angiotensin-aldosterone system and angiotensin blockade. Am J Cardiol 49:1385–1389

De Paolis P, Porcellini A, Gigante B, Giliberti R, Lombardi A, Savoia C, Rubattu S, Volpe M (1999) Modulation of the AT2 subtype receptor gene activation and expression by the AT1 receptor in endothelial cells. J Hypertens 17:1873–1877

Decarie A, Raymond P, Gervais N, Couture R, Adam A (1996) Serum interspecies differences in metabolic pathways of bradykinin and [des-Arg9]BK: influence of enalaprilat. Am J Physiol 271:H1340–1347

Deddish PA, Marcic BM, Tan F, Jackman HL, Chen Z, Erdos EG (2002) Neprilysin inhibitors potentiate effects of bradykinin on B2 receptor. Hypertension 39:619–623

Dell'Italia LJ, Husain A (2002) Dissecting the role of chymase in angiotensin II formation and heart and blood vessel diseases. Curr Opin Cardiol 17:374–379

Delles C, Michelson G, Harazny J, Oehmer S, Hilgers KF, Schmieder RE (2004) Impaired endothelial function of the retinal vasculature in hypertensive patients. Stroke 35:1289–1293

Dendorfer A, Wolfrum S, Dominiak P (1999) Pharmacology and cardiovascular implications of the kinin-kallikrein system. Jpn J Pharmacol 79:403–426

Derian CK, Moskowitz MA (1986) Polyphosphoinositide hydrolysis in endothelial cells and carotid artery segments. Bradykinin-2 receptor stimulation is calcium-independent. J Biol Chem 261:3831–3837

Desideri G, Bravi MC, Tucci M, Croce G, Marinucci MC, Santucci A, Alesse E, Ferri C (2003) Angiotensin II inhibits endothelial cell motility through an AT1-dependent oxidant-sensitive decrement of nitric oxide availability. Arterioscler Thromb Vasc Biol 23:1218–1223

Diep QN, El Mabrouk M, Cohn JS, Endemann D, Amiri F, Virdis A, Neves MF, Schiffrin EL (2002) Structure, endothelial function, cell growth, and inflammation in blood vessels of angiotensin II-infused rats: role of peroxisome proliferator-activated receptor-gamma. Circulation 105:2296–2302

Dietze G, Wicklmayr M, Bottger I, Schifmann R, Geiger R, Fritz H, Mehnert H (1980) The kallikrein-kinin system and muscle metabolism: biochemical aspects. Agents Actions 10:335–338

Dimmeler S, Zeiher AM (2000) Reactive oxygen species and vascular cell apoptosis in response to angiotensin II and pro-atherosclerotic factors. Regul Pept 90:19–25

Dimmeler S, Rippmann V, Weiland U, Haendeler J, Zeiher AM (1997) Angiotensin II induces apoptosis of human endothelial cells. Protective effect of nitric oxide. Circ Res 81:970–976

Dobuler KJ, Catravas JD, Gillis CN (1982) Early detection of oxygen-induced lung injury in conscious rabbits. Reduced in vivo activity of angiotensin converting enzyme and removal of 5-hydroxytryptamine. Am Rev Respir Dis 126:534–539

Donoghue M, Hsieh F, Baronas E, Godbout K, Gosselin M, Stagliano N, Donovan M, Woolf B, Robison K, Jeyaseelan R, Breitbart RE, Acton S (2000) A novel angiotensin-converting enzyme-related carboxypeptidase (ACE2) converts angiotensin I to angiotensin 1-9. Circ Res 87:E1–9

Dzau VJ (2001) Theodore Cooper Lecture: tissue angiotensin and pathobiology of vascular disease: a unifying hypothesis. Hypertension 37:1047–1052

Dzau VJ, Bernstein K, Celermajer D, Cohen J, Dahlof B, Deanfield J, Diez J, Drexler H, Ferrari R, Van Gilst W, Hansson L, Hornig B, Husain A, Johnston C, Lazar H, Lonn E, Luscher T, Mancini J, Mimran A, Pepine C, Rabelink T, Remme W, Ruilope L, Ruzicka M, Schunkert H, Swedberg K, Unger T, Vaughan D, Weber M (2002) Pathophysiologic and therapeutic importance of tissue ACE: a consensus report. Cardiovasc Drugs Ther 16:149–160

Ehlers MR, Fox EA, Strydom DJ, Riordan JF (1989) Molecular cloning of human testicular angiotensin-converting enzyme: the testis isozyme is identical to the C-terminal half of endothelial angiotensin-converting enzyme. Proc Natl Acad Sci U S A 86:7741–7745

Eiskjaer H, Sorensen SS, Danielsen H, Pedersen EB (1992) Glomerular and tubular antinatriuretic actions of low-dose angiotensin II infusion in man. J Hypertens 10:1033–1040

El-Dahr SS, Harrison-Bernard LM, Dipp S, Yosipiv IV, Meleg-Smith S (2000) Bradykinin B2 null mice are prone to renal dysplasia: gene-environment interactions in kidney development. Physiol Genomics 3:121–131

Emanueli C, Maestri R, Corradi D, Marchione R, Minasi A, Tozzi MG, Salis MB, Straino S, Capogrossi MC, Olivetti G, Madeddu P (1999) Dilated and failing cardiomyopathy in bradykinin B(2) receptor knockout mice. Circulation 100:2359–2365

Erdos EG (1990a) Some old and some new ideas on kinin metabolism. J Cardiovasc Pharmacol 15 Suppl 6:S20–24

Erdos EG (1990b) Angiotensin I converting enzyme and the changes in our concepts through the years. Lewis K. Dahl memorial lecture. Hypertension 16:363–370

Erdos EG, Yang HY (1967) An enzyme in microsomal fraction of kidney that inactivates bradykinin. Life Sci 6:569–574

Erdos EG, Deddish PA, Marcic BM (1999) Potentiation of bradykinin actions by ACE inhibitors. Trends Endocrinol Metab 10:223–229

Esther CR, Marino EM, Howard TE, Machaud A, Corvol P, Capecchi MR, Bernstein KE (1997) The critical role of tissue angiotensin-converting enzyme as revealed by gene targeting in mice. J Clin Invest 99:2375–2385

Esther CR Jr, Howard TE, Marino EM, Goddard JM, Capecchi MR, Bernstein KE (1996) Mice lacking angiotensin-converting enzyme have low blood pressure, renal pathology, and reduced male fertility. Lab Invest 74:953–965

Fein AM, Bernard GR, Criner GJ, Fletcher EC, Good JT Jr, Knaus WA, Levy H, Matuschak GM, Shanies HM, Taylor RW, Rodell TC (1997) Treatment of severe systemic inflammatory response syndrome and sepsis with a novel bradykinin antagonist, deltibant (CP-0127). Results of a randomized, double-blind, placebo-controlled trial. CP-0127 SIRS and Sepsis Study Group. JAMA 277:482–487

Fernandes L, Fortes ZB, Nigro D, Tostes RCA, Santos RAS, Catelli de Carvalho MH (2001) Potentiation of bradykinin by angiotensin-(1–7) on arterioles of spontaneously hypertensive rats studied in vivo. Hypertension 37:703–709

Ferrario CM, Flack JM (1996) Pathologic consequences of increased angiotensin II activity. Cardiovasc Drugs Ther 10:511–518

Ferrario CM, Brosnihan KB, Diz DI, Jaiswal N, Khosla MC, Milsted A, Tallant EA (1991) Angiotensin-(1–7): a new hormone of the angiotensin system. Hypertension 18:III126–133

Ferrario CM, Chappell MC, Tallant EA, Brosnihan KB, Diz DI (1997) Counterregulatory actions of angiotensin-(1–7). Hypertension 30:535–541

Ferreira SH, Bartelt DC, Greene LJ (1970) Isolation of bradykinin-potentiating peptides from *Bothrops jararaca* venom. Biochemistry 9:2583–2593

Ferri C, Desideri G, Baldoncini R, Bellini C, Valenti M, Santucci A, De Mattia G (1999) Angiotensin II increases the release of endothelin-1 from human cultured endothelial cells but does not regulate its circulating levels. Clin Sci (Lond) 96:261–270

Frossard N, Stretton CD, Barnes PJ (1990) Modulation of bradykinin responses in airway smooth muscle by epithelial enzymes. Agents Actions 31:204–209

Gans RO, Bilo HJ, Nauta JJ, Popp-Snijders C, Heine RJ, Donker AJ (1991) The effect of angiotensin-I converting enzyme inhibition on insulin action in healthy volunteers. Eur J Clin Invest 21:527–533

Gohlke P, Pees C, Unger T (1998) AT2 receptor stimulation increases aortic cyclic GMP in SHRSP by a kinin-dependent mechanism. Hypertension 31:349–355

Grafe M, Auch-Schwelk W, Zakrzewicz A, Regitz-Zagrosek V, Bartsch P, Graf K, Loebe M, Gaehtgens P, Fleck E (1997) Angiotensin II-induced leukocyte adhesion on human coronary endothelial cells is mediated by E-selectin. Circ Res 81:804–811

Griendling KK, Minieri CA, Ollerenshaw JD, Alexander RW (1994) Angiotensin II stimulates NADH and NADPH oxidase activity in cultured vascular smooth muscle cells. Circ Res 74:1141–1148

Griendling KK, Ushio-Fukai M, Lassegue B, Alexander RW (1997) Angiotensin II signaling in vascular smooth muscle. New concepts. Hypertension 29:366–373

Griesbacher T, Legat FJ (2000) Effects of the non-peptide B2 receptor antagonist FR173657 in models of visceral and cutaneous inflammation. Inflamm Res 49:535–540

Guimaraes S, Paiva MQ, Moura D (1998) Different receptors for angiotensin II at pre- and postjunctional level of the canine mesenteric and pulmonary arteries. Br J Pharmacol 124:1207–1212

Hall KL, Venkateswaran S, Hanesworth JM, Schelling ME, Harding JW (1995) Characterization of a functional angiotensin IV receptor on coronary microvascular endothelial cells. Regul Pept 58:107–115

Harada K, Komuro I, Sugaya T, Murakami K, Yazaki Y (1999) Vascular injury causes neointimal formation in angiotensin II type 1a receptor knockout mice. Circ Res 84:179–185

Harding JW, Cook VI, Miller-Wing AV, Hanesworth JM, Sardinia MF, Hall KL, Stobb JW, Swanson GN, Coleman JK, Wright JW, et al (1992) Identification of an AII(3–8) [AIV] binding site in guinea pig hippocampus. Brain Res 583:340–343

Heitsch H, Brovkovych S, Malinski T, Wiemer G (2001) Angiotensin-(1–7)-stimulated nitric oxide and superoxide release from endothelial cells. Hypertension 37:72–76

Henrion D, Kubis N, Levy BI (2001) Physiological and pathophysiological functions of the AT2 subtype receptor of angiotensin II: from large arteries to the microcirculation. Hypertension 38:1150–1157

Hilgenfeldt U, Kienapfel G, Kellermann W, Schott R, Schmidt M (1987) Renin-angiotensin system in sepsis. Clin Exp Hypertens A 9:1493–1504

Hill-Kapturczak N, Kapturczak MH, Block ER, Patel JM, Malinski T, Madsen KM, Tisher CC (1999) Angiotensin II-stimulated nitric oxide release from porcine pulmonary endothelium is mediated by angiotensin IV. J Am Soc Nephrol 10:481–491

Hornig B, Kohler C, Schlink D, Tatge H, Drexler H (2003) AT1-receptor antagonism improves endothelial function in coronary artery disease by a bradykinin/B2-receptor-dependent mechanism. Hypertension 41:1092–1095

Hu DE, Fan TP (1993) [Leu8]des-Arg9-bradykinin inhibits the angiogenic effect of bradykinin and interleukin-1 in rats. Br J Pharmacol 109:14–17

Hubert C, Houot AM, Corvol P, Soubrier F (1991) Structure of the angiotensin I-converting enzyme gene. Two alternate promoters correspond to evolutionary steps of a duplicated gene. J Biol Chem 266:15377–15383

Imai T, Hirata Y, Emori T, Yanagisawa M, Masaki T, Marumo F (1992) Induction of endothelin-1 gene by angiotensin and vasopressin in endothelial cells. Hypertension 19:753–757

Imanishi T, Hano T, Nishio I (2004) Angiotensin II potentiates vascular endothelial growth factor-induced proliferation and network formation of endothelial progenitor cells. Hypertens Res 27:101–108

Isami S, Kishikawa H, Araki E, Uehara M, Kaneko K, Shirotani T, Todaka M, Ura S, Motoyoshi S, Matsumoto K, Miyamura N, Shichiri M (1996) Bradykinin enhances GLUT4 translocation through the increase of insulin receptor tyrosine kinase in primary adipocytes: evidence that bradykinin stimulates the insulin signalling pathway. Diabetologia 39:412–420

Ito H, Takemori K, Suzuki T (2001) Role of angiotensin II type 1 receptor in the leucocytes and endothelial cells of brain microvessels in the pathogenesis of hypertensive cerebral injury. J Hypertens 19:591–597

Iwai M, Liu HW, Chen R, Ide A, Okamoto S, Hata R, Sakanaka M, Shiuchi T, Horiuchi M (2004) Possible inhibition of focal cerebral ischemia by angiotensin II type 2 receptor stimulation. Circulation 110:843–848

Iyer SN, Averill DB, Chappell MC, Yamada K, Allred AJ, Ferrario CM (2000) Contribution of angiotensin-(1–7) to blood pressure regulation in salt-depleted hypertensive rats. Hypertension 36:417–422

Jaiswal N, Diz DI, Chappell MC, Khosla MC, Ferrario CM (1992) Stimulation of endothelial cell prostaglandin production by angiotensin peptides. Characterization of receptors. Hypertension 19:II49–55

Jalowy A, Schulz R, Dorge H, Behrends M, Heusch G (1998) Infarct size reduction by AT1-receptor blockade through a signal cascade of AT2-receptor activation, bradykinin and prostaglandins in pigs. J Am Coll Cardiol 32:1787–1796

Jose PJ, Page DA, Wolstenholme BE, Williams TJ, Dumonde DC (1981) Bradykinin-stimulated prostaglandin E2 production by endothelial cells and its modulation by antiinflammatory compounds. Inflammation 5:363–378

Ju H, Venema VJ, Marrero MB, Venema RC (1998) Inhibitory interactions of the bradykinin B2 receptor with endothelial nitric-oxide synthase. J Biol Chem 273:24025–24029

Kansui Y, Fujii K, Goto K, Abe I, Iida M (2002) Angiotensin II receptor antagonist improves age-related endothelial dysfunction. J Hypertens 20:439–446

Kaplan AP, Joseph K, Shibayama Y, Nakazawa Y, Ghebrehiwet B, Reddigari S, Silverberg M (1998) Bradykinin formation. Plasma and tissue pathways and cellular interactions. Clin Rev Allergy Immunol 16:403–429

Katada J, Majima M (2002) AT2 receptor-dependent vasodilation is mediated by activation of vascular kinin generation under flow conditions. Br J Pharmacol 136:484–491

Kerins DM, Hao Q, Vaughan DE (1995) Angiotensin induction of PAI-1 expression in endothelial cells is mediated by the hexapeptide angiotensin IV. J Clin Invest 96:2515–2520

Kintsurashvili E, Duka I, Gavras I, Johns C, Farmakiotis D, Gavras H (2001) Effects of ANG II on bradykinin receptor gene expression in cardiomyocytes and vascular smooth muscle cells. Am J Physiol Heart Circ Physiol 281:H1778–1783

Kiron M, Soffer R (1989) Purification and properties of a soluble angiotensin II-binding protein from rabbit liver. J Biol Chem 264:4138–4142

Kishi K, Muromoto N, Nakaya Y, Miyata I, Hagi A, Hayashi H, Ebina Y (1998) Bradykinin directly triggers GLUT4 translocation via an insulin-independent pathway. Diabetes 47:550–558

Kohlman O Jr, Neves Fde A, Ginoza M, Tavares A, Cezaretti ML, Zanella MT, Ribeiro AB, Gavras I, Gavras H (1995) Role of bradykinin in insulin sensitivity and blood pressure regulation during hyperinsulinemia. Hypertension 25:1003–1007

Kramer C, Sunkomat J, Witte J, Luchtefeld M, Walden M, Schmidt B, Tsikas D, Boger RH, Forssmann WG, Drexler H, Schieffer B (2002) Angiotensin II receptor-independent antiinflammatory and antiaggregatory properties of losartan: role of the active metabolite EXP3179. Circ Res 90:770–776

Krege JH, John SW, Langenbach LL, Hodgin JB, Hagaman JR, Bachman ES, Jennette JC, O'Brien DA, Smithies O (1995) Male-female differences in fertility and blood pressure in ACE-deficient mice. Nature 375:146–148

Kuno A, Miura T, Tsuchida A, Hasegawa T, Miki T, Nishino Y, Shimamoto K (2002) Blockade of angiotensin II type 1 receptors suppressed free radical production and preserved coronary endothelial function in the rabbit heart after myocardial infarction. J Cardiovasc Pharmacol 39:49–57

Kurisu S, Ozono R, Oshima T, Kambe M, Ishida T, Sugino H, Matsuura H, Chayama K, Teranishi Y, Iba O, Amano K, Matsubara H (2003) Cardiac angiotensin II type 2 receptor Activates the kinin/NO system and inhibits fibrosis. Hypertension 41:99–107

Lassegue B, Sorescu D, Szocs K, Yin Q, Akers M, Zhang Y, Grant SL, Lambeth JD, Griendling KK (2001) Novel gp91phox homologues in vascular smooth muscle cells: nox1 mediates angiotensin II-induced superoxide formation and redox-sensitive signaling pathways. Circ Res 88:888–894

Lehtonen JYA, Horiuchi M, Daviet L, Akishita M, Dzau VJ (1999) Activation of the de novo biosynthesis of sphingolipids mediates angiotensin II type 2 receptor-induced apoptosis. J Biol Chem 274:16901–16906

Lemos VS, Cortes SF, Silva DM, Campagnole-Santos MJ, Santos RA (2002) Angiotensin-(1–7) is involved in the endothelium-dependent modulation of phenylephrine-induced contraction in the aorta of mRen-2 transgenic rats. Br J Pharmacol 135:1743–1748

Li D, Tomson K, Yang B, Mehta P, Croker BP, Mehta JL (1999a) Modulation of constitutive nitric oxide synthase, bcl-2 and Fas expression in cultured human coronary endothelial cells exposed to anoxia-reoxygenation and angiotensin II: role of AT1 receptor activation. Cardiovasc Res 41:109–115

Li D, Yang B, Philips MI, Mehta JL (1999b) Proapoptotic effects of ANG II in human coronary artery endothelial cells: role of AT1 receptor and PKC activation. Am J Physiol 276:H786–792

Li DY, Zhang YC, Philips MI, Sawamura T, Mehta JL (1999c) Upregulation of endothelial receptor for oxidized low-density lipoprotein (LOX-1) in cultured human coronary artery endothelial cells by angiotensin II type 1 receptor activation. Circ Res 84:1043–1049

Li JS, Sharifi AM, Schiffrin EL (1997b) Effect of AT1 angiotensin-receptor blockade on structure and function of small arteries in SHR. J Cardiovasc Pharmacol 30:75–83

Li P, Chappell MC, Ferrario CM, Brosnihan KB (1997c) Angiotensin-(1–7) augments bradykinin-induced vasodilation by competing with ACE and releasing nitric oxide. Hypertension 29:394–398

Li P, Fukuhara M, Diz DI, Ferrario CM, Brosnihan KB (2000) Novel angiotensin II AT(1) receptor antagonist irbesartan prevents thromboxane A(2)-induced vasoconstriction in canine coronary arteries and human platelet aggregation. J Pharmacol Exp Ther 292:238–246

Li Q, Feenstra M, Pfaffendorf M, Eijsman L, van Zwieten PA (1997a) Comparative vasoconstrictor effects of angiotensin II, III, and IV in human isolated saphenous vein. J Cardiovasc Pharmacol 29:451–456

Lima CV, Paula RD, Resende FL, Khosla MC, Santos RAS (1997) Potentiation of the hypotensive effect of bradykinin by short-term infusion of angiotensin-(1–7) in normotensive and hypertensive rats. Hypertension 30:542–548

Liu X, Fernandez M, Wouters MA, Heyberger S, Husain A (2001) Arg(1098) is critical for the chloride dependence of human angiotensin I-converting enzyme C-domain catalytic activity. J Biol Chem 276:33518–33525

Liu YH, Yang XP, Sharov VG, Nass O, Sabbah HN, Peterson E, Carretero OA (1997) Effects of angiotensin-converting enzyme inhibitors and angiotensin II type 1 receptor antagonists in rats with heart failure. Role of kinins and angiotensin II type 2 receptors. J Clin Invest 99:1926–1935

Liu YH, Xu J, Yang XP, Yang F, Shesely E, Carretero OA (2002) Effect of ACE inhibitors and angiotensin II type 1 receptor antagonists on endothelial NO synthase knockout mice with heart failure. Hypertension 39:375–381

Luft FC, Wilcox CS, Unger T, Kuhn R, Demmert G, Rohmeiss P, Ganten D, Sterzel RB (1989) Angiotensin-induced hypertension in the rat. Sympathetic nerve activity and prostaglandins. Hypertension 14:396–403

Luscher TF, Boulanger CM, Dohi Y, Yang ZH (1992) Endothelium-derived contracting factors. Hypertension 19:117–130

Madeddu P, Emanueli C, Maestri R, Salis MB, Minasi A, Capogrossi MC, Olivetti G (2000) Angiotensin II type 1 receptor blockade prevents cardiac remodeling in bradykinin B(2) receptor knockout mice. Hypertension 35:391–396

Maeso R, Rodrigo E, Munoz-Garcia R, Navarro-Cid J, Ruilope LM, Cachofeiro V, Lahera V (1998) Factors involved in the effects of losartan on endothelial dysfunction induced by aging in SHR. Kidney Int Suppl 68:S30–35

Maia LG, Ramos MC, Fernandes L, de Carvalho MH, Campagnole-Santos MJ, Souza dos Santos RA (2004) Angiotensin-(1–7) antagonist A-779 attenuates the potentiation of bradykinin by captopril in rats. J Cardiovasc Pharmacol 43:685–691

Makino I, Shibata K, Ohgami Y, Fujiwara M, Furukawa T (1996) Transient upregulation of the AT2 receptor mRNA level after global ischemia in the rat brain. Neuropeptides 30:596–601

Maly J, Karasova L, Simova M, Vitko S, El-Dahr SS (2001) Angiotensin II-induced hypertension in bradykinin B2 receptor knockout mice. Hypertension 37:967–973

Marceau F, Barabe J, St-Pierre S, Regoli D (1980) Kinin receptors in experimental inflammation. Can J Physiol Pharmacol 58:536–542

Marceau F, Hess JF, Bachvarov DR (1998) The B1 receptors for kinins. Pharmacol Rev 50:357–386

Marcic B, Deddish PA, Jackman HL, Erdos EG (1999) Enhancement of bradykinin and resensitization of its B2 receptor. Hypertension 33:835–843

Margolius HS (1998) Tissue kallikreins structure, regulation, and participation in mammalian physiology and disease. Clin Rev Allergy Immunol 16:337–349

McCloud LL, Parkerson JB, Freant L, Hoffman WH, Catravas JD (2004a) β-Hydroxybutyrate induces acute pulmonary endothelial dysfunction in rabbits. Exp Lung Res 30:193–206

McCloud LL, Parkerson JB, Zou L, Rao RN, Catravas JD (2004b) Reduced pulmonary endothelium-bound angiotensin converting enzyme activity in diabetic rabbits. Vascul Pharmacol 41:159–165

McCormick JR, Chrzanowski R, Andreani J, Catravas JD (1987) Early pulmonary endothelial enzyme dysfunction after phorbol ester in conscious rabbits. J Appl Physiol 63:1972–1978

Mehta JL, Li DY, Yang H, Raizada MK (2002) Angiotensin II and IV stimulate expression and release of plasminogen activator inhibitor-1 in cultured human coronary artery endothelial cells. J Cardiovasc Pharmacol 39:789-794

Meng W, Busija D (1993) Comparative effects of angiotensin-(1-7) and angiotensin II on piglet pial arterioles. Stroke 24:2041-2044

Mervaala EMA, Cheng ZJ, Tikkanen I, Lapatto R, Nurminen K, Vapaatalo H, Muller DN, Fiebeler A, Ganten U, Ganten D, Luft FC (2001) Endothelial dysfunction and xanthine oxidoreductase activity in rats with human renin and angiotensinogen genes. Hypertension 37:414-418

Mikrut K, Paluszak J, Kozlik J, Sosnowski P, Krauss H, Grzeskowiak E (2001) The effect of bradykinin on the oxidative state of rats with acute hyperglycaemia. Diabetes Res Clin Pract 51:79-85

Minshall RD, Tan F, Nakamura F, Rabito SF, Becker RP, Marcic B, Erdos EG (1997) Potentiation of the actions of bradykinin by angiotensin I-converting enzyme inhibitors: the role of expressed human bradykinin B2 receptors and angiotensin I-converting enzyme in CHO cells. Circ Res 81:848-856

Mitchell KD, Navar LG (1989) The renin-angiotensin-aldosterone system in volume control. Baillieres Clin Endocrinol Metab 3:393-430

Miura S, Matsuo Y, Saku K (2003) Transactivation of KDR/Flk-1 by the B2 receptor induces tube formation in human coronary endothelial cells. Hypertension 41:1118-1123

Moeller I, Clune EF, Fennessy PA, Bingley JA, Albiston AL, Mendelsohn FA, Chai SY (1999) Up regulation of AT4 receptor levels in carotid arteries following balloon injury. Regul Pept 83:25-30

Morbidelli L, Parenti A, Giovannelli L, Granger HJ, Ledda F, Ziche M (1998) B1 receptor involvement in the effect of bradykinin on venular endothelial cell proliferation and potentiation of FGF-2 effects. Br J Pharmacol 124:1286-1292

Mukae S, Aoki S, Itoh S, Iwata T, Ueda H, Katagiri T (2000) Bradykinin B(2) receptor gene polymorphism is associated with angiotensin-converting enzyme inhibitor-related cough. Hypertension 36:127-131

Muller-Esterl W, Rauth G, Lottspeich F, Kellermann J, Henschen A (1985) Limited proteolysis of human low-molecular-mass kininogen by tissue kallikrein. Isolation and characterization of the heavy and the light chains. Eur J Biochem 149:15-22

Muscella A, Marsigliante S, Vilella S, Jimenez E, Storelli C (1999) Angiotensin II stimulates the Na+/H+ exchanger in human umbilical vein endothelial cells via AT1 receptor. Life Sci 65:2385-2394

Nakamoto H, Ferrario CM, Fuller SB, Robaczewski DL, Winicov E, Dean RH (1995) Angiotensin-(1-7) and nitric oxide interaction in renovascular hypertension. Hypertension 25:796-802

Nakanishi S (1987) Substance P precursor and kininogen: their structures, gene organizations, and regulation. Physiol Rev 67:1117-1142

Nickenig G, Harrison DG (2002) The AT(1)-type angiotensin receptor in oxidative stress and atherogenesis: part I: oxidative stress and atherogenesis. Circulation 105:393-396

Nora EH, Munzenmaier DH, Hansen-Smith FM, Lombard JH, Greene AS (1998) Localization of the ANG II type 2 receptor in the microcirculation of skeletal muscle. Am J Physiol 275:H1395-1403

Odya CE, Marinkovic DV, Hammon KJ, Stewart TA, Erdos EG (1978) Purification and properties of prolylcarboxypeptidase (angiotensinase C) from human kidney. J Biol Chem 253:5927-5931

Ogino Y, Costa T (1992) The epithelial phenotype of human neuroblastoma cells express bradykinin, endothelin, and angiotensin II receptors that stimulate phosphoinositide hydrolysis. J Neurochem 58:46–56

Ohashi H, Takagi H, Oh H, Suzuma K, Suzuma I, Miyamoto N, Uemura A, Watanabe D, Murakami T, Sugaya T, Fukamizu A, Honda Y (2004) Phosphatidylinositol 3-kinase/Akt regulates angiotensin II-induced inhibition of apoptosis in microvascular endothelial cells by governing survivin expression and suppression of caspase-3 activity. Circ Res 94:785–793

Oliveira MA, Carvalho MH, Nigro D, Passaglia Rde C, Fortes ZB (2002) Angiotensin-(1–7) and bradykinin interaction in diabetes mellitus: in vivo study. Peptides 23:1449–1455

Oliveira MA, Carvalho MH, Nigro D, Passaglia Rde C, Fortes ZB (2003) Elevated glucose blocks angiotensin-(1–7) and bradykinin interaction: the role of cyclooxygenase products. Peptides 24:449–454

Oliverio MI, Kim HS, Ito M, Le T, Audoly L, Best CF, Hiller S, Kluckman K, Maeda N, Smithies O, Coffman TM (1998) Reduced growth, abnormal kidney structure, and type 2 (AT2) angiotensin receptor-mediated blood pressure regulation in mice lacking both AT1A and AT1B receptors for angiotensin II. Proc Natl Acad Sci U S A 95:15496–15501

Orfanos SE, Chen XL, Ryan JW, Chung AY, Burch SE, Catravas JD (1994) Assay of pulmonary microvascular endothelial angiotensin-converting enzyme in vivo: comparison of three probes. Toxicol Appl Pharmacol 124:99–111

Orfanos SE, Langleben D, Khoury J, Schlesinger RD, Dragatakis L, Roussos C, Ryan JW, Catravas JD (1999) Pulmonary capillary endothelium-bound angiotensin-converting enzyme activity in humans. Circulation 99:1593–1599

Orfanos SE, Armaganidis A, Glynos C, Psevdi E, Kaltsas P, Sarafidou P, Catravas JD, Dafni UG, Langleben D, Roussos C (2000a) Pulmonary capillary endothelium-bound angiotensin-converting enzyme activity in acute lung injury. Circulation 102:2011–2018

Orfanos SE, Parkerson JB, Chen X, Fisher EL, Glynos C, Papapetropoulos A, Gerrity RG, Catravas JD (2000b) Reduced lung endothelial angiotensin-converting enzyme activity in Watanabe hyperlipidemic rabbits in vivo. Am J Physiol Lung Cell Mol Physiol 278:L1280–1288

Osei SY, Ahima RS, Minkes RK, Weaver JP, Khosla MC, Kadowitz PJ (1993) Differential responses to angiotensin-(1–7) in the feline mesenteric and hindquarters vascular beds. Eur J Pharmacol 234:35–42

Papademetriou V (2002) The potential role of AT(1)-receptor blockade in the prevention and reversal of atherosclerosis. J Hum Hypertens 16 Suppl 3:S34–41

Parenti A, Morbidelli L, Ledda F, Granger HJ, Ziche M (2001) The bradykinin/B1 receptor promotes angiogenesis by up-regulation of endogenous FGF-2 in endothelium via the nitric oxide synthase pathway. FASEB J 15:1487–1489

Passos GF, Fernandes ES, Campos MM, Araujo JG, Pesquero JL, Souza GE, Avellar MC, Teixeira MM, Calixto JB (2004) Kinin B1 receptor up-regulation after lipopolysaccharide administration: role of proinflammatory cytokines and neutrophil influx. J Immunol 172:1839–1847

Patchett AA (1984) The chemistry of enalapril. Br J Clin Pharmacol 18 Suppl 2:201S–207S

Patel JM, Martens JR, Li YD, Gelband CH, Raizada MK, Block ER (1998) Angiotensin IV receptor-mediated activation of lung endothelial NOS is associated with vasorelaxation. Am J Physiol 275:L1061–1068

Paula RD, Lima CV, Khosla MC, Santos RAS (1995) Angiotensin-(1–7) potentiates the hypotensive effect of bradykinin in conscious rats. Hypertension 26:1154–1159

Perron MS, Gobeil F Jr, Pelletier S, Regoli D, Sirois P (1999) Involvement of bradykinin B1 and B2 receptors in pulmonary leukocyte accumulation induced by Sephadex beads in guinea pigs. Eur J Pharmacol 376:83–89

Pesquero JB, Araujo RC, Heppenstall PA, Stucky CL, Silva JA Jr, Walther T, Oliveira SM, Pesquero JL, Paiva AC, Calixto JB, Lewin GR, Bader M (2000) Hypoalgesia and altered inflammatory responses in mice lacking kinin B1 receptors. Proc Natl Acad Sci U S A 97:8140–8145

Phillips MI, Kagiyama S (2002) Angiotensin II as a pro-inflammatory mediator. Curr Opin Investig Drugs 3:569–577

Pitt BR, Lister G (1983) Pulmonary metabolic function in the awake lamb: effect of development and hypoxia. J Appl Physiol 55:383–391

Porsti I, Bara A, Busse R, Hecker M (1994) Release of nitric oxide by angiotensin-(1–7) from porcine coronary endothelium: implications for a novel angiotensin receptor [published erratum appears in Br J Pharmacol 1996 Jan;117(1):231]. Br J Pharmacol 111:652–654

Pueyo ME, N'Diaye N, Michel JB (1996) Angiotensin II-elicited signal transduction via AT1 receptors in endothelial cells. Br J Pharmacol 118:79–84

Pueyo ME, Arnal JF, Rami J, Michel JB (1998) Angiotensin II stimulates the production of NO and peroxynitrite in endothelial cells. Am J Physiol 274:C214–220

Ramchandran R, Sen I (1995) Cleavage processing of angiotensin-converting enzyme by a membrane-associated metalloprotease. Biochemistry 34:12645–12652

Regitz-Zagrosek V, Fielitz J, Fleck E (1998) Myocardial angiotensin receptors in human hearts. Basic Res Cardiol 93 Suppl 2:37–42

Regoli D, Barabe J (1980) Pharmacology of bradykinin and related kinins. Pharmacol Rev 32:1–46

Ren Y, Garvin JL, Carretero OA (2002) Vasodilator action of angiotensin-(1–7) on isolated rabbit afferent arterioles. Hypertension 39:799–802

Rett K, Jauch KW, Wicklmayr M, Dietze G, Fink E, Mehnert H (1986) Angiotensin converting enzyme inhibitors in diabetes: experimental and human experience. Postgrad Med J 62 Suppl 1:59–64

Rett K, Wicklmayr M, Dietze GJ, Haring HU (1996) Insulin-induced glucose transporter (GLUT1 and GLUT4) translocation in cardiac muscle tissue is mimicked by bradykinin. Diabetes 45 Suppl 1:S66–69

Rhinehart K, Handelsman CA, Silldorff EP, Pallone TL (2003) ANG II AT2 receptor modulates AT1 receptor-mediated descending vasa recta endothelial Ca2+ signaling. Am J Physiol Heart Circ Physiol 284:H779–789

Rigat B, Hubert C, Alhenc-Gelas F, Cambien F, Corvol P, Soubrier F (1990) An insertion/deletion polymorphism in the angiotensin I-converting enzyme gene accounting for half the variance of serum enzyme levels. J Clin Invest 86:1343–1346

Riva L, Galzin AM (1996) Pharmacological characterization of a specific binding site for angiotensin IV in cultured porcine aortic endothelial cells. Eur J Pharmacol 305:193–199

Roks AJ, Nijholt J, van Buiten A, van Gilst WH, de Zeeuw D, Henning RH (2004) Low sodium diet inhibits the local counter-regulator effect of angiotensin-(1–7) on angiotensin II. J Hypertens 22:2355–2361

Rousseau A, Michaud A, Chauvet MT, Lenfant M, Corvol P (1995) The hemoregulatory peptide N-acetyl-Ser-Asp-Lys-Pro is a natural and specific substrate of the N-terminal active site of human angiotensin-converting enzyme. J Biol Chem 270:3656–3661

Rueckschloss U, Quinn MT, Holtz J, Morawietz H (2002) Dose-dependent regulation of NAD(P)H oxidase expression by angiotensin II in human endothelial cells: protective effect of angiotensin II type 1 receptor blockade in patients with coronary artery disease. Arterioscler Thromb Vasc Biol 22:1845–1851

Rupniak NM, Boyce S, Webb JK, Williams AR, Carlson EJ, Hill RG, Borkowski JA, Hess JF (1997) Effects of the bradykinin B1 receptor antagonist des-Arg9[Leu8]bradykinin and genetic disruption of the B2 receptor on nociception in rats and mice. Pain 71:89–97

Ryan JW, Chung A, Ammons C, Carlton ML (1977) A simple radioassay for angiotensin-converting enzyme. Biochem J 167:501–504

Ryan JW, Chung A, Martin LC, Ryan US (1978) New substrates for the radioassay of angiotensin converting enzyme of endothelial cells in culture. Tissue Cell 10:555–562

Saito S, Hirata Y, Emori T, Imai T, Marumo F (1996) Angiotensin II activates endothelial constitutive nitric oxide synthase via AT1 receptors. Hypertens Res 19:201–206

Santhamma KR, Sadhukhan R, Kinter M, Chattopadhyay S, McCue B, Sen I (2004) Role of tyrosine phosphorylation in the regulation of cleavage secretion of angiotensin-converting enzyme. J Biol Chem 279:40227–40236

Sato M, Engelman RM, Otani H, Maulik N, Rousou JA, Flack JE III, Deaton DW, Das DK (2000) Myocardial protection by preconditioning of heart with losartan, an angiotensin II type 1-receptor blocker: implication of bradykinin-dependent and bradykinin-independent mechanisms. Circulation 102:346III–351

Schiffrin EL (2001) Small artery remodeling in hypertension: can it be corrected? Am J Med Sci 322:7–11

Schiffrin EL, Park JB, Intengan HD, Touyz RM (2000) Correction of arterial structure and endothelial dysfunction in human essential hypertension by the angiotensin receptor antagonist losartan. Circulation 101:1653–1659

Schiffrin EL, Park JB, Pu Q (2002) Effect of crossing over hypertensive patients from a beta-blocker to an angiotensin receptor antagonist on resistance artery structure and on endothelial function. J Hypertens 20:71–78

Schmaier AH (2003) The kallikrein-kinin and the renin-angiotensin systems have a multi-layered interaction. Am J Physiol Regul Integr Comp Physiol 285:R1–13

Segar JL, Barna TJ, Acarregui MJ, Lamb FS (2001) Responses of fetal ovine systemic and umbilical arteries to angiotensin II. Pediatr Res 49:826–833

Seyedi N, Xu X, Nasjletti A, Hintze TH (1995) Coronary kinin generation mediates nitric oxide release after angiotensin receptor stimulation. Hypertension 26:164–170

Shigematsu S, Ishida S, Gute DC, Korthuis RJ (2002) Bradykinin-induced proinflammatory signaling mechanisms. Am J Physiol Heart Circ Physiol 283:H2676–2686

Shinozaki K, Ayajiki K, Nishio Y, Sugaya T, Kashiwagi A, Okamura T (2004) Evidence for a causal role of the renin-angiotensin system in vascular dysfunction associated with insulin resistance. Hypertension 43:255–262

Shiuchi T, Cui TX, Wu L, Nakagami H, Takeda-Matsubara Y, Iwai M, Horiuchi M (2002) ACE inhibitor improves insulin resistance in diabetic mouse via bradykinin and NO. Hypertension 40:329–334

Shivakumar BR, Wang Z, Hammond TG, Harris RC (2004) EP24.15 interacts with the angiotensin II type I receptor and bradykinin B(2) receptor. Cell Biochem Funct 23:195–204

Silvestre JS, Bergaya S, Tamarat R, Duriez M, Boulanger CM, Levy BI (2001) Proangiogenic effect of angiotensin-converting enzyme inhibition is mediated by the bradykinin B(2) receptor pathway. Circ Res 89:678–683

Siragy HM, Jaffa AA, Margolius HS, Carey RM (1996) Renin-angiotensin system modulates renal bradykinin production. Am J Physiol Regul Integr Comp Physiol 271:R1090–1095

Skeggs LT Jr, Kahn JR, Shumway NP (1956) The preparation and function of the hypertensin-converting enzyme. J Exp Med 103:295–299

Soffer RL, Reza R, Caldwell PR (1974) Angiotensin-converting enzyme from rabbit pulmonary particles. Proc Natl Acad Sci U S A 71:1720–1724

Sohn HY, Raff U, Hoffmann A, Gloe T, Heermeier K, Galle J, Pohl U (2000) Differential role of angiotensin II receptor subtypes on endothelial superoxide formation. Br J Pharmacol 131:667–672

Soubrier F, Alhenc-Gelas F, Hubert C, Allegrini J, John M, Tregear G, Corvol P (1988) Two putative active centers in human angiotensin I-converting enzyme revealed by molecular cloning. Proc Natl Acad Sci U S A 85:9386–9390

Sowers JR (2002) Hypertension, angiotensin II, and oxidative stress. N Engl J Med 346:1999–2001

Starke K, Schumann HJ (1972) Interactions of angiotensin, phenoxybenzamine and propranolol on noradrenaline release during sympathetic nerve stimulation. Eur J Pharmacol 18:27–30

Stewart JM (2003) Bradykinin antagonists as anti-cancer agents. Curr Pharm Des 9:2036–2042

Stoll M, Meffert S, Stroth U, Unger T (1995a) Growth or antigrowth: angiotensin and the endothelium. J Hypertens 13:1529–1534

Stoll M, Steckelings UM, Paul M, Bottari SP, Metzger R, Unger T (1995b) The angiotensin AT2-receptor mediates inhibition of cell proliferation in coronary endothelial cells. J Clin Invest 95:651–657

Strawn WB, Ferrario CM (2002) Mechanisms linking angiotensin II and atherogenesis. Curr Opin Lipidol 13:505–512

Swanson GN, Hanesworth JM, Sardinia MF, Coleman JK, Wright JW, Hall KL, Miller-Wing AV, Stobb JW, Cook VI, Harding EC, et al (1992) Discovery of a distinct binding site for angiotensin II (3–8), a putative angiotensin IV receptor. Regul Pept 40:409–419

Sweet CS (1983) Pharmacological properties of the converting enzyme inhibitor, enalapril maleate (MK-421). Fed Proc 42:167–170

Takeuchi K (1999) Signal transduction systems of angiotensin II receptors. Nippon Rinsho 57:1070–1077

Tallant EA, Lu X, Weiss RB, Chappell MC, Ferrario CM (1997) Bovine aortic endothelial cells contain an angiotensin-(1–7) receptor. Hypertension 29:388–392

Tamarat R, Silvestre JS, Durie M, Levy BI (2002) Angiotensin II angiogenic effect in vivo involves vascular endothelial growth factor- and inflammation-related pathways. Lab Invest 82:747–756

Tan Y, Hutchison FN, Jaffa AA (2004) Mechanisms of angiotensin II-induced expression of B2 kinin receptors. Am J Physiol Heart Circ Physiol 286:H926–932

Tanimoto K, Sugiyama F, Goto Y, Ishida J, Takimoto E, Yagami K, Fukamizu A, Murakami K (1994) Angiotensinogen-deficient mice with hypotension. J Biol Chem 269:31334–31337

Tatemoto K, Takayama K, Zou MX, Kumaki I, Zhang W, Kumano K, Fujimiya M (2001) The novel peptide apelin lowers blood pressure via a nitric oxide-dependent mechanism. Regul Pept 99:87–92

Tayeh MA, Scicli AG (1998) Angiotensin II and bradykinin regulate the expression of P-selectin on the surface of endothelial cells in culture. Proc Assoc Am Physicians 110:412–421

Tipnis SR, Hooper NM, Hyde R, Karran E, Christie G, Turner AJ (2000) A human homolog of angiotensin-converting enzyme. Cloning and functional expression as a captopril-insensitive carboxypeptidase. J Biol Chem 275:33238–33243

Todd PA, Heel RC (1986) Enalapril. A review of its pharmacodynamic and pharmacokinetic properties, and therapeutic use in hypertension and congestive heart failure. Drugs 31:198–248

Tom B, de Vries R, Saxena PR, Danser AHJ (2001) Bradykinin potentiation by angiotensin-(1–7) and ACE inhibitors correlates with ACE C- and N-domain blockade. Hypertension 38:95–99

Tom B, Dendorfer A, Danser AH (2003) Bradykinin, angiotensin-(1–7), and ACE inhibitors: how do they interact? Int J Biochem Cell Biol 35:792–801

Tomita H, Egashira K, Ohara Y, Takemoto M, Koyanagi M, Katoh M, Yamamoto H, Tamaki K, Shimokawa H, Takeshita A (1998) Early induction of transforming growth factor-beta via angiotensin II type 1 receptors contributes to cardiac fibrosis induced by long-term blockade of nitric oxide synthesis in rats. Hypertension 32:273–279

Torlone E, Rambotti AM, Perriello G, Botta G, Santeusanio F, Brunetti P, Bolli GB (1991) ACE-inhibition increases hepatic and extrahepatic sensitivity to insulin in patients with type 2 (non-insulin-dependent) diabetes mellitus and arterial hypertension. Diabetologia 34:119–125

Tsuchida S, Matsusaka T, Chen X, Okubo S, Niimura F, Nishimura H, Fogo A, Utsunomiya H, Inagami T, Ichikawa I (1998) Murine double nullizygotes of the angiotensin type 1A and 1B receptor genes duplicate severe abnormal phenotypes of angiotensinogen nullizygotes. J Clin Invest 101:755–760

Tsuchida S, Miyazaki Y, Matsusaka T, Hunley TE, Inagami T, Fogo A, Ichikawa I (1999) Potent antihypertrophic effect of the bradykinin B2 receptor system on the renal vasculature. Kidney Int 56:509–516

Tsutsumi Y, Matsubara H, Masaki H, Kurihara H, Murasawa S, Takai S, Miyazaki M, Nozawa Y, Ozono R, Nakagawa K, Miwa T, Kawada N, Mori Y, Shibasaki Y, Tanaka Y, Fujiyama S, Koyama Y, Fujiyama A, Takahashi H, Iwasaka T (1999) Angiotensin II type 2 receptor overexpression activates the vascular kinin system and causes vasodilation. J Clin Invest 104:925–935

Turner AJ, Hooper NM (2002) The angiotensin-converting enzyme gene family: genomics and pharmacology. Trends Pharmacol Sci 23:177–183

Ullian ME, Walsh LG, Morinelli TA (1996) Potentiation of angiotensin II action by corticosteroids in vascular tissue. Cardiovasc Res 32:266–273

Vaughan D (2000) Pharmacology of ACE inhibitors versus AT1 blockers. Can J Cardiol 16 Suppl E:36E–40E

Vavrek RJ, Stewart JM (1985) Competitive antagonists of bradykinin. Peptides 6:161–164

Vaziri ND, Ding Y, Ni Z, Barton CH (2005) Bradykinin down-regulates whereas arginine analogs up-regulate eNOS expression in coronary endothelial cells. J Pharmacol Exp Ther 313:121–126

Venema RC (2002) Post-translational mechanisms of endothelial nitric oxide synthase regulation by bradykinin. Int Immunopharmacol 2:1755–1762

Vickers C, Hales P, Kaushik V, Dick L, Gavin J, Tang J, Godbout K, Parsons T, Baronas E, Hsieh F, Acton S, Patane M, Nichols A, Tummino P (2002) Hydrolysis of biological peptides by human angiotensin-converting enzyme-related carboxypeptidase. J Biol Chem 277:14838–14843

Viswanathan M, Saavedra JM (1992) Expression of angiotensin II AT2 receptors in the rat skin during experimental wound healing. Peptides 13:783–786

Wang CH, Leung N, Lapointe N, Szeto L, Uffelman KD, Giacca A, Rouleau JL, Lewis GF (2003) Vasopeptidase inhibitor omapatrilat induces profound insulin sensitization and increases myocardial glucose uptake in Zucker fatty rats: studies comparing a vasopeptidase inhibitor, angiotensin-converting enzyme inhibitor, and angiotensin II type I receptor blocker. Circulation 107:1923–1929

Warnholtz A, Nickenig G, Schulz E, Macharzina R, Brasen JH, Skatchkov M, Heitzer T, Stasch JP, Griendling KK, Harrison DG, Bohm M, Meinertz T, Munzel T (1999) Increased NADH-oxidase-mediated superoxide production in the early stages of atherosclerosis: evidence for involvement of the renin-angiotensin system. Circulation 99:2027–2033

Wassmann S, Hilgers S, Laufs U, Bohm M, Nickenig G (2002) Angiotensin II type 1 receptor antagonism improves hypercholesterolemia-associated endothelial dysfunction. Arterioscler Thromb Vasc Biol 22:1208–1212

Watanabe T, Suzuki J, Yamawaki H, Sharma VK, Sheu SS, Berk BC (2005) Losartan metabolite EXP3179 activates Akt and endothelial nitric oxide synthase via vascular endothelial growth factor receptor-2 in endothelial cells: angiotensin II type 1 receptor-independent effects of EXP3179. Circulation 112:1798–1805

Wei CC, Ferrario CM, Brosnihan KB, Farrell DM, Bradley WE, Jaffa AA, Dell'Italia LJ (2002) Angiotensin peptides modulate bradykinin levels in the interstitium of the dog heart in vivo. J Pharmacol Exp Ther 300:324–329

Williams TA, Michaud A, Houard X, Chauvet MT, Soubrier F, Corvol P (1996) Drosophila melanogaster angiotensin I-converting enzyme expressed in Pichia pastoris resembles the C domain of the mammalian homologue and does not require glycosylation for secretion and enzymic activity. Biochem J 318:125–131

Witherow FN, Helmy A, Webb DJ, Fox KAA, Newby DE (2001) Bradykinin contributes to the vasodilator effects of chronic angiotensin-converting enzyme inhibition in patients with heart failure. Circulation 104:2177–2181

Wright JW, Harding JW (1995) Brain angiotensin receptor subtypes AT1, AT2, and AT4 and their functions. Regul Pept 59:269–295

Yan C, Kim D, Aizawa T, Berk BC (2003) Functional interplay between angiotensin II and nitric oxide: cyclic GMP as a key mediator. Arterioscler Thromb Vasc Biol 23:26–36

Yanai K, Saito T, Kakinuma Y, Kon Y, Hirota K, Taniguchi-Yanai K, Nishijo N, Shigematsu Y, Horiguchi H, Kasuya Y, Sugiyama F, Yagami K, Murakami K, Fukamizu A (2000) Renin-dependent cardiovascular functions and renin-independent blood-brain barrier functions revealed by renin-deficient mice. J Biol Chem 275:5–8

Yang C, Hsu WH (1995) Stimulatory effect of bradykinin on insulin release from the perfused rat pancreas. Am J Physiol 268:E1027–1030

Yang HY, Erdos EG (1967) Second kininase in human blood plasma. Nature 215:1402–1403

Yang HY, Erdos EG, Levin Y (1970) A dipeptidyl carboxypeptidase that converts angiotensin I and inactivates bradykinin. Biochim Biophys Acta 214:374–376

Yang HY, Erdos EG, Levin Y (1971) Characterization of a dipeptide hydrolase (kininase II: angiotensin I converting enzyme). J Pharmacol Exp Ther 177:291–300

Yoo KH, Thornhill BA, Wolstenholme JT, Chevalier RL (1998) Tissue-specific regulation of growth factors and clusterin by angiotensin II. Am J Hypertens 11:715–722

Yosipiv IV, Dipp S, El-Dahr SS (2001) Targeted disruption of the bradykinin B(2) receptor gene in mice alters the ontogeny of the renin-angiotensin system. Am J Physiol Renal Physiol 281:F795–801

Zuccollo A, Navarro M, Frontera M, Cueva F, Carattino M, Catanzaro OL (1999) The involvement of kallikrein-kinin system in diabetes type I (insulitis). Immunopharmacology 45:69–74

Endothelin

A. P. Davenport (✉) · J. J. Maguire

Clinical Pharmacology Unit, University of Cambridge, Addenbrooke's Hospital,
Cambridge CB2 2QQ, UK
apd10@medschl.cam.ac.uk

1	Introduction	296
2	Endothelins and Sarafotoxins	297
2.1	ET-1 and Big ET-1	297
2.2	ET-2 and Big ET-2	298
2.3	ET-3 and Big ET-3	298
2.4	Sarafotoxins	299
3	Endothelin Synthesis	299
3.1	Endothelin Converting Enzyme-1 (ECE-1)	299
3.1.1	Distribution of ECE-1 in Human Endothelium	301
3.1.2	Localisation of ECE-1 in Endothelial Cells	302
3.2	Endothelin Converting Enzyme-2 (ECE-2)	303
3.3	Alternative Pathways for ET Synthesis: ET-1_{1-31} and Chymase	304
4	Endothelin Receptors	305
4.1	ET_A and ET_B Subtypes	305
4.2	Receptor Mutations	306
4.2.1	ET_A/ET-1 Mutations and Knockouts	306
4.2.2	ET_B/ET-3 Mutations and Knockouts	306
4.3	Splice Variants of ET Receptors	307
4.3.1	Splice Variants of ET_A Receptors	307
4.3.2	Splice Variants of ET_B Receptors	307
4.4	ET Ligands	307
4.4.1	Endogenous and Synthetic Agonists	307
4.4.2	Peptide and Non-peptide ET_A Antagonists	308
4.4.3	Peptide and Non-peptide ET_B Antagonists	309
4.4.4	Mixed ET_A/ET_B Antagonists	310
5	Physiological and Pathophysiological Role	310
5.1	ET-1: The Universal Vasoconstrictor?	310
5.1.1	Smooth Muscle ET_A Receptors	311
5.1.2	Endothelial ET_B Receptors	312
5.1.3	ET_B Clearing Receptors	313
5.2	Pulmonary Arterial Hypertension (PAH)	314
5.3	Essential Hypertension	315
5.4	Atherosclerosis	315
6	Conclusions	317
	References	318

Abstract In humans, the endothelins (ETs) comprise a family of three 21-amino-acid peptides, ET-1, ET-2 and ET-3. ET-1 is synthesised from a biologically inactive precursor, Big ET-1, by an unusual hydrolysis of the Trp^{21}-Val^{22} bond by the endothelin converting enzyme (ECE-1). In humans, there are four isoforms (ECE-1a-d) derived from a single gene by the action of alternative promoters. Structurally, they differ only in the amino acid sequence of the extreme N-terminus. A second enzyme, ECE-2, also exists as four isoforms and differs from ECE-1 in requiring an acidic pH for optimal activity. Human chymase can also cleave Big ET-1 to $ET-1_{1-31}$, which is cleaved, in turn, to the mature peptide as an alternative pathway. ET-1 is the principal isoform in the human cardiovascular system and remains one of the most potent constrictors of human vessels discovered. ET-1 is unusual in being released from a dual secretory pathway. The peptide is continuously released from vascular endothelial cells by the constitutive pathway, producing intense constriction of the underlying smooth muscle and contributing to the maintenance of endogenous vascular tone. ET-1 is also released from endothelial cell-specific storage granules (Weibel-Palade bodies) in response to external stimuli. ETs mediate their action by activating two G protein-coupled receptor sub-types, ET_A and ET_B. Two therapeutic strategies have emerged to oppose the actions of ET-1, namely inhibition of the synthetic enzyme by combined ECE/neutral endopeptidase inhibitors such as SLV306, and receptor antagonists such as bosentan. The ET system is up-regulated in atherosclerosis, and ET antagonists may be of benefit in reducing blood pressure in essential hypertension. Bosentan, the first ET antagonist approved for clinical use, represents a significant new therapeutic strategy in the treatment of pulmonary arterial hypertension (PAH).

Keywords Endothelin converting enzyme · Receptors · Atherosclerosis · Essential hypertension · Pulmonary arterial hypertension

1
Introduction

The existence of a peptidic endothelium-derived constricting factor was proposed 20 years ago by Hickey et al. (1985). A trypsin-sensitive factor from cultured bovine endothelial cells was isolated, but the structure was not determined. In 1988, Yanagisawa and colleagues identified the structure of endothelin (now called endothelin-1 or ET-1) as a 21-amino-acid peptide (Fig. 1). In a remarkable paper in *Nature*, they showed that the synthetic peptide had potent constrictor activity (Yanagisawa et al. 1988), which stimulated a considerable amount of interest, with over 18,000 papers on the subject published to date. By analysis of the ET-1 gene, two further members of the family, endothelin-2 (ET-2) and endothelin-3 (ET-3), were identified (Inoue et al. 1989), together with two receptor sub-types, ET_A (Arai et al. 1990) and ET_B (Sakurai et al. 1990). Subsequently, novel enzymes responsible for ET synthesis from its precursor-those enzymes being endothelin converting enzyme-1 (ECE-1) (Takahashi et al. 1993; Xu et al. 1994) and ECE-2 (Emoto and Yanagisawa 1995)-were identified.

The aim of this chapter is to focus on ET peptides, receptors and converting enzymes in the human vascular endothelium and their role in the pathophysiology of atherosclerosis, pulmonary arterial and essential hypertension.

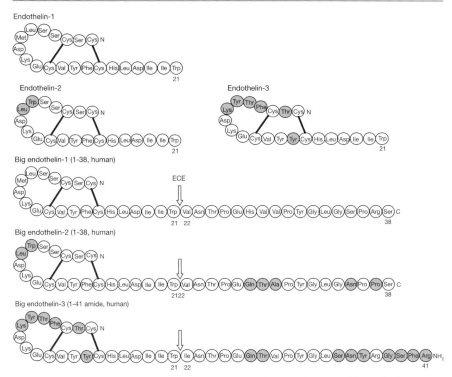

Fig. 1 Structure of ET peptides and their precursors. The site of action of the endothelin converting enzyme (*ECE*) is indicated with an *arrow*. (Modified from Davenport and Maguire 2002)

2
Endothelins and Sarafotoxins

2.1
ET-1 and Big ET-1

The structure of ET-1 is unique amongst the mammalian bioactive peptides in possessing not one but two intramolecular disulphide bonds between cysteine residues cross-linked at positions 1 and 15 and 3 and 11 (Fig. 1). ET-1 is one of the few peptides in which the crystal structure has been solved. Residues at positions 10, 17, 18 and 21 are crucial for binding (Janes et al. 1994).

ET-1 is the principal isoform in the human cardiovascular system and remains the most potent constrictor of human vessels discovered. ET-1 is unusual amongst the mammalian bioactive peptides in being released from a dual secretory pathway (Russell et al. 1998a, b; Davenport and Russell 2001). The peptide is continuously released from vascular endothelial cells by the constitutive pathway, producing intense constriction of the underlying smooth muscle and contributing to the maintenance of endogenous vascular tone (Haynes and Webb 1994). The peptide is also released from endothelial cell-specific storage granules (Weibel-Palade bodies) in response to external physiological or perhaps pathophysiological stimuli, producing further vasoconstriction (Russell et al. 1998a, b; Davenport and Russell 2001). Thus, ET-1 functions as a locally released, rather than circulating, hormone and concentrations are comparatively low in plasma and other tissues.

2.2
ET-2 and Big ET-2

ET-2 differs by only two amino acids from ET-1, and despite the relatively large Leu-to-Trp substitution at position 6, this has little or no effect on the binding affinity (Fig. 1). Although ET-2 is as potent a vasoconstrictor as ET-1 (Maguire and Davenport 1995), the peptide has been less extensively studied than ET-1. ET-2 messenger RNA (mRNA) (O'Reilly et al. 1992, 1993) and ET-2 peptide (Plumpton et al. 1993, 1996a) have been detected in the human cardiovascular system. Both ET-2 mRNA (O'Reilly et al. 1993) and the precursor Big ET-2 have been detected in the cytoplasm of endothelial cells (Howard et al. 1992), suggesting that the peptide may also be released locally from endothelial cells and contributes to maintaining tone. In support of this hypothesis, Big ET-2 levels are higher in normal human plasma than Big ET-1 (Matsumoto et al. 1994). Using a specific enzyme-linked immunosorbent assay (ELISA) that does not cross-react with ET-1, plasma levels of ET-2 are detectable that give an average value in 40 volunteers of 0.9±0.03 pmol/l. ET-2 has also been identified in failing hearts from humans (Plumpton et al. 1993). However, the precise physiological or pathophysiological role of this isoform remains to be discovered.

2.3
ET-3 and Big ET-3

Endothelial cells do not synthesise ET-3, but the mature peptide and Big ET-3 are detectable in plasma (Matsumoto et al. 1994) and other tissues including heart (Plumpton et al. 1996b) and brain (Takahashi et al. 1991). The adrenal gland may also be a source of Big ET-3 (Davenport et al. 1996). Antisera to this precursor stained secretory cells of the medulla, although mature ET-3 was not detected within homogenates of adrenal tissue. If released, further processing of Big ET-3 could occur within the vasculature by smooth muscle cells (Davenport et al. 1998a), and the adrenals may be a source of the ET-3 that can be detected in human plasma.

ET-3 is unique in that it is the only endogenous isoform that distinguishes between the two endothelin receptors. It has the same affinity at the ET_B receptor as ET-1 but, at physiological concentrations, has little or no affinity for the ET_A sub-type. In humans, ET_A receptors predominate in the human vasculature, and the low density of ET_B receptors (<15%) present on the smooth muscle of the vasculature contribute little to vasoconstriction (Maguire and Davenport 1995). ET_B receptors are the principal sub-type in the kidney, localising to non-vascular tissues. Evidence is emerging that the ET_B sub-type functions as a clearing receptor to remove ET from the circulation. Blockade of the ET_B receptor results in a rise in circulating immunoreactive ET. Blockade of the ET_B receptor by receptor antagonists results in a corresponding rise in circulating levels of ET-3 (Plumpton et al. 1996b). ET-3 may play a beneficial role in human disease by activating endothelial ET_B receptors to release opposing vasodilators, thus limiting unwanted vasoconstriction.

2.4
Sarafotoxins

The only peptides with a high degree of sequence similarity to the endothelins are the sarafotoxins, a family of four (S6a, S6b, S6c, S6d) 21-amino-acid peptides that was discovered in the venom of a snake, *Atractaspis engaddensis*, that has evolved to immobilise larger mammalian prey. In humans, symptoms of envenomation include a rapid rise in blood pressure consistent with systemic vasoconstriction, with changes in ECG consistent with coronary vasoconstriction or direct inotropic actions on the heart (Kurnik et al. 1999). Sarafotoxin S6c is used as a moderately selective ET_B agonist.

3
Endothelin Synthesis

3.1
Endothelin Converting Enzyme-1 (ECE-1)

Following the removal of the signal sequence from pre-proendothelin-1 (a 212-amino-acid peptide, the initial product of the ET-1 gene), the resulting proendothelin is cleaved by the enzyme furin to yield the 38-amino-acid peptide Big ET-1. ET-1 is synthesised from Big ET-1 by an unusual hydrolysis of Trp^{21}-Val^{22} (Fig. 1) catalysed by ECEs, rather than the more frequent Arg-Arg or Arg-Lys as in other peptide precursors (Turner and Murphy 1996).

mRNA encoding ECE-1 is widely distributed in homogenates of human tissue (Rossi et al. 1995; Valdenaire et al. 1995; Schweizer et al. 1997). In humans and other mammals, there are four isoforms of ECE-1 (ECE-1a-d), derived from a single gene by the action of alternative promoters. Structurally, they differ only in the amino acid sequence of the extreme N-terminus (Shimada et al. 1995a, b; Valdenaire et al. 1995; Turner et al. 1998). mRNA encoding all four isoforms has been detected in cultured human umbilical vein endothelial cells, whereas ECE-1a was the only isoform not detected in cultured human smooth muscle cells (Valdenaire et al. 1999). These sequence differences have been exploited to generate site-directed antisera to the deduced amino acids in the N-terminus of human ECE-1a (ECE-1β_{2-16}), ECE-1b$_{1-16}$, ECE-1c (ECE-1$\alpha_{(2-16)}$) and ECE-1d$_{1-14}$. These antisera have been extensively characterised and used to compare their cellular distribution in human tissues (Mockridge et al. 1998; Russell et al. 1998a, c).

ECE-1c (also called ECE-1α) consists of 754 amino acids in man, and mRNA encoding the protein has been shown to predominate in human tissues (Schweizer et al. 1997). These studies revealed unexpected anomalies, so that levels of mRNA encoding ECE-1 were relatively low in human brain compared with peripheral tissues such as the lungs. In agreement with the molecular studies, measurement of protein levels showed ECE-1 to be the most abundant isoform in microsomal fractions prepared from homogenates of a number of human tissues (Mockridge et al. 1998). In the heart, levels of ECE-1 measured by competition ELISAn were 0.9±0.3 and 0.4±0.1 pmol/g wet weight in the atria and ventricles, respectively. These levels are comparatively low, reflecting the localisation of the enzyme to the endothelium, which represents only a small proportion of the cell type within the heart.

ECE-1a (also called ECE-1β) is a 758-amino-acid enzyme in humans and, with ECE-1c, has been detected in human umbilical vein and coronary artery endothelial cells (Russell et al. 1998a, c). However, the concentration of ECE-1a in these tissues was below the level for detection by competition ELISA, suggesting that ECE-1c was the predominant isoform. ECE-1b is a 770-amino-acid protein that is identical to ECE-1c, except for an additional 17 amino acids at the N-terminus, replacing the first methionine of ECE-1c. ECE-1b complementary DNA (cDNA) has only been identified in humans (Schweizer et al. 1997). Intense immunoreactivity was localised within renal and pulmonary epithelial cells with lower levels of staining displayed by perivascular astrocytes and neuronal processes in the cerebral cortex from the brain. In diseased vessels, ECE-1b antisera stained macrophages infiltrating atherosclerotic plaques within coronary arteries. These results suggest that ECE-1b may also be expressed in normal and diseased human tissue (Davenport and Kuc 2000). ECE-1d comprises 767 amino acids, and mRNA encoding it was detected in all human tissues examined (Valdenaire et al. 1999).

The physiological significance of multiple ECE isoforms in human tissue is unclear. All isoforms have the same kinetic rate constants for cleaving Big ET-1 when expressed in cell lines and would be expected to synthesise comparable amounts of the mature peptide. It is possible that the isoforms may occupy different compartments within the same cell. When artificially expressed in CHO cells, all four isoforms are present intracellularly but with varying degrees of expression on the cell surface, with ECE-1d not expressed (Valdenaire et al. 1999). A second possibility is that expression may vary according to cell type: This may account for the particularly intense staining with antisera to ECE-1b in epithelial cells, whereas endothelial cell staining was difficult to detect (Davenport and Kuc 2000).

3.1.1
Distribution of ECE-1 in Human Endothelium

ET-1, together with its precursor Big ET-1, is the predominant isoform synthesised and released from the human endothelium (Fig. 2). The mature peptide has been localised in endothelial cells of all human vessels examined, includ-

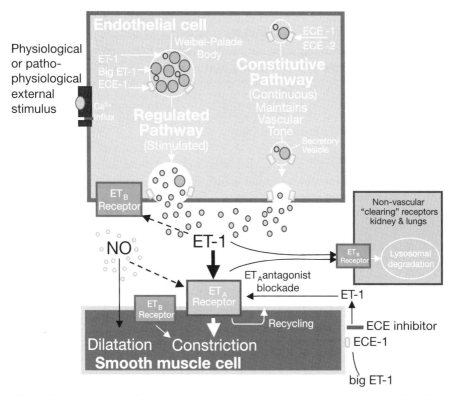

Fig. 2 Schematic model of ET-1 in the human cardiovascular system. Within endothelial cells, two distinct exocytic pathways transport ET-1 to the cell surface. ET-1, synthesised by ECE-1/ECE-2, is continuously released via the constitutive pathway, contributing to vascular tone. ET-1 is also synthesised by ECE-1 and stored in Weibel-Palade bodies until released following an external physiological or pathophysiological stimulus (regulated pathway) to produce further vasoconstriction. Following release, ET-1 interacts with ET_A receptors that predominate on the smooth muscle. In some, but not all, human vessels, a small population of ET_B receptors can also mediate constriction. Activation of endothelial ET_B receptors by ET-1 limits the constrictor response by the release of vasodilators (*NO*). Non-vascular ET_B receptors in, for example, kidney and lungs may remove ET-1 from the circulation, as well as having a beneficial role in limiting any rise in ET-1 resulting from ET_A receptor blockade. Some Big ET-1 escapes conversion by endothelial cell ECE. This circulating precursor is converted to ET-1 at target sites by smooth muscle ECE that can be blocked by peptidase inhibitors

ing large conduit and small resistance vessels (Hemsen et al. 1991; Howard et al. 1992; Ashby et al. 1995; Plumpton et al. 1996b). Conservation of the C-terminus has permitted the development of antisera which can cross-react with all ECE enzymes discovered to date in human tissue. Davenport et al. (1998a, b) used site-directed antisera raised against the C-terminus of mammalian ECE-1 (bECE-$1_{744-758}$; XU et al. 1994), which also cross-reacted with the C-terminus of the deduced amino-acid sequence of bovine ECE-2 (Emoto and Yanagisawa 1995) that has four identical amino acids to ECE-1 at the extreme C-terminus. Using these antisera, Davenport et al. (1998b) showed that immunoreactive ECE had a ubiquitous distribution in human endothelial cells lining large conduit and smaller resistance vessels within cardiac, adrenal, respiratory and brain tissue. This pattern of staining in the vascular endothelium paralleled that of its substrate, Big ET-1, and its product, the mature ET peptide.

3.1.2
Localisation of ECE-1 in Endothelial Cells

The processing of Big ET-1 to ET-1 has been attributed to activity of one or more converting enzymes that are located mainly on the plasma membrane or within intracellular compartments. Initial studies using endothelial cells isolated from animal tissues (Harrison et al. 1995; Takahashi et al. 1995; Barnes et al. 1996; 1998) or transformed endothelial cell lines (Waxman et al. 1994) suggested that ECE activity is localised to the cell surface and the enzyme acts mainly in a post-secretory processing role. However, the co-localisation of the mature peptide and Big ET-1 within endothelial cells implies that at least some ECE activity is located intracellularly. Evidence from a number of different studies demonstrated ECE is either primarily expressed or has predominant activity within intracellular compartments (Gui et al. 1993; Xu et al. 1994; Davenport et al. 1998a; Russell et al. 1998a, b, c). Davenport et al. (1998a) compared the ability of permeabilised and non-permeabilised human endothelial cells to convert Big ET-1 to the mature peptide and found that about 85% of ECE activity was located in intracellular compartments. ECE-like immunoreactivity was visualised by scanning electron microscopy on the surface of the plasma membrane of cultured endothelial cells (Russell et al. 1998a) as well as *enface* preparations of human coronary artery.

The sub-cellular expression of Big ET-1, ECE-1c and ECE-1a was compared with von Willebrand factor, a marker of Weibel-Palade bodies, in human endothelial cells that had been permeabilised to allow access of antisera to subcellular structures. The resulting cells were optically sectioned using confocal microscopy. In agreement with the results of the scanning electron microscopy, only moderate levels of ECE-1c were detected over the plasma membrane. ECE-1c and ECE-1a, together with Big ET-1, were found to co-localise with von Willebrand factor in the Weibel-Palade bodies. Co-localisation of ECE isoforms to Weibel-Palade bodies was confirmed by immunoelectron microscopy in ultra-thin sections of human coronary artery. These numerous rod-shaped structures, about 0.2 μm in diameter and 2–3 μm in length, are located beneath the plasma membrane and are specific to endothelial cells. Stimulation by the calcium ionophore released ET-1 from cultured human umbilical vein endothelial cells (HUVECs). These results suggest that ET-1 is synthesised by the regulated pathway and released in response to external stimuli (Fig. 2; Russell et al. 1998a).

Intense staining with antisera to ECE was also discovered in smaller punctate vesicles, establishing that ET is also synthesised via the constitutive secretory pathway (Fig. 2). These results are in agreement with the ultrastructural localisation of the mature peptide in human coronary artery. Quantitative immunoelectron microscopy revealed the presence of ET-like immunoreactivity in the secretory vesicles as well as the Weibel-Palade bodies (Russell et al. 1998b). The combined results demonstrate that ET is released from human endothelial cells via two distinct pathways. Thus, ET is continuously transported in and released from secretory vesicles by the constitutive secretory pathway, contributing to the maintenance of normal vascular tone. Continuous release from this pathway accounts for the rise in the concentration of plasma ET following systemic administration of ET receptor antagonists in volunteers (Plumpton 1996a). In addition, ET stored in Weibel-Palade bodies may be released following a physiological or pathophysiological stimulus by the regulated pathway, to cause additional local vasoconstriction (Russell and Davenport 1999b).

3.2
Endothelin Converting Enzyme-2 (ECE-2)

ECE-2 is a membrane-bound metalloprotease with 59% homology with bovine ECE-1 (Emoto and Yanagisawa 1995). However, the enzyme has distinct biochemical properties: The optimum pH for ECE-2 activity in cleaving Big ET-1 to the mature peptide is acidic (5.5) compared with a neutral range for ECE-1. ECE-2 is 250-fold more sensitive to phosphoramidon than ECE-1. Like ECE-1, four isoforms have been identified that differ in their N-terminus and may reflect differences in the types of cell expressing the protein. In bovine tissues, mRNA encoding ECE-2a-1 and ECE-2a-2 isoforms is abundant in the periphery, including liver, kidney, adrenal gland and endothelial cells, whereas ECE-2b-1 and ECE-2b-2 are restricted to the brain, perhaps functioning in neural tissue (Ikeda et al. 2002).

In humans, ECE-2 as well as ECE-1 is present in endothelial cells, including coronary arteries (Davenport and Kuc 2000; Russell and Davenport 1999b). Confocal microscopy, using antisera that would cross-react with all ECE-2 isoforms, revealed staining within secretory vesicles (Fig. 2), suggesting a role in processing Big ET-1 whilst in transit to the cell surface via the constitutive secretory pathway. No staining was detected in storage granules. In agreement with this intracellular localization, ECE activity with an acid pH optimum in sub-cellular fractions of endothelial cells was inhibited by low concentrations of phosphoramidon (Russell et al. 1998b, c).

The precise physiological or pathophysiological role of ECE-2 in human endothelium remains to be established but may be related to the enzyme requiring an acidic pH for activity. Emoto and Yanagisawa (1995) predicted that the enzyme would be restricted to the acidified environment of the trans-Golgi network or vesicles of the secretory pathway. In human endothelial cells, ECE-2-like immunoreactivity is localised to secretory vesicles (Russell and Davenport 1999a), suggesting that ECE-2 could contribute to synthesis of ET under physiological conditions. Alternatively, synthesis of ET-1 by ECE-2 may become more important under pathophysiological conditions in which the cellular pH is reduced, such as ischaemic heart disease where intracellular pH values of 5.8 have been detected in hearts subjected to global ischaemia (Docherty et al. 1997), and a correlation between myocardial ischaemia and increased plasma levels of ET is now well established (Tonnessen et al. 1993; Cohn 1996). The increased severity of developmental defects observed when both ECE-2 and ECE-1 are knocked out implies a role in synthesising ET-1 during development (Yanagisawa et al. 2000). ECE-2 also cleaves other peptides, including the vasodilator bradykinin. The brain of ECE-2 knockout mice has significantly higher levels of beta amyloid but the significance of this to humans is not yet clear (Eckman et al. 2003).

3.3
Alternative Pathways for ET Synthesis: ET-1$_{1\text{-}31}$ and Chymase

ECE isoforms may not be the only enzymes synthesising ET-1. Human chymase, a chymotrypsin-like serine protease, can cleave Big ET-1 to yield a novel 31-amino-acid peptide, ET-1$_{1\text{-}31}$ (Nakano et al. 1997). In human vessels in vitro, including coronary arteries (Maguire et al. 2001; Maguire and Davenport 2004), ET-1$_{1\text{-}31}$ does not bind to ET receptors at physiological concentrations but is converted by enzymatic activity to ET-1, measured by radioimmunoassay in the bathing medium, to cause potent vasoconstriction. Whilst the selective ECE inhibitor PD159790 blocks the conversion of Big ET-1 in human vessels as expected, the compound has no effect on ET-1$_{1\text{-}31}$ vasoconstriction, indicating that ET-1 formation is via an alternative pathway. Thus, in human vessels, Big ET-1 can be converted directly to ET-1 by ECE or to ET-1$_{1\text{-}31}$ by chymase, with the resulting ET-1$_{1\text{-}31}$ subsequently converted to ET-1 by uncharacterised enzymes that could include neutral endopeptidase (NEP) (D'Orleans-Juste et al. 2003).

At present, there are no specific chymase inhibitors to prove conclusively that the chymostatin-sensitive enzyme is chymase and not another serine protease such as cathepsin G. However, a similar role for chymase has been proposed in the processing of angiotensin I to angiotensin II in human arteries (Takai et al.

1999) and heart (Katugampola and Davenport 2002). Mast cells are a major source of chymase and are found, for example, in umbilical cords in close proximity to the vessels that are particularly responsive to Big ET-1 (Takeji et al. 2000; Maguire and Davenport 2004). They are present in human atherosclerotic lesions, with the number and degree of degranulation increasing as the lesions develop. If chymase proves to be an alternative synthetic enzyme for ET-1 in vivo then, in those cardiovascular diseases in which plasma ET levels are raised, an alternative therapeutic strategy to ET receptor antagonism may require the dual inhibition of both ECE and chymase.

4
Endothelin Receptors

4.1
ET_A and ET_B Subtypes

Endothelins mediate their action by two sub-types of receptor (Davenport 2002) isolated and cloned from mammalian tissues, ET_A (Arai et al. 1990) and ET_B (Sakurai et al. 1990). Both sub-types belong to class 1 (Family A or rhodopsin-like), the most numerous of the G protein-coupled (GPC) seven-transmembrane-spanning family of receptors, which are also the major targets for nearly half of all currently available drugs, including many cardiovascular agents such as β-blockers and angiotensin II receptor antagonists. The ET_B receptor is characterised by an unusually long N-terminus that can be cleaved by a metalloprotease to remove the first 64 amino acids while still retaining ET-1 binding. There are two separate ligand interaction sub-domains on each endothelin receptor. The extracellular loops, particularly between transmembrane-spanning domains 4–6, determine selectivity. The amino acid sequences of ET_A receptors also differ between humans and other species, for example by 9% between human and rat ET_A receptors and by 12% for the ET_B. These may contribute to differences in efficacy and potency of selective agonists and antagonists (Davenport 2002).

The existence of further sub-types in mammals is unlikely. Following completion of 99% of the human genome, bioinformatics has been applied to identify most, if not all, of the remaining genes that potentially could encode the remaining unliganded receptors (Foord et al. 2005; Maguire and Davenport 2005). It is accepted that these have all been artificially expressed in artificial cell lines and screened against libraries of existing transmitters but no further receptors have been identified that might bind endothelin peptides.

Previous studies have suggested that different ET_B receptors may be present on endothelial versus smooth muscle cells. In detailed studies in ET_B receptor knockout mice, both the direct constrictor responses and indirect vasodilatation by the ET_B agonist sarafotoxin S6c were abolished as expected (Mizuguchi et al. 1997). In agreement, a highly detailed binding study was unable to distinguish between ET_B receptors expressed by human isolated endothelial cells compared with smooth muscle cells in culture (Flynn et al. 1998). In human tissue, both ET_A- and ET_B-selective radiolabelled ligands bound with a single affinity and Hill slopes close to unity (Molenaar et al. 1992, 1993; Davenport 1997; Davenport et al. 1994, 1998c). Similarly, competition studies using unlabelled ligands provided no evidence for further sub-types (Peter and Davenport 1995, 1996; Kuc et al. 1995; Russell and Davenport 1996).

4.2
Receptor Mutations

Disruption of genes encoding ET-1, ET-3, ET_A, ET_B, ECE-1 and ECE-2 have shown that, in addition to a role in cardiovascular regulation, the ET system is essential for correct embryonic neural crest development, a completely novel finding for GPC receptors (Kurihara et al. 2001).

4.2.1
ET_A/ET-1 Mutations and Knockouts

ET-1-deficient homozygous mice die at birth of respiratory failure secondary to severe craniofacial and cardiovascular abnormalities. Surprisingly, ET-1$^{+/-}$ heterozygous mice, which produce lower levels of ET-1 than wild-type mice, develop an elevated blood pressure (Kurihara et al. 1994). One explanation is that lower circulating ET levels may result in reduced activation of vasodilator ET_B receptors on endothelial cells. Remarkably, ET_A receptor and ECE-1 knockout mice have similar morphological abnormalities (Clouthier et al. 1998; Hosoda et al. 1994; Yanagisawa et al. 1998), implying the ET_A/ET-1 signalling system is essential for cardiovascular and craniofacial development.

4.2.2
ET_B/ET-3 Mutations and Knockouts

Homozygote ET_B knockout mice exhibit a different and non-overlapping phenotype to ET_A-deficient animals; they are viable at birth, and can survive for up to 8 weeks but display aganglionic megacolon as a result of absence of ganglion neurons, together with a pigmentary disorder in their coats (Kurihara et al. 2001). This is a result of the failure of enteric nervous system precursors and neural crest-derived epidermal melanoblasts to colonise the intestine and skin. ET-3 knockouts display an identical phenotype (Kurihara et al. 2001). Intriguingly, heterozygous knockout of ET_B (but not ET_A) receptors causes hypertension, consistent with a role in clearing ETs from the circulation.

A similar phenotype is observed in 'spotting lethal' rats that have a naturally occurring 301-bp deletion of the ET_B gene, resulting in a lack of ET_B

expression, elevation of plasma ET levels and aganglionic megacolon. ET_B deficiency caused early onset of renal impairment characterised by reduced sodium excretion and decreased glomerular filtration rate (Hocher et al. 2001; Taylor et al. 2003). This animal is used as a model of Hirschsprung disease, a multigenetic disorder, where one of the causative genes includes mutations in ET_B receptor expression (Tanaka et al. 1998).

4.3
Splice Variants of ET Receptors

4.3.1
Splice Variants of ET_A Receptors

The human ET_B receptor gene has been proposed to give rise to at least three alternatively spliced ET_A receptor transcripts, corresponding to deletion of exon 3 (producing a protein with two membrane-spanning domains), exon 4 (producing a protein with three membrane-spanning domains) and exon 3 plus exon 4 (producing a protein lacking the third and fourth domain; Miyamoto et al. 1996; Bourgeois et al. 1997). Although alternative transcripts were identified in human tissues including lung, aorta and atrium, the truncated receptors when expressed in COS cell lines did not bind ET-1 (Miyamoto et al. 1996), suggesting a mechanism for limiting ET_A receptor expression. For example, mRNA encoding the putative truncated receptor with the deletion of exon 3 plus 4 was more abundant than the wild-type in human melanoma cell lines and melanoma tissue (Zhang et al. 1998).

4.3.2
Splice Variants of ET_B Receptors

Alternative splice variants of ET_B receptors have been reported, but to date these variants show little or no change in binding characteristics and their physiological or pathophysiological significance is unclear.

4.4
ET Ligands

4.4.1
Endogenous and Synthetic Agonists

ET receptors are unusual in being isolated and cloned before the discovery of sub-type selective antagonists. The two sub-types were originally distinguished and continue to be classified by their rank order of affinity for the endogenous peptides: ET-3 typically displays at least two orders of magnitude lower affinity for the ET_A receptor than ET-1, whereas both peptides are equipotent at the ET_B receptor (Tables 1 and 2).

Table 1 Properties of ET$_A$ receptors, agonists and antagonists

Receptor	ET$_A$	
Structural information	7TM	
	Human, 427 aa	Adachi et al. (1991)
	Rat 426 aa	Lin et al. (1991)
	Mouse 427 aa	
Agonists	Selective: none with high affinity	
Agonist potencies	ET-1 = ET-2 > S6b >> ET-3 (human coronary artery)	
Antagonist potencies	BQ123 (pA$_2$ 6.9–7.4)	Ihara et al. (1992a)
	FR139317 (7.3–7.9)	Aramori et al. (1993)
	PD156707 (8–8.7) [CI1020]	Doherty et al. (1995)
	SB234551 (9)	Ohlstein et al. (1998)
	L754142 (7.7–8.7)	Williams et al. (1995)
	BMS182874 (6.2)	Stein et al. (1994)
	A127722 (9–10.5) [Atrasentan]	Opgenorth et al. (1996)
	TBC11251 (8.0) [Sitaxsentan]	Wu et al. (1997)
	LU127043 (7.3)	Raschack et al. (1995)
	LU135252 [Darusentan]	Münter et al. (1996)
Radioligand assays	Human, rat and porcine heart; A10 smooth muscle cells	
Radioligands	[^{125}I]-ET-1 (K_d = 0.01–5 nM)	Davenport (1997)
	[^{125}I]-PD151242 (0.5 nM)	Davenport et al. (1994)
	[^{125}I]-PD164333 (0.2 nM)	Davenport et al. (1998c)
	[^{3}H]-BQ123 (3.2 nM)	Ihara et al. (1995)

Names of antagonists that have undergone clinical trials are given in square brackets

4.4.2
Peptide and Non-peptide ET$_A$ Antagonists

A selective ET$_A$ receptor agonist with comparable potency to ET-1 has not been discovered, although a peptide agonist with two orders of magnitude lower potency has been reported (Langlois et al. 2003). Antagonists are currently classified as ET$_A$-selective, ET$_B$-selective or mixed antagonists that display similar affinity for both receptor sub-types. The most highly selective (by 4–5 orders of magnitude) peptide antagonists for the ET$_A$ receptors are the cyclic pentapeptide, BQ123 (Ihara et al. 1992a) and the modified linear peptide FR139317 (Aramori et al. 1993). A linear tetrapeptide analogue of FR139317, [^{125}I]-PD151242 binds with sub-nanomolar affinity to the ET$_A$ receptor and

Table 2 Properties of ET_B receptors, agonists and antagonists

Receptor	ET_B	
Structural information	7TM	
	Human 442 aa	Nakamuta et al. (1991)
	Rat 441 aa	Sakurai et al. (1990)
	Mouse 442 aa	Baynash et al. (1994)
Agonists	Selective	
	[Ala^{1,3,11,15}]ET-1	Saeki et al. (1991)
	BQ3020	Ihara et al. (1992b)
	IRL1620	Takai et al. (1992)
	S6c	Williams et al. (1991)
Agonist potencies	ET-1 = ET-2 = ET-3 = S6b (rat glomeruli)	
Antagonist potencies	IRL2500 (pA$_2$ 7.8)	Balwierczak et al. (1995)
	RES7011 (6.0)	Tanaka et al. (1994)
	BQ788 (6.9)	Ishikawa et al. (1994)
	Ro468443 (pA$_2$ 8.1)	Clozel and Breu (1996)
	A192621 (8.1)	Von Geldern et al. (1999)
Radioligand assays	Brain, lung, placenta and kidney	
Radioligands	[^{125}I]-ET-1 (K_d = 0.01–5 nM)	Davenport (1997)
	[^{125}I]-BQ3020 (0.1 nM)	Ihara et al. (1992b)
	[^{125}I]-[Ala^{1,3,11,15}]ET-1 (0.2 nM)	Molenaar et al. (1992)
	[^{125}I]-IRL1620 (0.02 nM)	Watakabe et al. (1992)

has about 10,000-fold selectivity for this sub-type in human and animal tissues. A non-peptide ET_A-selective ligand, [^{125}I]-PD164333 (Davenport et al. 1998c) also binds with comparable affinity. A number of non-peptide ET_A antagonists (Table 1; Davenport and Battistini 2002) are in clinical development with good oral bioavailability and some may cross the blood-brain barrier. The majority of these are more potent, with pA$_2$ values of up to 10 compared with 7–8 for the peptides BQ123 or FR139317, but are less selective for the ET_A versus the ET_B receptor (Table 1).

4.4.3
Peptide and Non-peptide ET_B Antagonists

Sarafotoxin S6c is widely used as an ET_B selective agonist, displaying over 200,000-fold selectivity in rat tissues (Williams et al. 1991), but is much less selective in human tissues, reflecting species differences in the receptors (Russell and Davenport 1996). The truncated, linear synthetic analogues BQ3020 ([Ala^{11,15}]Ac-ET-1$_{(6-21)}$) and IRL1620 [Suc-(Glu^9, Ala^{11,15})-ET-1$_{(8-21)}$] are the most widely used selective synthetic agonists to characterise ET_B receptors. Both peptides can be radiolabelled to produce [^{125}I]-BQ3020 (Molenaar et al. 1992) and [^{125}I]-IRL1620 (Watakabe et al. 1992). Both bind with sub-nanomolar affinity, with at least 1,500-fold selectivity for this sub-type over the ET_A receptor (Table 2). Few peptide or non-peptide ET_B antagonists have been developed, reflecting the lack of clinical need for this type of compound. They are less potent than ET_A antagonists and display lower selectivity (usually only 1–2 orders of magnitude) for the ET_B sub-type (Table 2).

4.4.4
Mixed ET_A/ET_B Antagonists

The distinction between antagonists that are ET_A selective and those that block both ET_A and ET_B receptors is not precise, but generally the former display greater than 100-fold selectivity for the ET_A subtype and the latter less than 100-fold. Bosentan (Tracleer) is the only ET antagonist currently in the clinic and has been approved for pulmonary artery hypertension (Sect. 5.2). This remarkable milestone in ET biology was achieved within 12 years of the discovery of the peptide.

5
Physiological and Pathophysiological Role

5.1
ET-1: The Universal Vasoconstrictor?

ET receptors are widely expressed in all human vessels (Davenport and Russell 2001), consistent with the physiological role of ET-1 as a ubiquitous, potent, long-lasting, endothelium-derived vasoactive peptide, contributing to the maintenance of normal vascular tone. A number of these features are unusual, if not unique, to the ET system in the human vasculature. First, in contrast to other vasoconstrictors where responses can be variable with a number of individuals not responding, a large conduit or small resistance human vessel from either central or peripheral vascular beds that does not respond to ET-1 has yet to be reported. The maximal constrictor response in human vessels produced by ET-1 is unsurpassed by any other constrictor, including compounds with more recently discovered vasoactivity such as urotensin II (Maguire and Davenport 2002, 2005). The time course for ET-1-induced vasoconstriction is unusually long lasting and can be sustained for many hours, a profile consistent with vasospasm observed in a number of pathophysiological conditions. Importantly, however, ET antagonists are able to fully reverse an established constrictor response (Pierre and Davenport 1999). The decrease in vascular resistance produced by infusion of ET antagonists in normotensive volunteers has established that ET has a physiological role in humans, contributing to vascular tone (Haynes and Webb 1994; Haynes et al. 1996; Plumpton et al. 1995). In contrast, antagonists to other vasoconstrictors, such as angiotensin II, have little or no effect on blood pressure in normotensive individuals.

5.1.1
Smooth Muscle ET_A Receptors

In human vessels (Fig. 2), the ET receptors located on vascular smooth muscle cells are mainly (>85%) of the ET_A sub-type (Davenport et al. 1995a, b; Russell et al. 1997) and are the principal sub-type mediating vasoconstriction (Davenport and Maguire 1994). A small population (<15%) of ET_B receptors are present in some human vessels (Davenport et al. 1993, 1995a, b, c; Bacon and Davenport 1996); this has been confirmed by electron microscope autoradiography (Russell et al. 1997). Sarafotoxin S6c (an ET_B agonist in animals) does cause vasoconstriction in a small number of human vessels but these responses are variable, occurring in less than 50% of individuals and, while potent, the magnitude of the response is much less than that to ET-1 (Davenport and Maguire 1994). However, little or no response to the endogenous agonist ET-3 has been detected in human vessels. Furthermore, ET_A-selective antagonists cause parallel and rightward shifts of the ET-1 concentration response curves in these vessels, with no portion of the curve resistant to ET_A blockade (Davenport and Maguire 1994; Maguire and Davenport 1995; Maguire et al. 1997a).

While ET_A receptors present on smooth muscle cells are mainly responsible for constriction in humans, in other animals this can vary depending on the species and vascular bed. For example, ET-1 mediates contraction only via ET_A receptors in rat aorta, by ET_B receptors in rabbit saphenous vein, but by both sub-types in porcine coronary artery (Davenport and Maguire 1994).

In the human brain (cortex) about 90% of the ET receptors are of the ET_B sub-type (Harland et al. 1995, 1998) and are localised to neural regions predominately on glial cells and to a lesser extent on neurons. ET_A receptors are present in high densities, localised to the cerebral vasculature and leptomeninges with lower but detectable expression in grey and white matter. Smooth muscle cells in both large arteries and small cerebral vessels only express the ET_A sub-type (Adner et al. 1994; Yu et al. 1995; Lucas et al. 1996; Harland et al. 1995, 1998; Pierre and Davenport 1995, 1998a, 1999). ET-1 potently constricts basilar arteries (Papadopoulos et al. 1990). The small pial arteries are exceptionally sensitive to ET-1 (Hardebo et al. 1989; Thorin et al. 1998; Pierre and Davenport 1998a, 1999) and, together with arterioles penetrating into the brain, play a major role in the maintenance of cerebral blood flow (autoregulation). Immunoreactive ET and ECE are present in the vascular endothelium of these vessels (Davenport et al. 1998b), which in the brain are also regulated by the release of vasoactive agents released from astrocytes that send processes to terminate upon the smooth muscle.

In normal human brain, while intense ECE staining was also detected in astrocytes including astrocytic processes (Davenport et al. 1998b), staining for ET was not detected (Giaid et al. 1991). However, intense ET staining was detected in reactive astrocytes surrounding metastases (Zhang and Olsson 1995) and following viral infections (Ma et al. 1994) as well as in rat perivascular astrocytic processes in an animal model of ischaemia (Gajkowska and Mossakowski 1995). These results suggest that ET-1 released from endothelial and reactive perivascular astrocytes may be involved in the genesis or maintenance of cerebrovascular disorders, such as the delayed vasospasm leading to cerebral ischaemia seen after aneurysmal subarachnoid haemorrhage, and could contribute to ischaemic core volume in stroke. Importantly, ET-1 does not normally cross the blood-brain barrier (Johnström et al. 2005). However, in these conditions the barrier may be compromised, and ET-1 synthesised in the periphery could be an additional source affecting both the (1) vascular receptors mediating cerebrovasospasm and (2) neural receptors mediating the increase in intracellular free calcium (Morton and Davenport 1992) that initiates the pathophysiological processes leading to neuronal death. ET_A receptors may also have a role at the blood brain-barrier. Ligand binding (Yamaga et al. 1995) and functional evidence suggest that human brain endothelial cells isolated from the capillaries (diameter $\sim$10 µm) that form the blood-brain barrier and larger microvessels, express ET_A receptors linked to phospholipase C and inositol trisphosphate accumulation (Stanimirovic et al. 1994; Spatz et al. 1997). ET-1 acting via this sub-type has been proposed to increase capillary permeability leading to oedema (Purkiss et al. 1994).

5.1.2
Endothelial ET_B Receptors

ET_B receptors are present on the endothelial cells (Fig. 2). Some of the ET-1 released from the endothelium may feed back onto these receptors to release endothelium-derived relaxing factors such as nitric oxide, prostacyclin or an endothelium-derived hyperpolarizing factor, opposing the constrictor response. In humans, infusion of low doses of ET-1 (Kiowski et al. 1991) into the brachial artery in vivo causes an initial reduction in forearm blood flow consistent with the peptides binding to ET_B receptors to cause vasodilatation. High concentrations of ET-3 also cause vasodilatation (Haynes et al. 1995). In agreement, blocking ET_B receptors with a selective antagonist, BQ788, causes vasoconstriction (Verhaar et al. 1998), since the constrictor actions of ET-1 on the underlying smooth muscle are unopposed. In vitro studies have examined a wider range of vascular beds. ET_B-mediated relaxation was reported in isolated preconstricted temporal and cerebral arteries (Lucas et al. 1996; Nilsson et al. 1997) but not in some peripheral vessels including internal mammary (Seo et al. 1994), radial (Liu et al. 1996), conduit or resistance coronary arteries (Pierre and Davenport 1995, 1998a, b) and small omental arteries (Riezebos et al. 1994). It is unclear if these results reflect heterogeneity in ET_B dilator responses within vessels from different vascular beds.

Staining for ECE was also detected within endocardial endothelial cells lining the ventricle, a second major source of cardiac ET-1 (Plumpton 1996a). ET-1 is a potent positive inotropic agent, acting directly on heart muscle (Moravec et al. 1989; Davenport et al. 1989). Synthesis of ET-1 by ECE within the endocardial endothelial cells may not only modulate the inotropic state of the heart but also exert effects on the conducting system in close proximity to endocardial cells.

5.1.3
ET_B Clearing Receptors

Systemic blockade of ET_B receptors results in a significant rise in circulating ET-1 (Plumpton et al. 1996b). This is not simply the result of occupancy of vascular receptors by the antagonist and displacement of ET-1, since ET-3 levels are also significantly elevated. Human lungs contain one of the highest densities of ET receptors, with a high proportion of the ET_B sub-type (McKay et al. 1991; Henry et al. 1990; Marciniak et al. 1992; Knott et al. 1995; Russell and Davenport 1996). The human kidney is also rich in ET_B receptors (comprising 70% in both cortex and medulla) expressed by endothelial cells, tubules and collecting ducts, whereas most ET_A receptors are localised to vascular smooth muscle of arteries and veins as well as intra-renal resistance vessels (Karet et al. 1993; Davenport et al. 1994; Maguire et al. 1994). The presence of mRNA encoding ET-1 and the detection of the peptide and its precursor by high-pressure liquid chromatography and radioimmunoassay (Karet and Davenport 1993, 1996) established that ET-1 was synthesised within the kidney. ET-1 functions as a locally acting renal peptide with two main actions, vasoconstriction via ET_A receptors and natriuresis, via the ET_B sub-type (Nambi et al. 1992).

In addition to these roles, the ET_B sub-type in rat lung and kidney have been proposed to function as clearing receptors (Fig. 2), removing ET-1 from the circulation (Fukuroda et al. 1994; Gasic et al. 1992). Dynamic imaging using positron emission tomography in the living animal showed that [^{18}F]-ET-1 rapidly accumulated in the lung, kidney and liver but only low levels were detected in the heart. Infusion of an ET_B-selective antagonist, BQ788, just before administration of the [^{18}F]-ET-1, blocked binding of the radioligand to ET_B receptors in the kidney and lungs as expected. Under ET_B blockade, [^{18}F]-ET-1 was now able to bind to ET_A receptors in the heart, revealing the importance of the ET_B sub-type in clearing ET-1 from the circulation, thus protecting cardiac tissue from the potentially deleterious action of the circulating peptide. Infusion of the antagonist 30 min after injecting [^{18}F]-ET-1 did not displace the ligand, consistent with internalisation of the ligand-receptor complex (Johnström et al. 2005). It is likely that ET_B receptors present on endothelial cells in the lungs and kidney are responsible for removing circulating

ET-1. In agreement, mice where the ET_B receptor has been selectively knocked out in endothelial cells (but not, for example, in epithelial cells), ET_B receptor density in ET_B-rich tissues such as lungs was significantly reduced but plasma concentrations of ET were elevated fourfold (Kelland et al. 2004).

5.2
Pulmonary Arterial Hypertension (PAH)

The first ET antagonist to receive FDA approval for clinical use was bosentan, and this compound was also the first orally active drug treatment for pulmonary arterial hypertension (PAH). This condition is characterised by hypertrophy of the small pulmonary arterioles, increasing vascular resistance and ultimately right ventricular heart failure. Overexpression of ET-1 in this condition leads to endothelial cell dysfunction and inflammation, with the peptide acting as a co-mitogen for smooth muscle cells contributing to vascular hypertrophy as well as fibrosis mediated via transforming growth factor-β (Clozel and Salloukh 2005).

In human lungs, ET_A receptors are present on resistance vessels and predominate in conduit arteries, comprising 90% of the ET receptors expressed by the medial layer from the main pulmonary artery. ET_A receptors are also present on lung parenchyma, submucosal glands, airway smooth muscle and epithelial cells. Synthesis and release of ET-1 from the pulmonary vascular endothelium is thought to cause constriction of pulmonary arteries, predominantly via the ET_A sub-type (Hay et al. 1993; Maguire and Davenport 1995), although McCulloch et al. (1996) have proposed a significant contribution of ET_B receptors in resistance arteries (150–200 μm in diameter). ET_B receptors are localised to airway smooth muscle, with lower levels in parenchyma, airway submucosal glands and small conduit arteries (Russell and Davenport 1995). In the bronchus, the constrictor action of the peptide released from epithelia and diffusing onto the underlying airway smooth muscle is via ET_B receptors (Adner et al. 1996; Takahashi et al. 1997; Hay et al. 1998) although Fukuroda et al. (1994) detected an ET_A component. Thus PAH may require blockade of both sub-types by a mixed antagonist such as bosentan.

The initial clinical trials with bosentan of 3–7 months duration reported improvements in patients with idiopathic PAH in exercise capacity and haemodynamics, and delayed clinical worsening (Channick et al. 2001; Rubin et al. 2002). Longer-term studies are currently following up these patients. Importantly, the survival of WHO class III patients treated with bosentan after 3 years was at least as good as those treated with intravenous epoprostenol (prostacyclin). The median predicted survival for these patients is under 3 years, but with bosentan treatment, 3-year survival was nearly 90% (Sitbon et al. 2003, 2004). Bosentan represents a significant new therapeutic strategy in PAH. Other emerging clinical indications include connective tissue disease, particularly systemic sclerosis (included as a subset in the original trials) and associated digital ulcers, HIV-associated PAH and, more speculatively, in liver disease, including portal hypertension (Clozel and Salloukh 2005).

5.3
Essential Hypertension

The potent constrictor actions of ET-1 in humans (Sect. 5.1), combined with endothelial cell dysfunction and a reduction in nitric oxide, suggest a role for ET-1 in essential hypertension. Lowering blood pressure by about 10 mmHg in such patients has proven benefits in reducing the risk of cardiovascular disease. In vitro, ET antagonists display desirable properties of anti-hypertensive drugs in dilating human isolated arteriole resistance and venous capacitance vessels. In vivo, systemic infusion of ET-1 to produce 30- to 50-fold rises in circulating peptide in healthy volunteers causes a 5–10 mmHg increase in mean blood pressure (Vierhapper et al. 1990; Pernow et al. 1996). Hypertensive patients display increased venoconstrictor responses to ET-1 (Haynes et al. 1994). In two cases of malignant haemangioendothelioma, a condition in which endothelial cell proliferation occurs, blood pressure and ET-1 levels were elevated. Both of these parameters were reduced towards normal levels on removal of the tumour (Yokokawa et al. 1991).

As proof of principle that ET-1 may contribute to hypertension in some individuals, in a study with nearly 300 hypertensive patients, the ET_A/ET_B antagonist bosentan at the highest dose tested resulted in a 10-mmHg reduction in systolic blood pressure, although a number of side-effects were reported, including liver function abnormalities. The ET_A selective antagonist darusentan (LU135252), in a multicentre trial of about 400 patients, produced similar reductions in diastolic (8.3 mmHg) and systolic (11.3 mmHg) pressure (Nakov et al. 2002). Salt sensitivity, in which hypertension is exacerbated by high salt intake is common in African-Americans. Intriguingly, hypertensive African-Americans have enhanced ET_A-dependent vascular tone compared with white patients, suggesting that ET_A antagonists would be particularly beneficial in the former ethnic group (Campia et al. 2004). ET_A antagonists are also highly effective in lowering blood pressure in hypertensive patients with chronic renal failure (Goddard et al. 2004). Targeting the ET system has potential in the treatment of hypertension, particularly when associated with salt-sensitivity and target organ damage.

5.4
Atherosclerosis

In patients with atherosclerosis, studies have shown a consistent pattern of raised plasma levels of immunoreactive ET (Table 3). Tissue levels of ET-1 mRNA (Winkles et al. 1993), the mature peptide and Big ET-1 are significantly increased within the wall of human vessels containing atherosclerotic lesions

Table 3 Changes in the ET system in human atherosclerosis

Coronary arteries	Cell type	Pathology	Reference
Advanced plaques	MSMC	ET_A	Bacon et al. (1995)
	ISMC	$ET_A\downarrow$	
	MAC	ET_B	
Advanced plaques		ET-1↑, Big ET-1↑	Bacon et al. (1996)
Advanced plaques	MSMC	ET_A	Katugampola et al. (2002)
	ISMC	$ET_A\downarrow$	Kuc and Davenport (2000)
Early lesions	ISMC, EC	ECE↑	Hai et al. (2004)
Advanced plaques	MAC, EC	ECE↑	Hai et al. (2004)
Carotid atherosclerosis +type II diabetes	ISMC/MSMC	Intimal: medial thickness↑	Migdalis et al. (2000)
Plasma			
Atherosclerosis Systemic		ET↑	Lerman et al. (1991)
Coronary atherosclerosis	Coronary	ET↑	Lerman et al. (1995)
	Systemic	ET↑	
Carotid plaques +type II diabetes		ET↑	Kalogeropoulou et al. (2002)
Carotid plaques Systemic +type II diabetes		ET↑	Migdalis et al. (2000)
Carotid plaques Systemic +essential hypertension		ET↑	Minami et al. (2001)
Atherosclerosis Systemic +NIDDM		ET↑	Perfetto et al. (1988)

EC, endothelial cell; ISMC, intimal smooth muscle cell; MAC, macrophage; MSMC, medial smooth muscle cell; ;NIDDM, non-insulin-dependent diabetes mellitus

(Bacon et al. 1996). Most of the immunoreactive ET-1 is confined to infiltrating macrophages of the lesion and not the smooth muscle.

About 1 in 4 molecules of Big ET-1 synthesised within endothelial cells escapes conversion and is released to circulate in the plasma, where it may be cleaved to ET-1 by smooth muscle ECE. Big ET-1 could function as a long-range signalling molecule. Infusion of Big ET-1 into volunteers causes pronounced vasoconstriction by local conversion to ET-1 by a phosphoramidon-sensitive ECE (Plumpton et al. 1995). Since most endothelial ECE is intra-cellular, conversion is predominantly via smooth muscle ECE. In agreement with the presence of ECE on smooth muscle, Big ET-1 constricts human vessels denuded of endothelium (Mombouli et al. 1993; Maguire et al. 1997b; Maguire and Davenport 1998a) which can be inhibited by phosphoramidon but not thiorphan (Maguire et al. 1997a; Rizzi et al. 1998) and PD159790, a selective inhibitor of ECE-1 (Maguire et al. 1999). In atherosclerosis, smooth muscle ECE activity is up-regulated, increasing the amount of ET-1 synthesised at the site of the lesion. In endothelium-denuded human coronary arteries, the response to Big ET-1 was significantly enhanced in vessels containing atherosclerotic lesions with a corresponding increase in mature ET formed in the bathing medium, compared with non-diseased arteries. There were no differences in responses of arteries from either group to ET-1, demonstrating up-regulation of ECE activity rather than an augmented response of the arteries to ET-1 (Maguire and Davenport 1998a).

Minamino et al. (1997) reported that ECE-1 immunoreactivity was present in both smooth muscle cells and macrophages in two human coronary atherectomy samples. Particularly intense staining for ECE-1b, ECE-1c and ECE-2 isoforms was detected within infiltrating macrophages of atherosclerotic plaques from human coronary arteries. Lower levels of staining were also visualised in smooth muscle within the intimal thickening and thinned medial layer (Davenport and Kuc 2000). Macrophages, in addition to smooth muscle ECE, may locally increase conversion of Big ET-1 in the vessel wall.

ET_A receptors predominate in the media of atherosclerotic coronary arteries but are down-regulated (together with ET_B receptors) in the intimal smooth muscle (Bacon et al. 1996; Katugampola et al. 2002). In agreement with ET_A receptors being the major sub-type present on smooth muscle, ET_A antagonists also fully reverse ET-1 induced vasoconstriction in diseased vessels (Maguire and Davenport 1998b). ET_B receptor expression is increased in atherosclerotic coronary arteries (Dagassan et al. 1996), but this is due to ET_B receptors localised to infiltrating macrophages and the increase in endothelial cells associated with neovascularisation. Smooth muscle ET_B receptors may have limited physiological or pathophysiological importance in mediating vasoconstriction.

6
Conclusions

Endothelin-1 remains one of the most powerful vasoconstrictors discovered, and overproduction of this peptide following endothelial cell dysfunction contributes to pathophysiological processes including vascular hypertrophy, cell proliferation, fibrosis and inflammation, making this system a particularly attractive drug target. The first ET antagonist approved for clinical use, bosentan, has proved successful in the treatment of PAH. Considered together, bosentan, ET_A-selective antagonists, and combined neutral endopeptidase/endothelin converting enzyme (NEP/ECE) inhibitors-which continue in clinical trials-represent a new therapeutic strategy to the clinicians' armamentarium and may find other clinical applications.

Acknowledgements Supported by grants from the British Heart Foundation

References

Adachi M, Yang YY, Furuichi Y, Miyamoto C (1991) Cloning and characterization of cDNA encoding human A-type endothelin receptor. Biochem Biophys Res Commun 180:1265–1272

Adner M, Jansen I, Edvinsson L (1994) Endothelin-A receptors mediate contraction in human cerebral, meningeal and temporal arteries. J Auton Nerv Syst 49 Suppl:S117–121

Adner M, Cardell LO, Sjoberg T, Ottosson A, Edvinsson L (1996) Contractile endothelin-B (ETB) receptors in human small bronchi. Eur Respir J 9:351–355

Arai H, Hori S, Aramori I, Ohkubo H, Nakanishi S (1990) Cloning and expression of a cDNA encoding an endothelin receptor. Nature 348:730–7322

Aramori I, Nirei H, Shoubo M, Sogabe K, Nakamura K, Kojo H, Notsu Y, Ono T, Nakanishi S (1993) Subtype selectivity of a novel endothelin antagonist, FR139317, for the two endothelin receptors in transfected Chinese hamster ovary cells. Mol Pharmacol 43:127–131

Ashby MJ, Plumpton C, Teale P, Kuc RE, Houghton E, Davenport AP (1995) Analysis of endogenous human endothelin peptides by high-performance liquid chromatography and mass spectrometry. J Cardiovasc Pharmacol 26 Suppl 3:S247–249

Bacon CR, Davenport AP (1996) Endothelin receptors in human coronary artery and aorta. Br J Pharmacol 117:986–992

Bacon CR, Cary NR, Davenport AP (1995) Distribution of endothelin receptors in atherosclerotic human coronary arteries. J Cardiovasc Pharmacol 26 Suppl 3:439–444

Bacon CR, Cary NR, Davenport AP (1996) Endothelin peptide and receptors in human atherosclerotic coronary artery and aorta. Circ Res 79:794–801

Balwierczak JL, Bruseo CW, DelGrande D, Jeng AY, Savage P, Shetty SS (1995) Characterization of a potent and selective endothelin-B receptor antagonist, IRL 2500. J Cardiovasc Pharmacol 26 Suppl 3:S393–396

Barnes K, Shimada K, Takahashi M, Tanzawa K, Turner AJ (1996) Metallopeptidase inhibitors induce an up-regulation of endothelin-converting enzyme levels and its redistribution from the plasma membrane to an intracellular compartment. J Cell Sci 109:919–928

Barnes K, Brown C, Turner AJ (1998) Endothelin-converting enzyme: ultrastructural localization and its recycling from the cell surface. Hypertension 31:3–9

Baynash AG, Hosoda K, Giaid A, Richardson JA, Emoto N, Hammer RE, Yanagisawa M (1994) Interaction of endothelin-3 with endothelin-B receptor is essential for development of epidermal melanocytes and enteric neurons. Cell 79:1277–1285

Bourgeois C, Robert B, Rebourcet R, Mondon F, Mignot TM, Duc-Goiran P, Ferre F (1997) Endothelin-1 and ETA receptor expression in vascular smooth muscle cells from human placenta: a new ETA receptor messenger ribonucleic acid is generated by alternative splicing of exon 3. J Clin Endocrinol Metab 82:3116–3123

Campia U, Cardillo C, Panza JA (2004) Ethnic differences in the vasoconstrictor activity of endogenous endothelin-1 in hypertensive patients. Circulation 109:3191–3195

Channick RN, Simonneau G, Sitbon O, Robbins IM, Frost A, Tapson VF, Badesch DB, Roux S, Rainisio M, Bodin F, Rubin LJ (2001) Effects of the dual endothelin-receptor antagonist bosentan in patients with pulmonary hypertension: a randomised placebo-controlled study. Lancet 358:1119–1123

Clouthier DE, Hosoda K, Richardson JA, Williams SC, Yanagisawa H, Kuwaki T, Kumada M, Hammer RE, Yanagisawa M (1998) Cranial and cardiac neural crest defects in endothelin-A receptor-deficient mice. Development 125:813–824

Clozel M, Breu V (1996) The role of ETB receptors in normotensive and hypertensive rats as revealed by the non-peptide selective ETB receptor antagonist Ro 46–8443. FEBS Lett 383:42–45

Clozel M, Salloukh H (2005) Role of endothelin in fibrosis and anti-fibrotic potential of bosentan. Ann Med 37:2–12

Cohn JN (1996) Is there a role for endothelin in the natural history of heart failure? Circulation 94:604–606

D'Orleans-Juste P, Plante M, Honore JC, Carrier E, Labonte J (2003) Synthesis and degradation of endothelin-1. Can J Physiol Pharmacol 81:503–510

Dagassan PH, Breu V, Clozel M, Kunzli A, Vogt P, Turina M, Kiowski W, Clozel JP (1996) Up-regulation of endothelin-B receptors in atherosclerotic human coronary arteries. J Cardiovasc Pharmacol 27:147–153

Davenport A (1997) Distribution of endothelin receptors. In: Huggins JP, Pelton JT (eds) Endothelin in biology and medicine. CRC Press, New York, pp 45–68

Davenport AP (2002) International Union of Pharmacology. XXIX. Update on endothelin receptor nomenclature. Pharmacol Rev 54:219–226

Davenport AP, Battistini B (2002) Classification of endothelin receptors and antagonists in clinical development. Clin Sci (Lond) 103 Suppl 48:1S–3S

Davenport AP, Kuc RE (2000) Cellular expression of isoforms of endothelin-converting enzyme-1 (ECE-1c, ECE-1b and ECE-1a) and endothelin-converting enzyme-2. J Cardiovasc Pharmacol 36:S12–14

Davenport AP, Maguire JJ (1994) Is endothelin-induced vasoconstriction mediated only by ET_A receptors in humans? Trends Pharmacol Sci 15:9–11

Davenport AP, Maguire JJ (2002) Of mice and men: advances in endothelin research and first antagonist gains FDA approval. Trends Pharmacol Sci 23:155–157

Davenport AP, Russell FD (2001) Endothelin converting enzymes and endothelin receptor localisation in human tissues. In: Warner TD (ed) Endothelin and its inhibitors. (Handbook of experimental pharmacology, vol 152) Springer, Heidelberg Berlin New York, pp 209–237

Davenport AP, Nunez DJ, Hall JA, Kaumann AJ, Brown MJ (1989) Autoradiographical localization of binding sites for porcine [125I]endothelin-1 in humans, pigs, and rats: functional relevance in humans. J Cardiovasc Pharmacol 13 Suppl 5:S166–S170

Davenport AP, O'Reilly G, Molenaar P, Maguire JJ, Kuc RE, Sharkey A, Bacon CR, Ferro A (1993) Human endothelin receptors characterized using reverse transcriptase-polymerase chain reaction, in situ hybridization, and subtype-selective ligands BQ123 and BQ3020: evidence for expression of ETB receptors in human vascular smooth muscle. J Cardiovasc Pharmacol 22 Suppl 8:S22–25

Davenport AP, Kuc RE, Fitzgerald F, Maguire JJ, Berryman K, Doherty AM (1994) [125I]-PD151242: a selective radioligand for human ETA receptors. Br J Pharmacol 111:4–6

Davenport AP, Kuc RE, Maguire JJ, Harland SP (1995a) ETA receptors predominate in the human vasculature and mediate constriction. J Cardiovasc Pharmacol 26 Suppl 3:S265–267

Davenport AP, O'Reilly G, Kuc RE (1995b) Endothelin ETA and ETB mRNA and receptors expressed by smooth muscle in the human vasculature: majority of the ETA sub-type. Br J Pharmacol 114:1110–1116

Davenport AP, Hoskins SL, Kuc RE, Plumpton C (1996) Differential distribution of endothelin peptides and receptors in human adrenal gland. Histochem J 28:779–789

Davenport AP, Kuc RE, Mockridge JW (1998a) Endothelin-converting enzyme in the human vasculature: evidence for differential conversion of big endothelin-3 by endothelial and smooth-muscle cells. J Cardiovasc Pharmacol 31 Suppl 1:S1–3

Davenport AP, Kuc RE, Plumpton C, Mockridge JW, Barker PJ, Huskisson NS (1998b) Endothelin-converting enzyme in human tissues. Histochem J 30:359–374

Davenport AP, Kuc RE, Ashby MJ, Patt WC, Doherty AM (1998c) Characterization of [125I]-PD164333, an ETA selective non-peptide radiolabelled antagonist, in normal and diseased human tissues. Br J Pharmacol 123:223–230

Docherty JC, Gunter HE, Kuzio B, Shoemaker L, Yang L, Deslauriers R (1997) Effects of cromakalim and glibenclamide on myocardial high energy phosphates and intracellular pH during ischemia-reperfusion: 31P NMR studies. J Mol Cell Cardiol 29:1665–1673

Doherty AM, Patt WC, Edmunds JJ, Berryman KA, Reisdorph BR, Plummer MS, Shahripour A, Lee C, Cheng XM, Walker DM, et al (1995) Discovery of a novel series of orally active non-peptide endothelin-A (ETA) receptor-selective antagonists. J Med Chem 38:1259–1263

Eckman EA, Watson M, Marlow L, Sambamurti K, Eckman CB (2003) Alzheimer's disease beta-amyloid peptide is increased in mice deficient in endothelin-converting enzyme. J Biol Chem 278:2081–2084

Emoto N, Yanagisawa M (1995) Endothelin-converting enzyme-2 is a membrane-bound, phosphoramidon-sensitive metalloprotease with acidic pH optimum. J Biol Chem 270:15262–15268

Flynn MA, Haleen SJ, Welch KM, Cheng XM, Reynolds EE (1998) Endothelin B receptors on human endothelial and smooth-muscle cells show equivalent binding pharmacology. J Cardiovasc Pharmacol 32:106–116

Foord SM, Bonner TI, Neubig RR, Rosser EM, Pin JP, Davenport AP, Spedding M, Harmar AJ (2005) International Union of Pharmacology. XLVI. G protein-coupled receptor list. Pharmacol Rev 57:279–288

Fukuroda T, Kobayashi M, Ozaki S, Yano M, Miyauchi T, Onizuka M, Sugishita Y, Goto K, Nishikibe M (1994) Endothelin receptor subtypes in human versus rabbit pulmonary arteries. J Appl Physiol 76:1976–1982

Gajkowska B, Mossakowski MJ (1995) Localization of endothelin in the blood-brain interphase in rat hippocampus after global cerebral ischemia. Folia Neuropathol 33:221–230

Gasic S, Wagner OF, Vierhapper H, Nowotny P, Waldhausl W (1992) Regional hemodynamic effects and clearance of endothelin-1 in humans: renal and peripheral tissues may contribute to the overall disposal of the peptide. J Cardiovasc Pharmacol 19:176–180

Giaid A, Gibson SJ, Herrero MT, Gentleman S, Legon S, Yanagisawa M, Masaki T, Ibrahim NB, Roberts GW, Rossi ML, et al (1991) Topographical localisation of endothelin mRNA and peptide immunoreactivity in neurones of the human brain. Histochemistry 95:303–314

Goddard J, Johnston NR, Hand MF, Cumming AD, Rabelink TJ, Rankin AJ, Webb DJ (2004) Endothelin-A receptor antagonism reduces blood pressure and increases renal blood flow in hypertensive patients with chronic renal failure: a comparison of selective and combined endothelin receptor blockade. Circulation 109:1186–1193

Gui G, Xu D, Emoto N, Yanagisawa M (1993) Intracellular localization of membrane-bound endothelin-converting enzyme from rat lung. J Cardiovasc Pharmacol 22 Suppl 8:S53–56

Hai E, Ikura Y, Naruko T, Shirai N, Yoshimi N, Kayo S, Sugama Y, Fujino H, Ohsawa M, Tanzawa K, Yokota T, Ueda M (2004) Alterations of endothelin-converting enzyme expression in early and advanced stages of human coronary atherosclerosis. Int J Mol Med 13:649–654

Hardebo JE, Kahrstrom J, Owman C, Salford LG (1989) Endothelin is a potent constrictor of human intracranial arteries and veins. Blood Vessels 26:249–253

Harland SP, Kuc RE, Pickard JD, Davenport AP (1995) Characterization of endothelin receptors in human brain cortex, gliomas, and meningiomas. J Cardiovasc Pharmacol 26 Suppl 3:S408–411

Harland SP, Kuc RE, Pickard JD, Davenport AP (1998) Expression of endothelin(A) receptors in human gliomas and meningiomas, with high affinity for the selective antagonist PD156707. Neurosurgery 43:890–898; discussion 898–899

Harrison VJ, Barnes K, Turner AJ, Wood E, Corder R, Vane JR (1995) Identification of endothelin 1 and big endothelin 1 in secretory vesicles isolated from bovine aortic endothelial cells. Proc Natl Acad Sci U S A 92:6344–6348

Hay DW, Henry PJ, Goldie RG (1993) Endothelin and the respiratory system. Trends Pharmacol Sci 14:29–32

Hay DW, Luttmann MA, Pullen MA, Nambi P (1998) Functional and binding characterization of endothelin receptors in human bronchus: evidence for a novel endothelin B receptor subtype? J Pharmacol Exp Ther 284:669–677

Haynes WG, Webb DJ (1994) Contribution of endogenous generation of endothelin-1 to basal vascular tone. Lancet 344:852–854

Haynes WG, Hand MF, Johnstone HA, Padfield PL, Webb DJ (1994) Direct and sympathetically mediated venoconstriction in essential hypertension. Enhanced responses to endothelin-1. J Clin Invest 94:1359–1364

Haynes WG, Strachan FE, Webb DJ (1995) Endothelin ETA and ETB receptors cause vasoconstriction of human resistance and capacitance vessels in vivo. Circulation 92:357–363

Haynes WG, Ferro CJ, O'Kane KP, Somerville D, Lomax CC, Webb DJ (1996) Systemic endothelin receptor blockade decreases peripheral vascular resistance and blood pressure in humans. Circulation 93:1860–1870

Hemsen A, Gillis C, Larsson O, Haegerstrand A, Lundberg JM (1991) Characterization, localization and actions of endothelins in umbilical vessels and placenta of man. Acta Physiol Scand 143:395–404

Henry PJ, Rigby PJ, Self GJ, Preuss JM, Goldie RG (1990) Relationship between endothelin-1 binding site densities and constrictor activities in human and animal airway smooth muscle. Br J Pharmacol 100:786–792

Hickey KA, Rubanyi G, Paul RJ, Highsmith RF (1985) Characterization of a coronary vasoconstrictor produced by cultured endothelial cells. Am J Physiol 248:C550–556

Hocher B, Dembowski C, Slowinski T, Friese ST, Schwarz A, Siren AL, Neumayer HH, Thone-Reineke C, Bauer C, Nafz B, Ehrenreich H (2001) Impaired sodium excretion, decreased glomerular filtration rate and elevated blood pressure in endothelin receptor type B deficient rats. J Mol Med 78:633–641

Hosoda K, Hammer RE, Richardson JA, Baynash AG, Cheung JC, Giaid A, Yanagisawa M (1994) Targeted and natural (piebald-lethal) mutations of endothelin-B receptor gene produce megacolon associated with spotted coat color in mice. Cell 79:1267–1276

Howard PG, Plumpton C, Davenport AP (1992) Anatomical localization and pharmacological activity of mature endothelins and their precursors in human vascular tissue. J Hypertens 10:1379–1386

Ihara M, Noguchi K, Saeki T, Fukuroda T, Tsuchida S, Kimura S, Fukami T, Ishikawa K, Nishikibe M, Yano M (1992a) Biological profiles of highly potent novel endothelin antagonists selective for the ET$_A$ receptor. Life Sci 50:247–255

Ihara M, Saeki T, Fukuroda T, Kimura S, Ozaki S, Patel AC, Yano M (1992b) A novel radioligand [^{125}I]BQ-3020 selective for endothelin (ETB) receptors. Life Sci 51:PL47–52

Ihara M, Yamanaka R, Ohwaki K, Ozaki S, Fukami T, Ishikawa K, Towers P, Yano M (1995) [3H]BQ-123, a highly specific and reversible radioligand for the endothelin ETA receptor subtype. Eur J Pharmacol 274:1–6

Ikeda S, Emoto N, Alimsardjono H, Yokoyama M, Matsuo M (2002) Molecular isolation and characterization of novel four subisoforms of ECE-2. Biochem Biophys Res Commun 293:421–426

Inoue A, Yanagisawa M, Kimura S, Kasuya Y, Miyauchi T, Goto K, Masaki T (1989) The human endothelin family: three structurally and pharmacologically distinct isopeptides predicted by three separate genes. Proc Natl Acad Sci U S A 86:2863–2867

Ishikawa K, Ihara M, Noguchi K, Mase T, Mino N, Saeki T, Fukuroda T, Fukami T, Ozaki S, Nagase T, et al (1994) Biochemical and pharmacological profile of a potent and selective endothelin B-receptor antagonist, BQ-788. Proc Natl Acad Sci U S A 91:4892–4896

Janes RW, Peapus DH, Wallace BA (1994) The crystal structure of human endothelin. Nat Struct Biol 1:311–319

Johnström P, Fryer TD, Richards HK, Harris NG, Barret O, Clark JC, Pickard JD, Davenport AP (2005) Positron emission tomography using 18F-labelled endothelin-1 reveals prevention of binding to cardiac receptors owing to tissue-specific clearance by ET B receptors in vivo. Br J Pharmacol 144:115–122

Kalogeropoulou K, Mortzos G, Migdalis I, Velentzas C, Mikhailidis DP, Georgiadis E, Cordopatis P (2002) Carotid atherosclerosis in type 2 diabetes mellitus: potential role of endothelin-1, lipoperoxides, and prostacyclin. Angiology 53:279–285

Karet FE, Davenport AP (1993) Human kidney: endothelin isoforms detected by HPLC with radioimmunoassay and receptor subtypes detected using ligands BQ123 and BQ3020. J Cardiovasc Pharmacol 22 Suppl 8:S29–33

Karet FE, Davenport AP (1996) Localization of endothelin peptides in human kidney. Kidney Int 49:382–387

Karet FE, Kuc RE, Davenport AP (1993) Novel ligands BQ123 and BQ3020 characterize endothelin receptor subtypes ETA and ETB in human kidney. Kidney Int 44:36–42

Katugampola SD, Davenport AP (2002) Radioligand binding reveals chymase as the predominant enzyme for mediating tissue conversion of angiotensin I in the normal human heart. Clin Sci (Lond) 102:15–21

Katugampola SD, Kuc RE, Maguire JJ, Davenport AP (2002) G-protein-coupled receptors in human atherosclerosis: comparison of vasoconstrictors (endothelin and thromboxane) with recently de-orphanized. Clin Sci (Lond) 2002 103 Suppl 48:171S–175S

Kelland NF, Bagnall AJ, Gulliver-Sloan FH, Kuc RE, Maguire JJ, Davenport AP, Gray GA, Kotelevtsev YV, Webb DJ (2004) Clearance of circulating endothelin-1 is mediated by the endothelial cell endothelin B receptor. http://www.pa2online.org/Vol2Issue4abst022P.html.

Kiowski W, Luscher TF, Linder L, Buhler FR (1991) Endothelin-1-induced vasoconstriction in humans. Reversal by calcium channel blockade but not by nitrovasodilators or endothelium-derived relaxing factor. Circulation 83:469–475

Knott PG, D'Aprile AC, Henry PJ, Hay DW, Goldie RG (1995) Receptors for endothelin-1 in asthmatic human peripheral lung. Br J Pharmacol 114:1–3

Kuc RE, Davenport AP (2000) Endothelin-A-receptors in human aorta and pulmonary arteries are downregulated in patients with cardiovascular disease: an adaptive response to increased levels of endothelin-1? J Cardiovasc Pharmacol 36:S377–379

Kuc RE, Karet FE, Davenport AP (1995) Characterization of peptide and nonpeptide antagonists in human kidney. J Cardiovasc Pharmacol 26 Suppl 3:S373–375

Kurihara H, Kurihara Y Yazaki Y (2001) Lessons from gene deletion in the endothelin system. In: Warner TD (ed) Endothelin and its inhibitors. (Handbook of experimental pharmacology, vol 152) Springer, Heidelberg Berlin New York, pp 141–154

Kurihara Y, Kurihara H, Suzuki H, Kodama T, Maemura K, Nagai R, Oda H, Kuwaki T, Cao WH, Kamada N, et al (1994) Elevated blood pressure and craniofacial abnormalities in mice deficient in endothelin-1. Nature 368:703–710

Kurnik D, Haviv Y, Kochva E (1999) A snake bite by the burrowing asp, *Atractaspis engaddensis*. Toxicon 37:223–227

Langlois C, Letourneau M, Lampron P, St-Hilaire V, Fournier A (2003) Development of agonists of endothelin-1 exhibiting selectivity towards ETA receptors. Br J Pharmacol 139:616–622

Lerman A, Edwards BS, Hallett JW, Heublein DM, Sandberg SM, Burnett JC Jr (1991) Circulating and tissue endothelin immunoreactivity in advanced atherosclerosis. N Engl J Med 325:997–1001

Lerman A, Holmes DR Jr, Bell MR, Garratt KN, Nishimura RA, Burnett JC Jr (1995) Endothelin in coronary endothelial dysfunction and early atherosclerosis in humans. Circulation 92:2426–2431

Lin HY, Kaji EH, Winkel GK, Ives HE, Lodish HF (1991) Cloning and functional expression of a vascular smooth muscle endothelin 1 receptor. Proc Natl Acad Sci U S A 88:3185–3189

Liu JJ, Chen JR, Buxton BF (1996) Unique response of human arteries to endothelin B receptor agonist and antagonist. Clin Sci (Lond) 90:91–96

Lucas GA, White LR, Juul R, Cappelen J, Aasly J, Edvinsson L (1996) Relaxation of human temporal artery by endothelin ETB receptors. Peptides 17:1139–1144

Ma KC, Nie XJ, Hoog A, Olsson Y, Zhang WW (1994) Reactive astrocytes in viral infections of the human brain express endothelin-like immunoreactivity. J Neurol Sci 126:184–192

Maguire J, Davenport AP (2004) Alternative pathway to endothelin-converting enzyme for the synthesis of endothelin in human blood vessels. J Cardiovasc Pharmacol 44 Suppl 1:S27–29

Maguire J, Davenport AP (2005) Regulation of vascular reactivity by established and emerging GPC receptors. Trends Pharmacol Sci 26:448–454

Maguire JJ, Davenport AP (1995) ETA receptor-mediated constrictor responses to endothelin peptides in human blood vessels in vitro. Br J Pharmacol 115:191–197

Maguire JJ, Davenport AP (1998a) Increased response to big endothelin-1 in atherosclerotic human coronary artery: functional evidence for up-regulation of endothelin-converting enzyme activity in disease. Br J Pharmacol 125:238–240

Maguire JJ, Davenport AP (1998b) PD156707: a potent antagonist of endothelin-1 in human diseased coronary arteries and vein grafts. J Cardiovasc Pharmacol 31 Suppl 1:S239–240

Maguire JJ, Davenport AP (2002) Is urotensin-II the new endothelin? Br J Pharmacol 137:579–588

Maguire JJ, Kuc RE, O'Reilly G, Davenport AP (1994) Vasoconstrictor endothelin receptors characterized in human renal artery and vein in vitro. Br J Pharmacol 113:49–54

Maguire JJ, Kuc RE, Davenport AP (1997a) Affinity and selectivity of PD156707, a novel nonpeptide endothelin antagonist, for human ET(A) and ET(B) receptors. J Pharmacol Exp Ther 280:1102–1108

Maguire JJ, Johnson CM, Mockridge JW, Davenport AP (1997b) Endothelin converting enzyme (ECE) activity in human vascular smooth muscle. Br J Pharmacol 122:1647–1654

Maguire JJ, Ahn K, Davenport AP (1999) Inhibition of big endothelin-1 (BIG ET-1) responses in endothelium-denuded human coronary artery by the selective endothelin-converting enzyme-1 (ECE-1) inhibitor PD159790. Br J Clin Pharmacol 126:193P

Maguire JJ, Kuc RE, Davenport AP (2001) Vasoconstrictor activity of novel endothelin peptide, ET-1(1–31), in human mammary and coronary arteries in vitro. Br J Pharmacol 134:1360–1366

Marciniak SJ, Plumpton C, Barker PJ, Huskisson NS, Davenport AP (1992) Localization of immunoreactive endothelin and proendothelin in the human lung. Pulm Pharmacol 5:175–182

Matsumoto H, Suzuki N, Kitada C, Fujino M (1994) Endothelin family peptides in human plasma and urine: their molecular forms and concentrations. Peptides 15:505–510

McCulloch KM, Docherty CC, Morecroft I, MacLean MR (1996) EndothelinB receptor-mediated contraction in human pulmonary resistance arteries. Br J Pharmacol 119:1125–1130

McKay KO, Black JL, Diment LM, Armour CL (1991) Functional and autoradiographic studies of endothelin-1 and endothelin-2 in human bronchi, pulmonary arteries, and airway parasympathetic ganglia. J Cardiovasc Pharmacol 17 Suppl 7:S206–209

Migdalis IN, Kalogeropoulou K, Iiopoulou V, Varvarigos N, Karmaniolas KD, Mortzos G, Cordopatis P (2000) Progression of carotid atherosclerosis and the role of endothelin in diabetic patients. Res Commun Mol Pathol Pharmacol 108:27–37

Minami S, Yamano S, Yamamoto Y, Sasaki R, Nakashima T, Takaoka M, Hashimoto T (2001) Associations of plasma endothelin concentration with carotid atherosclerosis and asymptomatic cerebrovascular lesions in patients with essential hypertension. Hypertens Res 24:663–670

Minamino T, Kurihara H, Takahashi M, Shimada K, Maemura K, Oda H, Ishikawa T, Uchiyama T, Tanzawa K, Yazaki Y (1997) Endothelin-converting enzyme expression in the rat vascular injury model and human coronary atherosclerosis. Circulation 95:221–230

Miyamoto Y, Yoshimasa T, Arai H, Takaya K, Ogawa Y, Itoh H, Nakao K (1996) Alternative RNA splicing of the human endothelin-A receptor generates multiple transcripts. Biochem J 313:795–801

Mizuguchi T, Nishiyama M, Moroi K, Tanaka H, Saito T, Masuda Y, Masaki T, de Wit D, Yanagisawa M, Kimura S (1997) Analysis of two pharmacologically predicted endothelin B receptor subtypes by using the endothelin B receptor gene knockout mouse. Br J Pharmacol 120:1427–1430

Mockridge JW, Kuc RE, Huskisson NS, Barker PJ, Davenport AP (1998) Characterization of site-directed antisera against endothelin-converting enzymes. J Cardiovasc Pharmacol 31 Suppl 1:S35–37

Molenaar P, Kuc RE, Davenport AP (1992) Characterization of two new ETB selective radioligands, [125I]-BQ3020 and [125I]-[Ala1,3,11,15]ET-1 in human heart. Br J Pharmacol 107:637–639

Molenaar P, O'Reilly G, Sharkey A, Kuc RE, Harding DP, Plumpton C, Gresham GA, Davenport AP (1993) Characterization and localization of endothelin receptor subtypes in the human atrioventricular conducting system and myocardium. Circ Res 72:526–538

Mombouli JV, Le SQ, Wasserstrum N, Vanhoutte PM (1993) Endothelins 1 and 3 and big endothelin-1 contract isolated human placental veins. J Cardiovasc Pharmacol 22 Suppl 8:S278–281

Moravec CS, Reynolds EE, Stewart RW, Bond M (1989) Endothelin is a positive inotropic agent in human and rat heart in vitro. Biochem Biophys Res Commun 159:14–18

Morton AJ, Davenport AP (1992) Cerebellar neurons and glia respond differentially to endothelins and sarafotoxin S6b. Brain Res 581:299–306

Münter K, Hergenröder S, Unger L, Kirchengast M (1996) Oral treatment with an ETA receptor antagonist inhibits neointima formation induced endothelial injury. Pharm Pharmacol Lett 6:90–92

Nakamuta M, Takayanagi R, Sakai Y, Sakamoto S, Hagiwara H, Mizuno T, Saito Y, Hirose S, Yamamoto M, Nawata H (1991) Cloning and sequence analysis of a cDNA encoding human non-selective type of endothelin receptor. Biochem Biophys Res Commun 177:34–39

Nakano A, Kishi F, Minami K, Wakabayashi H, Nakaya Y, Kido H (1997) Selective conversion of big endothelins to tracheal smooth muscle-constricting 31-amino acid-length endothelins by chymase from human mast cells. J Immunol 159:1987–1992

Nakov R, Pfarr E, Eberle S (2002) Darusentan: an effective endothelinA receptor antagonist for treatment of hypertension. Am J Hypertens 15:583–589

Nambi P, Pullen M, Wu HL, Aiyar N, Ohlstein EH, Edwards RM (1992) Identification of endothelin receptor subtypes in human renal cortex and medulla using subtype-selective ligands. Endocrinology 131:1081–1086

Nilsson T, Cantera L, Adner M, Edvinsson L (1997) Presence of contractile endothelin-A and dilatory endothelin-B receptors in human cerebral arteries. Neurosurgery 40:346–351; discussion 351–353

O'Reilly G, Charnock-Jones DS, Davenport AP, Cameron IT, Smith SK (1992) Presence of messenger ribonucleic acid for endothelin-1, endothelin-2, and endothelin-3 in human endometrium and a change in the ratio of ETA and ETB receptor subtype across the menstrual cycle. J Clin Endocrinol Metab 75:1545–1549

O'Reilly G, Charnock-Jones DS, Morrison JJ, Cameron IT, Davenport AP, Smith SK (1993) Alternatively spliced mRNAs for human endothelin-2 and their tissue distribution. Biochem Biophys Res Commun 193:834–840

Ohlstein EH, Nambi P, Hay DW, Gellai M, Brooks DP, Luengo J, Xiang JN, Elliott JD (1998) Nonpeptide endothelin receptor antagonists. XI. Pharmacological characterization of SB 234551, a high-affinity and selective nonpeptide ETA receptor antagonist. J Pharmacol Exp Ther 286:650–656

Opgenorth TJ, Adler AL, Calzadilla SV, Chiou WJ, Dayton BD, Dixon DB, Gehrke LJ, Hernandez L, Magnuson SR, Marsh KC, Novosad EI, Von Geldern TW, Wessale JL, Winn M, Wu-Wong JR (1996) Pharmacological characterization of A-127722: an orally active and highly potent ETA-selective receptor antagonist. J Pharmacol Exp Ther 276:473–481

Papadopoulos SM, Gilbert LL, Webb RC, D'Amato CJ (1990) Characterization of contractile responses to endothelin in human cerebral arteries: implications for cerebral vasospasm. Neurosurgery 26:810–815

Perfetto F, Tarquini R, Tapparini L, Tarquini B (1998) Influence of non-insulin-dependent diabetes mellitus on plasma endothelin-1 levels in patients with advanced atherosclerosis. J Diabetes Complications 12:187–192

Pernow J, Kaijser L, Lundberg JM, Ahlborg G (1996) Comparable potent coronary constrictor effects of endothelin-1 and big endothelin-1 in humans. Circulation 94:2077–2082

Peter MG, Davenport AP (1995) Selectivity of [125I]-PD151242 for human, rat and porcine endothelin ETA receptors in the heart. Br J Pharmacol 114:297–302

Peter MG, Davenport AP (1996) Characterization of the endothelin receptor selective agonist, BQ3020 and antagonists BQ123, FR139317, BQ788, 50235, Ro462005 and bosentan in the heart. Br J Pharmacol 117:455–462

Pierre LN, Davenport AP (1995) Autoradiographic study of endothelin receptors in human cerebral arteries. J Cardiovasc Pharmacol 26 Suppl 3:S326–328

Pierre LN, Davenport AP (1998a) Relative contribution of endothelin A and endothelin B receptors to vasoconstriction in small arteries from human heart and brain. J Cardiovasc Pharmacol 31 Suppl 1:S74–76

Pierre LN, Davenport AP (1998b) Endothelin receptor subtypes and their functional relevance in human small coronary arteries. Br J Pharmacol 124:499–506

Pierre LN, Davenport AP (1999) Blockade and reversal of endothelin-induced constriction in pial arteries from human brain. Stroke 30:638–643

Plumpton C, Champeney R, Ashby MJ, Kuc RE, Davenport AP (1993) Characterization of endothelin isoforms in human heart: endothelin-2 demonstrated. J Cardiovasc Pharmacol 22 Suppl 8:S26–28

Plumpton C, Haynes WG, Webb DJ, Davenport AP (1995) Phosphoramidon inhibition of the in vivo conversion of big endothelin-1 to endothelin-1 in the human forearm. Br J Pharmacol 116:1821–1828

Plumpton C, Ashby MJ, Kuc RE, O'Reilly G, Davenport AP (1996a) Expression of endothelin peptides and mRNA in the human heart. Clin Sci (Lond) 90:37–46

Plumpton C, Ferro CJ, Haynes WG, Webb DJ, Davenport AP (1996b) The increase in human plasma immunoreactive endothelin but not big endothelin-1 or its C-terminal fragment induced by systemic administration of the endothelin antagonist TAK-044. Br J Pharmacol 119:311–314

Purkiss JR, West D, Wilkes LC, Scott C, Yarrow P, Wilkinson GF, Boarder MR (1994) Stimulation of phospholipase C in cultured microvascular endothelial cells from human frontal lobe by histamine, endothelin and purinoceptor agonists. Br J Pharmacol 111:1041–1046

Raschack M, Unger L, Riechers H, Klinge D (1995) Receptor selectivity of endothelin antagonists and prevention of vasoconstriction and endothelin-induced sudden death. J Cardiovasc Pharmacol 26 Suppl 3:S397–399

Riezebos J, Watts IS, Vallance PJ (1994) Endothelin receptors mediating functional responses in human small arteries and veins. Br J Pharmacol 111:609–615

Rizzi A, Calo G, Battistini B, Regoli D (1998) Contractile activity of endothelins and their precursors in human umbilical artery and vein: identification of distinct endothelin-converting enzyme activities. J Cardiovasc Pharmacol 31 Suppl 1:S58–61

Rossi GP, Albertin G, Franchin E, Sacchetto A, Cesari M, Palu G, Pessina AC (1995) Expression of the endothelin-converting enzyme gene in human tissues. Biochem Biophys Res Commun 211:249–253

Rubin LJ, Badesch DB, Barst RJ, Galie N, Black CM, Keogh A, Pulido T, Frost A, Roux S, Leconte I, Landzberg M, Simonneau G (2002) Bosentan therapy for pulmonary arterial hypertension. N Engl J Med 346:896–903

Russell FD, Davenport AP (1995) Characterization of endothelin receptors in the human pulmonary vasculature using bosentan, SB209670, and 97–139. J Cardiovasc Pharmacol 26 Suppl 3:S346–347

Russell FD, Davenport AP (1996) Characterization of the binding of endothelin ETB selective ligands in human and rat heart. Br J Pharmacol 119:631–636

Russell FD, Davenport AP (1999a) Secretory pathways in endothelin synthesis. Br J Pharmacol 126:391–398

Russell FD, Davenport AP (1999b) Evidence for intracellular endothelin-converting enzyme-2 expression in cultured human vascular endothelial cells. Circ Res 84:891–896

Russell FD, Skepper JN, Davenport AP (1997) Detection of endothelin receptors in human coronary artery vascular smooth muscle cells but not endothelial cells by using electron microscope autoradiography. J Cardiovasc Pharmacol 29:820–826

Russell FD, Skepper JN, Davenport AP (1998a) Human endothelial cell storage granules: a novel intracellular site for isoforms of the endothelin-converting enzyme. Circ Res 83:314–321

Russell FD, Skepper JN, Davenport AP (1998b) Evidence using immunoelectron microscopy for regulated and constitutive pathways in the transport and release of endothelin. J Cardiovasc Pharmacol 31:424–430

Russell FD, Skepper JN, Davenport AP (1998c) Endothelin peptide and converting enzymes in human endothelium. J Cardiovasc Pharmacol 31 Suppl 1:S19–S21

Saeki T, Ihara M, Fukuroda T, Yamagiwa M, Yano M (1991) [Ala1,3,11,15]endothelin-1 analogs with ETB agonistic activity. Biochem Biophys Res Commun 179:286–292

Sakurai T, Yanagisawa M, Takuwa Y, Miyazaki H, Kimura S, Goto K, Masaki T (1990) Cloning of a cDNA encoding a non-isopeptide-selective subtype of the endothelin receptor. Nature 348:732–735

Schweizer A, Valdenaire O, Nelbock P, Deuschle U, Dumas Milne Edwards JB, Stumpf JG, Loffler BM (1997) Human endothelin-converting enzyme (ECE-1): three isoforms with distinct subcellular localizations. Biochem J 328:871–877

Seo B, Oemar BS, Siebenmann R, von Segesser L, Luscher TF (1994) Both ETA and ETB receptors mediate contraction to endothelin-1 in human blood vessels. Circulation 89:1203–1208

Shimada K, Matsushita Y, Wakabayashi K, Takahashi M, Matsubara A, Iijima Y, Tanzawa K (1995a) Cloning and functional expression of human endothelin-converting enzyme cDNA. Biochem Biophys Res Commun 207:807–812

Shimada K, Takahashi M, Ikeda M, Tanzawa K (1995b) Identification and characterization of two isoforms of an endothelin-converting enzyme-1. FEBS Lett 371:140–144

Sitbon O, Badesch DB, Channick RN, Frost A, Robbins IM, Simonneau G, Tapson VF, Rubin LJ (2003) Effects of the dual endothelin receptor antagonist bosentan in patients with pulmonary arterial hypertension: a 1-year follow-up study. Chest 124:247–254

Sitbon O, Gressin V, Speich R, Macdonald PS, Opravil M, Cooper DA, Fourme T, Humbert M, Delfraissy JF, Simonneau G (2004) Bosentan for the treatment of human immunodeficiency virus-associated pulmonary arterial hypertension. Am J Respir Crit Care Med 170:1212–1217

Spatz M, Kawai N, Merkel N, Bembry J, McCarron RM (1997) Functional properties of cultured endothelial cells derived from large microvessels of human brain. Am J Physiol 272:C231–239

Stanimirovic DB, Yamamoto T, Uematsu S, Spatz M (1994) Endothelin-1 receptor binding and cellular signal transduction in cultured human brain endothelial cells. J Neurochem 62:592–601

Stein PD, Hunt JT, Floyd DM, Moreland S, Dickinson KE, Mitchell C, Liu EC, Webb ML, Murugesan N, Dickey J, et al (1994) The discovery of sulfonamide endothelin antagonists and the development of the orally active ETA antagonist 5-(dimethylamino)-N-(3,4-dimethyl-5-isoxazolyl)-1-naphthalenesulf onamide. J Med Chem 37:329–331

Takahashi K, Ghatei MA, Jones PM, Murphy JK, Lam HC, O'Halloran DJ, Bloom SR (1991) Endothelin in human brain and pituitary gland: presence of immunoreactive endothelin, endothelin messenger ribonucleic acid, and endothelin receptors. J Clin Endocrinol Metab 72:693–699

Takahashi M, Matsushita Y, Iijima Y, Tanzawa K (1993) Purification and characterization of endothelin-converting enzyme from rat lung. J Biol Chem 268:21394–21398

Takahashi M, Fukuda K, Shimada K, Barnes K, Turner AJ, Ikeda M, Koike H, Yamamoto Y, Tanzawa K (1995) Localization of rat endothelin-converting enzyme to vascular endothelial cells and some secretory cells. Biochem J 311:657–665

Takahashi T, Barnes PJ, Kawikova I, Yacoub MH, Warner TD, Belvisi MG (1997) Contraction of human airway smooth muscle by endothelin-1 and IRL 1620: effect of bosentan. Eur J Pharmacol 324:219–222

Takai M, Umemura I, Yamasaki K, Watakabe T, Fujitani Y, Oda K, Urade Y, Inui T, Yamamura T, Okada T (1992) A potent and specific agonist, Suc-[Glu9,Ala11,15]-endothelin-1(8–21), IRL 1620, for the ETB receptor. Biochem Biophys Res Commun 184:953–959

Takai S, Jin D, Sakaguchi M, Miyazaki M (1999) Chymase-dependent angiotensin II formation in human vascular tissue. Circulation 100:654–658

Takeji T, Nakaya Y, Kamada M, Maeda K, Saijo Y, Mitani R, Irahara M, Aono T (2000) Effect of a novel vasoconstrictor endothelin-1 (1–31) on human umbilical artery. Biochem Biophys Res Commun 270:622–624

Tanaka H, Moroi K, Iwai J, Takahashi H, Ohnuma N, Hori S, Takimoto M, Nishiyama M, Masaki T, Yanagisawa M, Sekiya S, Kimura S (1998) Novel mutations of the endothelin B receptor gene in patients with Hirschsprung's disease and their characterization. J Biol Chem 273:11378–11383

Tanaka T, Tsukuda E, Nozawa M, Nonaka H, Ohno T, Kase H, Yamada K, Matsuda Y (1994) RES-701-1, a novel, potent, endothelin type B receptor-selective antagonist of microbial origin. Mol Pharmacol 45:724–730

Taylor TA, Gariepy CE, Pollock DM, Pollock JS (2003) Gender differences in ET and NOS systems in ETB receptor-deficient rats: effect of a high salt diet. Hypertension 41:657–662

Thorin E, Nguyen TD, Bouthillier A (1998) Control of vascular tone by endogenous endothelin-1 in human pial arteries. Stroke 29:175–180

Tonnessen T, Naess PA, Kirkeboen KA, Offstad J, Ilebekk A, Christensen G (1993) Endothelin is released from the porcine coronary circulation after short-term ischemia. J Cardiovasc Pharmacol 22 Suppl 8:S313–316

Turner AJ, Murphy LJ (1996) Molecular pharmacology of endothelin converting enzymes. Biochem Pharmacol 51:91–102

Turner AJ, Barnes K, Schweizer A, Valdenaire O (1998) Isoforms of endothelin-converting enzyme: why and where? Trends Pharmacol Sci 19:483–486

Valdenaire O, Rohrbacher E, Mattei MG (1995) Organization of the gene encoding the human endothelin-converting enzyme (ECE-1). J Biol Chem 270:29794–29798

Valdenaire O, Lepailleur-Enouf D, Egidy G, Thouard A, Barret A, Vranckx R, Tougard C, Michel JB (1999) A fourth isoform of endothelin-converting enzyme (ECE-1) is generated from an additional promoter molecular cloning and characterization. Eur J Biochem 264:341–349

Verhaar MC, Strachan FE, Newby DE, Cruden NL, Koomans HA, Rabelink TJ, Webb DJ (1998) Endothelin-A receptor antagonist-mediated vasodilatation is attenuated by inhibition of nitric oxide synthesis and by endothelin-B receptor blockade. Circulation 97:752–756

Vierhapper H, Wagner O, Nowotny P, Waldhausl W (1990) Effect of endothelin-1 in man. Circulation 81:1415–1418

von Geldern TW, Tasker AS, Sorensen BK, Winn M, Szczepankiewicz BG, Dixon DB, Chiou WJ, Wang L, Wessale JL, Adler A, Marsh KC, Nguyen B, Opgenorth TJ (1999) Pyrrolidine-3-carboxylic acids as endothelin antagonists. 4. Side chain conformational restriction leads to ET(B) selectivity. J Med Chem 42:3668–3678

Watakabe T, Urade Y, Takai M, Umemura I, Okada T (1992) A reversible radioligand specific for the ETB receptor: [125I]Tyr13-Suc-[Glu9,Ala11,15]-endothelin-1(8–21), [125I]IRL 1620. Biochem Biophys Res Commun 185:867–873

Waxman L, Doshi KP, Gaul SL, Wang S, Bednar RA, Stern AM (1994) Identification and characterization of endothelin converting activity from EAHY 926 cells: evidence for the physiologically relevant human enzyme. Arch Biochem Biophys 308:240–253

Williams DL Jr, Jones KL, Pettibone DJ, Lis EV, Clineschmidt BV (1991) Sarafotoxin S6c: an agonist which distinguishes between endothelin receptor subtypes. Biochem Biophys Res Commun 175:556–561

Williams DL Jr, Murphy KL, Nolan NA, O'Brien JA, Pettibone DJ, Kivlighn SD, Krause SM, Lis EV Jr, Zingaro GJ, Gabel RA, et al (1995) Pharmacology of L-754,142, a highly potent, orally active, nonpeptidyl endothelin antagonist. J Pharmacol Exp Ther 275:1518–1526

Winkles JA, Alberts GF, Brogi E, Libby P (1993) Endothelin-1 and endothelin receptor mRNA expression in normal and atherosclerotic human arteries. Biochem Biophys Res Commun 191:1081–1088

Wu C, Chan MF, Stavros F, Raju B, Okun I, Mong S, Keller KM, Brock T, Kogan TP, Dixon RA (1997) Discovery of TBC11251, a potent, long acting, orally active endothelin receptor-A selective antagonist. J Med Chem 40:1690–1697

Xu D, Emoto N, Giaid A, Slaughter C, Kaw S, deWit D, Yanagisawa M (1994) ECE-1: a membrane-bound metalloprotease that catalyzes the proteolytic activation of big endothelin-1. Cell 78:473–485

Yamaga S, Tsutsumi K, Niwa M, Kitagawa N, Anda T, Himeno A, Khalid H, Taniyama K, Shibata S (1995) Endothelin receptor in microvessels isolated from human meningiomas: quantification with radioluminography. Cell Mol Neurobiol 15:327–340

Yanagisawa H, Yanagisawa M, Kapur RP, Richardson JA, Williams SC, Clouthier DE, de Wit D, Emoto N, Hammer RE (1998) Dual genetic pathways of endothelin-mediated intercellular signaling revealed by targeted disruption of endothelin converting enzyme-1 gene. Development 125:825–836

Yanagisawa H, Hammer RE, Richardson JA, Emoto N, Williams SC, Takeda S, Clouthier DE, Yanagisawa M (2000) Disruption of ECE-1 and ECE-2 reveals a role for endothelin-converting enzyme-2 in murine cardiac development. J Clin Invest 105:1373–1382

Yanagisawa M, Kurihara H, Kimura S, Tomobe Y, Kobayashi M, Mitsui Y, Yazaki Y, Goto K, Masaki T (1988) A novel potent vasoconstrictor peptide produced by vascular endothelial cells. Nature 332:411–415

Yokokawa K, Tahara H, Kohno M, Murakawa K, Yasunari K, Nakagawa K, Hamada T, Otani S, Yanagisawa M, Takeda T (1991) Hypertension associated with endothelin-secreting malignant hemangioendothelioma. Ann Intern Med 114:213–215

Yu JC, Pickard JD, Davenport AP (1995) Endothelin ETA receptor expression in human cerebrovascular smooth muscle cells. Br J Pharmacol 116:2441–2446

Zhang M, Olsson Y (1995) Reactions of astrocytes and microglial cells around hematogenous metastases of the human brain. Expression of endothelin-like immunoreactivity in reactive astrocytes and activation of microglial cells. J Neurol Sci 134:26–32

Zhang YF, Jeffery S, Burchill SA, Berry PA, Kaski JC, Carter ND (1998) Truncated human endothelin receptor A produced by alternative splicing and its expression in melanoma. Br J Cancer 78:1141–1146

Subject Index

α catenin 122
β catenin 122
γ catenin 122

acetylcholine 223
actin
– α-SMA 83
– α-smooth muscle 83
adenylate cyclase 193
adhesion 118
adhesion molecules 225
adrenal gland 298, 302, 303
albumin 107
albumin-gold complex 110
aldosterone 256
ALI 128
allantois 75, 85
AMP-activated protein
  kinase 234
Ang-1 131
angioadaptation 26
angiogenesis 26, 73, 75, 227
angiopoietin 86
angiotensin I 304
angiotensin II 221, 304, 305
angiotensin converting enzyme
  (ACE) 274
angiotensinogen 257
anti-oxidant 195, 235
aorta-gonad-mesonephros
– AGM 76
apelin 93
arachidonic acid (AA) 190
artery 76
– aorta 76
– aortic primordia 75
– intra-aortic clusters 76
– omphalomesenteric 76
– subaortic patches 76

– umbilical 76
arthritis 196
aspirin 197
asthma 199
astrocytes 11, 300
asymmetric dimethylarginine
  (ADMA) 235
atherogenesis 195
atherosclerosis 197, 222, 231
– endothelin 296
Atractaspis engaddensis 299
autoinhibitory control element
  (ACE) 217

basal lamina 46, 54, 62, 63, 65, 66
– hyperplasic basal lamina 63
– modified and reassembled lipoproteins
  60, 63, 65
beta amyloid 304
bioavailability of NO 231
bleeding time 225
blood islands 74, 75, 78
blood pressure 196, 201
blood–endothelial interface 44
– α-2-macroglobulin 44
– albumin 44, 58, 59, 61
– angiotensin-converting enzyme 45
– associated plasma proteins 44
– fibrinogen 44
– immunoglobulin 44
– lipoprotein lipase 45
– N-acetylglucosaminyl and galactosyl
  residues 50
– proteoglycans 46, 50, 51, 54
– sialoconjugates 44
– strong anionic 46, 49, 51, 56, 61
body homeostasis 43, 44, 59
bone morphogenetic protein
– BMP 78

– BMP4   78
brachyury   74
bradykinin   219, 304
brain
– endothelin converting enzyme-1 (ECE-1)   300

$[Ca^{2+}]_i$ 120
Ca^{2+} 156
– agonists   147
– mechanical stress   147
– shear stress   147
– voltage-dependent   156
Ca^{2+} channel
– cation   157
– diltiazem   156
– monovalent   157
– non-selective   156
– purinoceptor   157
– verapamil   156
Ca^{2+} entry   146, 147, 157, 160, 169
– I_{ARC} 160
– I_{CRAC} 157
– arachidonate-regulated   160
– arachidonic acid   160
– Ca^{2+} release-activated   157
– Ca^{2+} uptake   169
– negative feedback   169
– NO   169
– non-capacitative   160
– non-selective   157
– store-operated   147, 157
– transient receptor potential (TRP)   157
– TRP-related   158
Ca^{2+} leak   154
– ATP concentration   154
– Ca^{2+} channel   155
– Ca^{2+}-free medium   154
– Ni^{2+} 154
Ca^{2+} oscillation   159
– agonist-induced   159
– Ca^{2+}-induced Ca^{2+} release   159
– IP_3 159
Ca^{2+} signalling   174
– Ca^{2+} depletion   174
– Ca^{2+} entry   174
Ca^{2+}-CaM   169
cadherin   79
– VE-cadherin   79

calcineurin   92
calcium (Ca^{2+})   219
– Ca^{2+} flux   227
calcium influx factor
– CIF   161
– CYP450   161
– econazole   161
calcium oscillations   15
caldesmon   83
calmodulin (CaM)   119, 170, 176, 216
– availability   173
– Ca^{2+}-binding protein   170
– CaM biosensor   172
– CaM-binding protein   172
– competition   172, 176
– eNOS   172
– kinetics   173
– limiting   172, 176
– MLCK   173
– phosphorylation   172
– PMCA   172
– SOCE   173
– transduction   176
calphostin C   123
calponin   83
cAMP   131
capillary endothelial cells   46, 49, 51, 56, 57, 65, 66
– continuous   50, 56, 58, 65
– discontinuous   56
– fenestrated   50, 51, 56
capillary types   4
cardiomyocytes   223
cardiovascular disease   231
cardiovascular dysfunction   195
cathepsin G   304
caveolae   217
caveolin-1   217
caveolin-3   217
CD11b/CD18   128
CD34   97
CD41   78
cdc42   131
cell respiration   229
cells   146
– non-excitable   146
cellular bioenergetics   229
cGMP-dependent protein kinases   227
– protein kinase GI (PKGI)   227

changes of the vascular endothelium in different pathologies 61
- Alzheimer's disease 61, 66
- atherosclerosis 57, 61–63
- diabetes 54, 60, 61, 64, 65
- lesional stage 63
- pre-lesional stage 62
channels 108, 125
- store-operated 108
- TRPC 125
cholera toxin 113
cholesterol 113
chordin 91
chymase 304, 305
claudin-5 10
coagulation 26
conductivity
- hydraulic 112
confocal
- imaging 112
confocal microscopy
- endothelin 303
conformational coupling 162
- Ca^{2+} entry 162
- Ca^{2+} release 162
- IP_3 receptors 162
contraction 108
- actin-myosin 108
coronary artery disease 232
coxibs 202
- celecoxib 202
- lumiracoxib 202
- rofecoxib 202
- valdecoxib 202
Csk 116
ct-βARK 114
CTD 122
cyclic adenosine monophosphate (cAMP) 196, 228
cyclic endoperoxides 191
cyclic guanosine monophosphate (cGMP) 227
cyclooxygenase (COX) 190, 260
- COX-1 191
- COX-2 191
cytochrome c oxidase 229
cytochrome P-450 reductase 215

DANCE 6

dephosphorylation 220
depletion Ca^{2+}-store 119
desmin 83
diabetes 231
dimethylarginine dimethylaminohydrolase (DDAH) 235
dynamin-2 113
dysfunction 118
- barrier 118

Edg 131
EDHF 262
endocytosis 108
endoglin 96
endoplasmic reticulum 150
- Ca^{2+} reserve 150
- Ca^{2+} store 150
- Ca^{2+} uptake 150
- IP_3 receptors 151
endothelial cell 42–44, 46, 49–53, 56–65
- coated/uncoated pits 45, 51, 56, 58–60
- coated/uncoated vesicles 45, 51, 58, 59
- cytoskeleton 52, 53, 57
- dysfunction 42, 61, 63, 64, 66
- endothelin converting enzyme-1 (ECE-1) 302
- endothelin converting enzyme-2 (ECE-2) 303
- fenestrae 45, 51, 56
- functions 42–44, 62, 66
- membrane microdomains 42
- organelles 42, 46, 63
- plasma membrane 42, 44, 46, 49
- plasmalemmal vesicles/caveolae 44
- transendothelial channels 42, 44, 45, 50, 57
- ultrastructure 42
- vesicles/caveolae 46
endothelial cell dysfunction 42, 61, 63, 64, 66
endothelial cell receptors for 57
- albumin 50, 58, 59, 61
- insulin 58, 60, 61
- lipoproteins 44, 58–60
- metalloproteins 57
- plasma proteins 49, 57, 61

- vasoactive mediators 57
endothelial nitric oxide synthase (eNOS) 113, 128, 147, 169, 172, 177, 215
- Ca^{2+}-CaM concentration 172
- Ca^{2+}-CaM network 147
- CaM availability 173
- CaM binding 172
- CaM buffering 172
- CaM network 177
- competition 172
- dominant affector 177
- $eNOS^{-/-}$ mice 223
- kinetics 173
- localisation 217
- major affector 147
- MLCK 173
- myristoylation 216
- nitric oxide 169
- palmitoylation 216
- phosphorylation 172
- polymorphisms 234
- reconstituted 172
- SOCE 173
- Thr^{497} 172
- time courses 173
- transfection 235
endothelin
- Big ET-1 297–305
- Big ET-2 298
- Big ET-3 298
- coronary artery 300, 302–304
- ELISA 298, 300
- endothelin-2 298
- endothelin-3 298
- $endothelin_{1-31}$ 304
- ET_A receptor 305
- ET_B receptor 305
- ischaemia 304
- macrophage 300
- renal artery 300
- resistance artery 302
endothelin converting enzyme-1 (ECE-1) 296, 299
- distribution 301
- ELISA 300
- immunocytochemistry 302
- mRNA 299
endothelin converting enzyme-2 (ECE-2) 296, 301, 303, 304
- inhibitor 303

- knock-out mouse 304
endothelium 107, 146
- multifunctional 146
endothelium-derived hyperpolarising factor 224
environmental factors 3
Eph 93
ephrin-B2 14
ephrins 93
epiblasts 77
ER
- Ca^{2+} release 152
- Ca^{2+}-ATPase 152
- osmotic swelling 152
- volume-sensitive 152
erythropoietin 80
exocytosis 108
extracellular matrix 46, 64
- hyperplasia 64
extraembryonic mesoderm 74

Fahraeus effect 21
fenestrae 117
fibrosis
- lung 117
filipin 113
flavin adenine dinucleotide 216
flavin mononucleotide 216
fluid
- interstitial 118
fluid phase 110
fluorescence imaging 15
forkhead transcription factors 87
free radicals 219
functional modifications 62
- hyperglycaemia 64, 65
- hyperlipaemia 60, 62, 63, 65
- inflammatory process 62, 63
fusion 108

G protein 147
- $G_{\alpha q}$ 148
- $G_{\alpha \gamma}$ 148
G proteins 113
- heterotrimeric 113
- monomeric 113
$G\beta\gamma$ 114
$G_{\alpha i}$ 114
$G_{12/13}$ 120
G_i 114

Subject Index

G_q 119
GAPs 124
GATA 77, 80
GDI 121
GEFs 124
gene 117
– caveolin-1 117
genetic factors 3
glucocorticoids 222
glycolysis 234
glycoprotein 17
glycosaminoglycan (GAG) 18
glypican 18
gp60 109
GST-PAK 133
GTPase 114

haem 216
haemangioblast 73–75, 77, 79
haematocrit 20
haematopoietically expressed homeobox
– (Hex) 77
haemogenic endothelium 75
haemoglobic endothelium
– intra-aortic clusters 76
– subaortic patches 76
heart 223
– endothelin converting enzyme-2 304
– endothelin-2 298
– endothelin-3 298
– endothelium converting enzyme-1 (ECE-1) 300
– sarafotoxin 299
heart attack 199
heart failure 231
heat shock protein 90 218
heparan sulphate (HS) 18
heparan sulphate proteoglycans (HSPGs) 18
heparinase 24
histamine 219
histamine receptors 54, 57
HMWK 264
homing 12
host-defence 108
hyaluronic acid 24
hydrogen peroxide (H_2O_2) 219
hypercholesterolaemia 222, 231
hyperhomocysteinaemia 231
hypertension 223, 231

hypoxia 80
hypoxia-inducible factor-1 234

^{125}I-albumin 120
ICAM-1 6, 127
icatibant 266
inducible nitric oxide synthase (iNOS) 215
inflammation 26, 108, 195
inflammatory mediators 108
– bradykinin 108
– histamine 108
– platelet activating factor 108
– thrombin 108
– vascular endothelial growth factor 108
influx 108
– Ca^{2+} 108
iNOS 128
– induction 222
insulin 109
– secretion 267
– sensitivity 267
integrins 18
interactions 113
– protein-protein 113
intercellular endothelial junctions 53, 56, 57
– adherent junctions (zonula adherens) 53
– gap (communicating) junctions (macula communicans) 53, 54
– tight junctions (zonula occludens) 44, 53, 54, 56, 66
intercellular junctions 42
internalisation 116
– SV40-induced 116
intimal hyperplasia 226
intracellular Ca^{2+} stores 148
IP deletion 197
IP knockout 196
IP_3 120, 148
– second messenger 148
IP_3 receptors 152, 153
– Ca^{2+} 153
– Golgi apparatus 153
– high-affinity 153
– IP_3-induced 153
– perinuclear 153
– secretory vesicles 153

– subunits   152
– tetramers   152

JMD   122
junctional complexes   108
– adherens   108
– tight   108

$K_{f,c}$ 127
kallikrein   264
kidney
– endothelin-1   299, 301, 303
kidney, glomerular capillaries   66
knock-out   224
knockout mice   113
– caveolin-1   113

L-arginine   215
L-selectin   12
leucocyte-endothelial interactions   13
lifespan   230
LMO-2   77
LMWK   264
losartan   258
low-density lipoprotein (LDL)   197, 232
lung microvascular bed   111
lymphatic tissues   12
lymphocytes   12

macrophage-derived foam cells   63, 64
macrophages   221
malformation
– cerebral cavernous   96
– venous   87
mast cells   54, 57, 64
master-regulator   113
matrix   118
– subendothelial   118
mechanical forces   151
– Ca^{2+} release   151
– Ca^{2+} stores   151
– osmotic   151
– shear stress   151
mechanism of endothelial sorting
  of molecules   59
– endocytosis   42, 58–61
– receptor-mediated and receptor-
  independent endocytosis   58, 60
– receptor-mediated and receptor-
  independent transcytosis
  42, 50, 57, 61

– receptor-mediated endocytosis   59
– transcytosis   42, 43, 47, 50, 57–61
– transport of plasma molecules
  42, 50, 61
membrane   112, 114
– apical   114
– basolateral membrane   112
membrane fission   112
– dynamin-dependent   112
membrane microdomains   112
– cholesterol-rich   112
– glycosphingolipid-rich   112
mesoangioblast   82
methyl-β-cyclodextrin   113
microdomains   113
microhaemodynamics   23
microparticle image velocimetry
  (μPIV)   23
microtubule   120
microvascular flow resistance   21
microvascular networks   21
microvessels   15
mitochondria   150, 223
– Ca^{2+} oscillation   151
– Ca^{2+} overload   150
– Ca^{2+} release   151
– Ca^{2+} reserve   150
– Ca^{2+} uptake   150
– excitable   150
– mitochondrial biogenesis   230
MLCK   118, 165
– barrier function   165
– bradykinin   165
– calcium signalling   165
– Cl^- influx   165
– cytoskeleton   166
– endothelial cells   165
– endothelium-derived relaxing
  factors   165
– HEK 293 cells   168
– ML-5   165
– ML-7   165
– ML-9   165
– MLC phosphorylation   166
– Mn^{2+} influx   168
– monocytes   166
– protein 4.1   167
– spectrin   167
– thapsigargin   165
– TRPC4   168

myocardial capillaries   56, 65
myocardin   84
myosin   83
myosin light chain   118

NADPH   216
NADPH oxidase   128
NADPH oxidases   233
netrin   92
neuronal (nNOS)   215
neuropilin   91
neuropilin 1   14
neutral endopeptidase   304
NF-κB   128
NG2 proteoglycan   83
nitric oxide   13, 259
nitric oxide (NO) synthase   13, 113
– endothelial   113
nitroglycerin   232
nitrosating species   228
nitrotyrosine   232
nitroxyl anion (NO$^-$)   215
noggin   91
non-steroidal anti-inflammatory drugs (NSAIDs)   201
– diclofenac   203
– flurbiprofen   203
– ibuprofen   194, 203
– indomethacin   194, 203
– meloxicam   203
– naproxen   203
NOS traffic inducer protein (NOSTRIN)   221
NOS-interacting protein (NOSIP)   221
Notch   95
notochord   91

occludin   8
oedema   126
– protein-rich   126
oestrogen   194, 218
oncotic pressure   108
oxidative phosphorylation chain   229
oxidative stress   225, 229
oxidised LDL   235

P-selectin   7
p115RhoGEF   121

p120-catenins   122
pain   200
PAR-1   119
pathology   57
pathway   107
– transcellular   107
pericytes   53, 54, 66, 72, 82, 84
peroxynitrite (ONOO$^-$)   194, 226
phase   110
phosphatase   114, 118
– MLC-associated   118
phosphatase   220
phosphorylation   108, 215
– myosin light chain   108
– Ser^{1177} 220
– Thr^{495} 220
PKA   131
PKCα   108
placenta   75
placenta growth factor
– PlGF   78
plasmalemma   114
platelet   224
– aggregation   201, 225
platelet endothelial cell adhesion molecule-1 (PECAM-1)   79
platelet-derived growth factors (PDGF)   86
platelets   193
PMN   128
polarized cells   44
pre-eclampsia   231
pregnancy   195
pressure   108
primary vascular plexus   75
primitive streak   74
prostacyclin   225
prostacyclin synthase   193
prostaglandins (PGs)   190, 224
prostanoid receptor   196
protein   113
– scaffolding   113
protein kinase C   168
– bradykinin   168
– endothelial cells   168
– SOCE   168
protein kinases   164
– Ca^{2+} entry   164
– genistein   164
– tyrosine kinase   164

proteoglycans 18, 24
pulmonary hypertension 231

Rac 131
Raynaud's syndrome 231
reactive oxygen species (ROS) 222, 259
renin 257
repeats 122
– cadherin 122
retinal capillary 66
retinoic acid 78
retraction 118
– endothelial cell 118
Rho GTPase 120
Rho kinase 108
RhoA 124
Runx1 80
ryanodine receptors 153
– endothelial 154
– isoforms 153
– non-endothelial 154

S-nitrosylation 228
S1P 131
sarafotoxin 299
– endothelin agonist 299
sarcoplasmic/endoplasmic reticulum Ca^{2+}-ATPase (SERCA)
– Ca^{2+}-binding sites 155
– calmidazolium 156
– CaM antagonists 155
– conformational changes 155
– high-energy state 155
– house-keeping 155
– isoforms 155
– low-affinity state 155
– non-muscle 155
– thapsigargin 155
– W-7 156
second messenger 149
– Ca^{2+}-releasing 149
– cADPR 149
– NAADP 149
selectins 18
semaphorin 91
septic shock 222
SERCA 155
serine kinase Akt 1 220
shear rates 13
shear stress 13, 148, 192, 197, 218

– IP$_3$ 148
signalling molecules 108
signalling pathways 108
– second messenger 108
Smad1 78
Smad5 78
smoking 231
smooth muscle cell (SMC) 43, 53, 54, 63, 65, 75, 83
smooth muscle proliferation 226
SOC 120
SOCE 147, 159
– Ca^{2+} release-activated Ca^{2+} current 159
– capacitative 159
– shear stress 159
– thapsigargin 159
– transient receptor potential channel 147
soluble guanylate cyclase 221
somitic tissue 75
sphingosine-1
– S1P 84
sprouting 90
Src 113
statins 219
stem cell leukaemia
– SCL 77, 80
– TAL-1 77
stem cells 78
– haematopoietic 76, 77, 97
store-operated Ca^{2+} entry 158
stress fibres 120
stroke 199
studies 121
– electron microscopy 121
superoxide anion (O$_2^-$) 218
syndecan 18

telangiectasia
– hereditary haemorrhagic 96
thapsigargin 123, 155, 159
thromboxane (TX) 190
thromboxane A$_2$ 193
thromboxane A$_2$ receptor (TP) 198
thromboxane synthase 193
thromboxane synthase inhibitors 200
Tie1 86
Tie2 86, 132

Subject Index

tight junctions 8
TLR2 128
TLR4 128
TNF-α 124
transendothelial electrical
 resistance 123
transferrin 80, 109
transforming growth factor
 (TGF)-β 227
transforming growth factor-β1
 – (TGF-β1) 78, 88
transport 107
TRP 158
– canonical 158
– TRP1 121
– TRP3 158
– TRPC 158
– TRPC4 120, 158
– TRPM 158
– vanilloid 158
tyrosine kinase 116

vascular barrier 108
– semi-permeable 108
vascular endothelial growth factor
 VEGF 72, 218
vascular endothelium 42, 44, 46, 56, 61, 66
– differentiated microdomains 56
– large vessel endothelial cells 45, 56, 58, 65
– phenotypic heterogeneity 56
vascular permeability 226
vascular smooth muscle cell 72, 195
– proliferation 201
vascular tone 26, 256
vasculogenesis 73, 75
vasoconstriction 222
VASP 131
VEGF-A receptor 2
– Flk1 74
– KDR 74
– VEGFR-2 74, 77, 78
vesicle carriers 107
vesicle fission 108
vesicle trafficking 108
vitelline vessels 75
vitronectin 7
von Willebrand factor 46, 56, 64
VVOs 117

Wnts 85

xanthine oxidase 233

yolk sac 73–75, 77